Springer Series in
Surface Sciences

Editor: Manuel Cardona 16

Springer Series in **Surface Sciences**

Editors: G. Ertl and R. Gomer   Managing Editor: H. K. V. Lotsch

Volume 1: **Physisorption Kinetics** By H. J. Kreuzer, Z. W. Gortel

Volume 2: **The Structure of Surfaces** Editors: M. A. Van Hove, S. Y. Tong

Volume 3: **Dynamical Phenomena at Surfaces, Interfaces and Superlattices**
Editors: F. Nizzoli, K.-H. Rieder, R. F. Willis

Volume 4: **Desorption Induced by Electronic Transitions, DIET II**
Editors: W. Brenig, D. Menzel

Volume 5: **Chemistry and Physics of Solid Surfaces VI**
Editors: R. Vanselow, R. Howe

Volume 6: **Low-Energy Electron Diffraction**
Experiment, Theory and Surface Structure Determination
By M. A. Van Hove, W. H. Weinberg, C.-M. Chan

Volume 7: **Electronic Phenomena in Adsorption and Catalysis**
By V. F. Kiselev, O. V. Krylov

Volume 8: **Kinetics of Interface Reactions** Editors: M. Grunze, H. J. Kreuzer

Volume 9: **Adsorption and Catalysis on Transition Metals and Their Oxides**
By V. F. Kiselev, O. V. Krylov

Volume 10: **Chemistry and Physics of Solid Surfaces VII**
Editors: R. Vanselow, R. Howe

Volume 11: **The Structure of Surfaces II**
Editors: J. F. van der Veen, M. A. Van Hove

Volume 12: **Diffusion at Interfaces: Microscopic Concepts**
Editors: M. Grunze, H. J. Kreuzer, J. J. Weimer

Volume 13: **Desorption Induced by Electronic Transitions, DIET III**
Editors: R. H. Stulen, M. L. Knotek

Volume 14: **Solvay Conference on Surface Science** Editor: F. W. de Wette

Volume 15: **Surfaces and Interfaces of Solids** By H. Lüth

Volume 16: **Theory of the Atomic and Electronic Structure of Surfaces**
By M. Lannoo, P. Friedel

Volume 17: **Adhesion and Friction** Editors: M. Grunze and H. J. Kreuzer

Volume 18: **Auger Spectroscopy and Electronic Structure**
Editors: G. Cubiotti, G. Mondio, K. Wandelt

Volume 19: **Desorption Induced by Electronic Transitions, DIET IV**
Editors: G. Betz, P. Varga

Volume 20: **Semiconductor Surfaces** By W. Mönch

Volume 21: **Surface Phonons** Editors: W. Kress, F. W. de Wette

Volume 22: **Chemistry and Physics of Solid Surfaces VIII**
Editors: R. Vanselow, R. Howe

M. Lannoo P. Friedel

Atomic and Electronic Structure of Surfaces

Theoretical Foundations

With 132 Figures

Springer-Verlag
Berlin Heidelberg New York
London Paris Tokyo
Hong Kong Barcelona
Budapest

Dr. Michel Lannoo

Laboratoire d'Etudes des Surfaces et Interfaces,
I.S.E.N. Institut Supérieur d'Électronique du Nord,
41, Boulevard Vauban, F-59046 Lille Cedex, France

Dr. Paul Friedel

Laboratoires D'Électronique Philips,
B.P. 15–22, Avenue Descartes, F-94453 Limeil-Brévannes Cedex, France

Guest Editor: Professor Dr., Dres. h. c. Manuel Cardona

Max-Planck-Institut für Festkörperforschung, Heisenbergstrasse 1,
D-7000 Stuttgart 80, Fed. Rep. of Germany

Series Editors

Professor Dr. Gerhard Ertl

Fritz-Haber-Institut der Max-Planck-Gesellschaft, Faradayweg 4–6,
D-1000 Berlin 33

Professor Robert Gomer, Ph. D.

The James Franck Institute, The University of Chicago, 5640 Ellis Avenue,
Chicago, IL 60637, USA

Managing Editor: Dr. Helmut K.V. Lotsch

Springer-Verlag, Tiergartenstrasse 17,
D-6900 Heidelberg, Fed. Rep. of Germany

ISBN 3-540-52682-X Springer-Verlag Berlin Heidelberg New York
ISBN 0-387-52682-X Springer-Verlag New York Berlin Heidelberg

Library of Congress Cataloging-in-Publication Data. Lannoo, M. (Michel), 1942– Atomic and electronic structure of surfaces : theoretical foundations / M. Lannoo, P. Friedel. p. cm.–(Springer series in surface sciences ; 16) Includes bibliographical references and index. ISBN 3-540-52682-X (acid-free paper).– ISBN 0-387-52682-X (acid-free paper) 1. Solids–Surfaces. 2. Semiconductors–Surfaces. 3. Atomic structure. 4. Electronic structure. I. Friedel, P. (Paul), 1955– . II. Title. III. Series. QC176.8.S8L36 1991 530.4'1–dc20 90-22640

54/3140-543210 – Printed on acid-free paper

To Ginette, Brigitte
and our children

Preface

The field of surface physics is presently extremely active and continues to evolve constantly. This is due in part to the improvement of experimental tools and techniques, which has enabled more information to be obtained about microscopic properties. A further incentive for development is the need to understand the mechanisms of growth, which themselves depend on the state of the free surface before adsorption of atoms occurs. In practical applications the influence of surfaces and near-surface regions is also very important: In the field of electronics, for instance, semiconductor heterojunctions and metal–semiconductor interfaces require extremely well-controlled growth conditions in order to minimize the number and influence of interface defects.

The results obtained during the last ten years have been extremely varied and in some cases quite surprising. For the free surfaces of semiconductors the essential results concern the important lattice reorganization resulting in surface relaxation or reconstruction. The details of the surface atomic structure are now well understood in many cases such as Si(111), Si(100), GaAs(110) and even some polar surfaces of compound semiconductors and insulators. Total energy calculations have been quite helpful in this context. Interesting information has also been obtained concerning the formation of interfaces, related to the bonding of adatoms at submonolayer coverages. This is especially true for alkali and noble metals on GaAs, which, however, remain less well understood than As monolayers on Si(111).

The motivation for writing this book can be summarized as follows: (i) Both of us have been working in the field for some time and have concentrated on complementary aspects. (ii) We have given several advanced lectures on this subject. (iii) One of us already had some experience of writing books in the same spirit but dealing with defects in semiconductors. (iv) We believe that the time is ripe in this fast moving field to present a coherent description of semiconductor surfaces ranging from the ideal case to interfaces with metals or other semiconductors. (v) The same set of simple concepts gives deep insight into both semiconductor and metal surface properties. As many research students are engaged in studying surface physics, we felt that a book incorporating the basic properties would be useful. Our central aim is thus to give these students an accurate description of what is going on, but at the same time to do this on the basis of simple physical models which allow a deeper understanding. This is apparent in all chapters of the book and at the end of each chapter we include some exercises for illustration and practice.

Finally we would like to thank all our collaborators. Although many have not been directly involved in the writing of the book, their scientific contributions mentioned as references have been essential in elaborating the material presented.

Lille
Limeil-Brévannes
February 1991

M. Lannoo
P. Friedel

Contents

1. Introduction

For the past few decades, surface physics has undergone a rapid and continuous development. Since the end of the 1970s the rate of increase of our knowledge has grown further due to the rapid development of theoretical and computational techniques. This book gives the reader an introduction to some of the theoretical techniques that have enabled us to understand basic phenomena such as surface reconstruction, chemisorption, heterojunction properties... A companion volume [1.1] stresses the experimental aspects.

Readers of this book should have a certain command of solid state physics at the level of the book of *Ashcroft* and *Mermin* [1.2]. The tools that are employed (e.g. the molecular model) here are quite simple and the intention is to convince the reader that simple arguments lead to the essentials of the physics. Only in a few cases are some deductions based on the results of numerical calculations that can be found in the literature. The numerical techniques used are in all cases based on simple methods and do not require large computer capacities. The didactic aspect is stressed by the inclusion of exercises at the end of each chapter.

Due to the common background of the authors, the text is somewhat biased towards the physics of semiconductors. The underlying theme of the main chapters of the book is the semiconductor surface, from the clean surface to the interface between two semiconductors (heterojunction). Intermediate steps include chemisorption and Schottky barriers on III-V compounds where recent progress has been made. All these aspects, as well as transition metal surfaces are studied using simple physico-chemical methods. However, a description of most recent advances in ab initio methods is included in the second chapter. No further use is made of these methods here since they are of limited help for extracting universal relations.

This first chapter presents the basic ideas necessary for an understanding of the book.

1.1 Energy Bands and Densities of States

1.1.1 Tight-Binding Approximation

The tight-binding approximation is a very attractive method for studying a wide range of problems in solid state physics. It gives a simple picture of the physics involved in complex systems. We shall describe it for the bulk of tetrahedrally coordinated compounds. It is a one-electron potential scheme. This means that all electrons feel an effective potential $V(\boldsymbol{r})$. This effective potential is written as a sum of atomic potentials:

$$V(\boldsymbol{r}) = \sum_{i} v(\boldsymbol{r} - \boldsymbol{R}_i) . \tag{1.1}$$

The wave function is expanded in terms of atomic orbitals:

$$\Phi_k(\boldsymbol{r}) = \sum_{i,\mu} C^k_{i\mu} \phi_{i\mu}(\boldsymbol{r} - \boldsymbol{R}_i) \tag{1.2}$$

where $\phi_{i\mu}$ is the μth orbital of the ith atom.

The eigenfunctions are solutions to the eigenvalue problem:

$$H|\psi\rangle = E|\psi\rangle , \tag{1.3}$$

where H is the Hamiltonian and E the eigenvalues. The atomic orbital basis that one chooses is the "minimal basis set". In the case of tetrahedral compounds like Si, GaAs, ..., the minimal basis set is built from ns and np shells. One takes only the corresponding free atom orbitals into account. In a first step one may treat these as an orthogonal set (the so-called tight-binding approximation). This means that the overlap integrals are neglected:

$$\langle \phi_{i\mu} | \phi_{j\nu} \rangle = \delta_{ij} \delta_{\mu\nu} . \tag{1.4}$$

Then the eigenvalue problem reduces to:

$$\det [M_{\mathrm{H}} - EI] = 0 , \tag{1.5}$$

where I is the unit matrix and M_{H} is the matrix of the Hamiltonian in the minimal basis set. Its values are:

$$M_{\mathrm{H}_{i\mu,j\nu}} = \langle \phi_{i\mu} | H | \phi_{j\nu} \rangle . \tag{1.6}$$

Two types of term can be distinguished:

$$\begin{aligned} M_{\mathrm{H}_{i\mu,j\nu}} &= \alpha_{i\mu} \delta_{\mu\nu} \\ M_{\mathrm{H}_{i\mu,j\nu}} &= \beta_{i\mu,j\nu} \quad \text{for} \quad i \neq j \end{aligned} \tag{1.7}$$

where the first (intra-atomic term) is the sum of the free atom eigenvalue and of a crystal field integral. It is diagonal in cubic symmetry. For a crystal containing only one chemical element, the $\alpha_{i\mu}$ do not depend on i. The $\beta_{i\mu,j\nu}$ are also called resonance or hopping integrals. They are inter-atomic terms. For non-monatomic solids, one should in principle solve the eigenvalue problem self-consistently. In this case, the intra-atomic term is a function of the net charge on the atoms; the net charge in turn is usually given by the filling of states of lower energy, at least for sp systems.

1.1.2 Molecular Model

For tetrahedrally coordinated compounds, a good deal of physical insight may be obtained by using a very much reduced number of parameters. We first start with the crudest approximation, which we demonstrate here for $A^N B^{8-N}$ compounds. In these crystals each atom is surrounded by four neighbors of the other type of atom. The local symmetry is that of a tetrahedron, so that the molecular bonds are built from sp^3 orbitals pointing towards each other. Instead of starting from the atomic orbitals, it is then possible to use as minimal basis set four such orbitals per atom. It is assumed that the overlap between these hybrids is negligible. The "molecular model" then consists in restricting the elements of the Hamiltonian matrix to:

$$\begin{aligned} \alpha_A &= \langle \phi_A | H | \phi_A \rangle , \\ \alpha_B &= \langle \phi_B | H | \phi_B \rangle , \\ \beta &= \langle \phi_A | H | \phi_B \rangle , \end{aligned} \tag{1.8}$$

where β is the resonance integral between sp^3 hybrids centered on nearest neighbors and pointing towards each other. Thus each pair of such hybrids is independent of the others. This leads to a set of identical diatomic molecules. The solutions of the eigenvalue problem are then a bonding E_b and an anti-bonding E_a level. Their energies are given by:

$$E_{\substack{a \\ b}} = \frac{\alpha_A + \alpha_B}{2} \pm \left[\left(\frac{\alpha_A - \alpha_B}{2} \right)^2 + \beta^2 \right]^{1/2} . \tag{1.9}$$

The eigenfunctions are

$$\begin{aligned} \Psi_b &= \frac{\phi_A + \lambda \phi_B}{\sqrt{1 + \lambda^2}} , \\ \Psi_a &= \frac{\lambda \phi_A + \phi_B}{\sqrt{1 + \lambda^2}} . \end{aligned} \tag{1.10}$$

If the number of atoms is $2M$, E_b and E_a are $4M$-fold degenerate, or $8M$-fold if one counts spin degeneracy. As there are $8M$ electrons in the solid, the bonding level will be completely filled and the anti-bonding level completely empty. This is a very crude first description of the band structure. The levels E_b and E_a can be thought of as the centers of mass of the valence and conduction bands respectively. More detailed information may be obtained from the net excess electrons on the atoms A and B, n_A^0 and n_B^0 respectively. From the wavefunctions of (1.10), one deduces:

$$n_A^0 = -n_B^0 = \frac{8}{1 + \lambda^2} - N . \tag{1.11}$$

The intra-atomic terms α, may be split into an excess charge-independent and a charge-dependent term:

$$\alpha_A = \alpha_A^0 + \left(J_A - \frac{\alpha e^2}{R}\right) n_A^0 \quad \text{and}$$

$$\alpha_B = \alpha_B^0 + \left(J_B - \frac{\alpha e^2}{R}\right) n_B^0 \, . \tag{1.12}$$

In this expression, α_A^0 and α_B^0 are the ionization potentials of the neutral atoms in their sp^3 valence state, J_A and J_B are the intra-atomic Coulomb terms and $|\alpha e^2/R|$ is a Madelung energy due to the excess charge on the sites other than the one considered. The A and B on-site energy difference is the important quantity, and is given by

$$\alpha_B - \alpha_A = \alpha_B^0 - \alpha_B^0 - \left(J_A + J_B - \frac{2\alpha e^2}{R}\right) n_A^0 \, . \tag{1.13}$$

This model may be compared to *Phillips'* spectroscopic model [1.3] where the average gap E_G is separated into a homopolar and a heteropolar part, E_H and C. These gaps are related by

$$E_G^2 = E_H^2 + C^2 \, . \tag{1.14}$$

In the molecular model, these quantities may be identified as follows:

$$\begin{aligned} E_G^2 &= 4\beta^2 + (\alpha_B - \alpha_A)^2 \\ E_H &= 2\beta \\ C &= |\alpha_B - \alpha_A| \, . \end{aligned} \tag{1.15}$$

The agreement is particularly good for the heteropolar gap and in the case of carbon for the homopolar gap. This shows that even a simple model may yield quite gratifying results. We shall often use the molecular model for the prediction of trends later on in this book.

1.1.3 Beyond the Molecular Model – The Formation of Bands

Starting from the molecular model, one may introduce further interactions to analyze the formation of bands. One may distinguish three classes of interactions: those between two bonding orbitals, between two anti-bonding, and finally between bonding and anti-bonding orbitals.

We shall initially neglect the last of these. Separate problems result for bonding and anti-bonding orbitals which are formally identical. We shall restrict the analytic description to the simplest model which only includes the interactions between orbitals of adjacent bonds. As defined in Fig. 1.1, one has four parameters $(\Delta_A)_{bb}$, $(\Delta_B)_{bb}$, $(\Delta_A)_{aa}$ and $(\Delta_B)_{aa}$. We shall treat both problems with two parameters Δ_A and Δ_B which can represent the resonance integrals either between bonding or between anti-bonding states. One can write the total wavefunction of the bonding (or anti-bonding) band as

$$\Psi = \sum_J a_J \phi_J \tag{1.16}$$

where J is the index of the bonds.

Taking the origin of energies at the bonding (E_b) (or anti-bonding, E_a) level one can write for instance (Fig. 1.2):

$$E a_{J_0} = \Delta_A \sum_{i=1}^{3} a_{J_i} + \Delta_B \sum_{\alpha=1}^{3} a_{J_{0,\alpha}} . \tag{1.17}$$

Introducing for all atoms i a quantity S_i equal to the sum of the coefficients a_J over the four bonds of this atom, one can rewrite (1.17) as

$$(E + \Delta_A) a_{J_0} = \Delta_A S_i + \Delta_B \sum_{\alpha=1}^{3} a_{J_{0,\alpha}} . \tag{1.18}$$

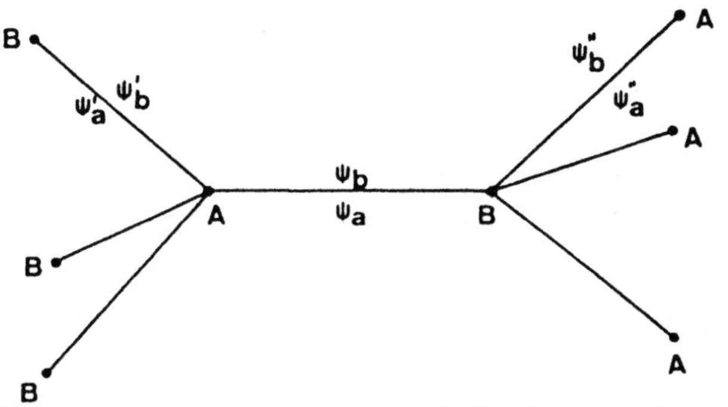

Fig. 1.1. Representation of one bond with its six adjacent bonds. The corresponding resonance integrals are: $(\Delta_A)_{bb} = \langle\Psi_b|H|\Psi'_b\rangle$, $(\Delta_B)_{bb} = \langle\Psi_b|H|\Psi''_b\rangle$, $(\Delta_A)_{aa} = \langle\Psi_a|H|\Psi'_a\rangle$, $(\Delta_B)_{aa} = \langle\Psi_a|H|\Psi''_a\rangle$

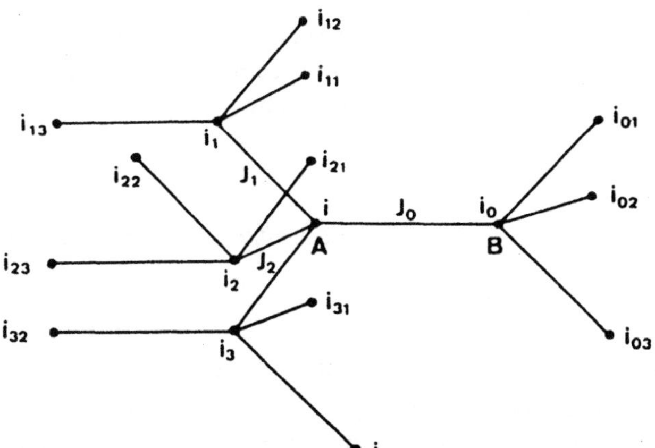

Fig. 1.2. Local environment of bond J_0. Atoms are labelled i, i_n, $i_{\alpha n}$, bonds J_n or $J_{\alpha n}$. For the latter type of bond, α, n is the same as for the corresponding terminal atom $i_{\alpha n}$

With the same arguments, it is easy to show that

$$(E+\Delta_A)\sum_{i=1}^{3} a_{J_{0,\alpha}} = \Delta_A \sum_{\alpha=1}^{3} S_{i_{0,\alpha}} + \Delta_B \left[3a_{J_0} + 2\sum_{\alpha=1}^{3} a_{J_{0,\alpha}}\right] . \tag{1.19}$$

One can now express $\sum_\alpha a_{J_{0,\alpha}}$ in (1.19) and put its value in (1.18). This leads to

$$(E+\Delta_A)(E+\Delta_A-2\Delta_B)a_{J_0} = \Delta_A(E+\Delta_A-2\Delta_B)S_i + \Delta_B\Delta_A\sum_{\alpha=1}^{3} S_{i_{0,\alpha}} + 3\Delta_B^2 a_{J_0} . \tag{1.20}$$

If one now writes the corresponding equations for J_1, J_2 and J_3, takes their sum and sets

$$\Delta = \frac{\Delta_A+\Delta_B}{2} , \qquad \delta = \frac{\Delta_A-\Delta_B}{2} , \tag{1.21}$$

one obtains

$$(E^2 - 4\Delta E - 12\delta^2)S_i = (\Delta^2-\delta^2)\sum_{j=0}^{3}\sum_{\alpha=0}^{3} S_{i_{j,\alpha}} . \tag{1.22}$$

This is the formal equivalent of the tight-binding equation of a one-band fcc structure with nearest neighbor interactions. Using the Bloch theorem, solutions to this problem are

$$E = 2\Delta \pm [4\Delta^2 + 12\delta^2 + (\Delta^2-\delta^2)\phi]^{1/2}$$
$$\phi = 4\left(\cos\frac{k_x a}{2}\cos\frac{k_y a}{2} + \cos\frac{k_y a}{2}\cos\frac{k_z a}{2} + \cos\frac{k_z a}{2}\cos\frac{k_x a}{2}\right) \tag{1.23}$$

where a is the lattice parameter, k_x, k_y, and k_z the components of the wavevector along the axes of the cubic cell. It is clear that ϕ varies continuously from 12 to −4. The extrema of the bands ($\phi = +12$) are then given by

$$E = \begin{matrix} 6\Delta \\ -2\Delta \end{matrix} \tag{1.24}$$

while for $\phi = -4$ a gap is opened in the band, whose upper and lower energy values are

$$E = 2\Delta \pm 4\delta . \tag{1.25}$$

This gap disappears for a homopolar crystal (C, Si or Ge for instance). This is why it is sometimes called the heteropolar gap. However, this nomenclature may cause some confusions since it is not the same quantity as the heteropolar gap defined by Phillips (see above). We shall thus not use this terminology here. Equation (1.23) gives only two bands although we started from four bonding (or anti-bonding) orbitals per atom. Two bands are thus missing and can be obtained by noticing that (1.20) has a trivial non-zero solution for which all the S_i are equal to zero ("pure p" solution). This requires that

$$\begin{aligned} &(E+\Delta_A)(E+\Delta_A-2\Delta_B)-3\Delta_B^2=0 \quad \text{i.e.} \\ &E=-(\Delta_A-\Delta_B)\pm 2\Delta_B \,. \end{aligned} \tag{1.26}$$

Only the second solution (with the minus sign) yields non-zero values of the a_J. One thus gets a flat two-fold degenerate band at

$$E=-2\Delta \,. \tag{1.27}$$

In summary, the inclusions of interactions between bonding orbitals has widened the degenerate bonding level E_b into a valence band of width $4|(\Delta_A)_{bb}+(\Delta_B)_{bb}|$ without changing the charge density or center of mass of the band. This remains true even if further interactions are included. The same may be said for the conduction band which is built from the anti-bonding level. A gap (forbidden band) of width 8δ separates these allowed bands. It is noteworthy that solutions to (1.22) with a purely imaginary wavevector $\boldsymbol{k}$ may be found in the gap. The interaction between the bonding and anti-bonding levels is in general sufficiently small to be treated in perturbation theory. If a gap already exists before this interaction is included, the effect is then merely to widen the gap.

1.1.4 Densities of States

With the inclusion of further interactions, the wavefunction is no longer localized as in the molecular model. One cannot speak of point charges any more. The evaluation of the charge on one atom now requires a determination of the density of states. For electronic systems, the density of states is taken with respect to energy. Different densities of states may be of interest.

The integral density of states $N(E)$ is given in terms of the density of states $n(E)$ by

$$\begin{aligned} N(E) &= \int^{E} n(E')dE' \,; \\ n(E) &= \frac{1}{N_0}\sum_{\boldsymbol{k},\nu}\delta(E-E_{\boldsymbol{k},\nu}) \\ &= \frac{V}{(2\pi)^3}\sum_{\nu}\iint_{S(E)}\frac{dS}{|\text{grad}_{\boldsymbol{k}} E_{\boldsymbol{k},\nu}|} \end{aligned} \tag{1.28}$$

where ν indexes the band, N_0 is the total number of atoms, V the volume per atom and $S(E)$ a constant energy surface. The sum over ν in (1.28) may be split to give the partial densities of states for a given band, $n_\nu(E)$:

$$n_\nu(E)=\frac{V}{(2\pi)^3}\iint_{S(E)}\frac{dS}{|\text{grad}_{\boldsymbol{k}} E_{\boldsymbol{k},\nu}|} \,, \tag{1.29}$$

which is of particular interest when discussing properties related to specific symmetries (optical properties for instance). In Chap. 2 we shall introduce a local partial density of states on one atom, which is formally equivalent to a point charge evaluation. It is most conveniently determined using a Green's function formalism. It may however be evaluated directly from the band dispersion relations $E(\boldsymbol{k})$ by a simple Fourier back-transform.

1.2 Two-Dimensional Lattices

We now introduce some basic notions necessary for understanding the properties of surfaces. The definition of the perfect two-dimensional lattice will be followed by a discussion of geometrical aspects associated with atomic relaxation of the surface. Surface-related electronic and acoustic properties will then be described.

1.2.1 The Two-Dimensional Lattice of a Perfect Surface

In this section, we first recall the general properties of a two-dimensional periodic system. Perfect surfaces have a translation symmetry: they are periodic in two dimensions. Atoms on a free surface are connected on only one side of the surface plane to atoms of the bulk. As a starting point we shall consider only a plane of atoms, forgetting the bonds to the bulk crystal. The primitive cell is defined by primitive translation vectors $(\boldsymbol{a}_i, i = 1,2)$. As an example, primitive cells of a square lattice and of a hexagonal lattice are shown in Fig. 1.3. This defines the direct space. The symmetries of the surface form the point group [1.4].

As is well known from the study of bulk crystals, the diffraction and electron conduction properties of periodic systems depend directly on the Fourier transformed space, the reciprocal space. The relations between direct and reciprocal space properties can be simply deduced from those known for three-dimensional solids. Let $\boldsymbol{a}_1$ and $\boldsymbol{a}_2$ be the surface primitive vectors in the plane, with $\boldsymbol{a}_3$ being a vector of arbitrary length t perpendicular to the plane of the surface. If

$$\boldsymbol{n} = \frac{\boldsymbol{a}_1 \wedge \boldsymbol{a}_2}{\|\boldsymbol{a}_1 \wedge \boldsymbol{a}_2\|}, \tag{1.30}$$

then $\boldsymbol{a}_3 = t\boldsymbol{n}$.

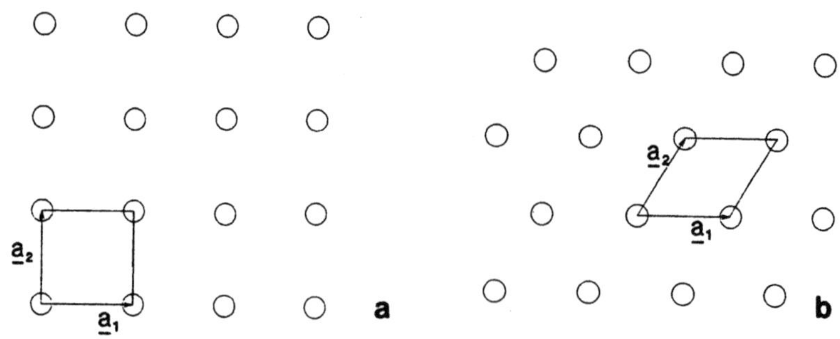

Fig. 1.3. Primitive cells in two dimensions. Primitive cells of a square (a) and a hexagonal (b) lattice

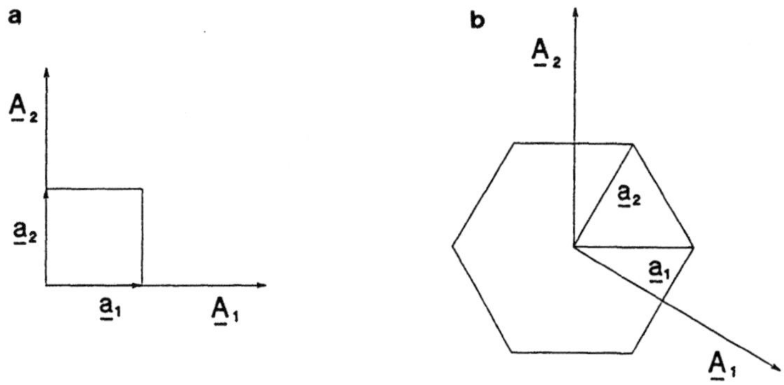

Fig. 1.4. Reciprocal space in two dimensions. Reciprocal spaces of the square (a) and hexagonal (b) lattices

One can express the reciprocal space primitive translations as in the three-dimensional case and then let t tend to infinity. Thus the reciprocal space primitive vectors defined by

$$\boldsymbol{A}_i \cdot \boldsymbol{a}_j = 2\pi\delta_{ij} \tag{1.31}$$

become

$$\begin{aligned}
\boldsymbol{A}_1 &= 2\pi\frac{\boldsymbol{a}_2 \wedge \boldsymbol{n}}{\boldsymbol{a}_1 \cdot (\boldsymbol{a}_2 \wedge \boldsymbol{n})}\,, \\
\boldsymbol{A}_2 &= 2\pi\frac{\boldsymbol{n} \wedge \boldsymbol{a}_1}{\boldsymbol{a}_1 \cdot (\boldsymbol{a}_2 \wedge \boldsymbol{n})}\,, \\
\boldsymbol{A}_3 &= \lim_{t\to\infty} \frac{2\pi}{t}\boldsymbol{n}\,.
\end{aligned} \tag{1.32}$$

This shows that the reciprocal lattice of a two-dimensional periodic lattice is a periodic array of straight lines all parallel to $\boldsymbol{n}$. The primitive vectors in the plane are $\boldsymbol{A}_1$ and $\boldsymbol{A}_2$. The projection of the reciprocal space on the surface plane is shown in Fig. 1.4 for the square and the hexagonal lattices. They are respectively a square and a hexagonal lattice.

The Hamiltonian of the surface then has the translation periodicity of the surface. Thus the eigenvalues obey Bloch's theorem and may be written:

$$\Psi(\boldsymbol{r}) = \mathrm{e}^{i\boldsymbol{k}\cdot\boldsymbol{r}} u_{\boldsymbol{k}}(\boldsymbol{r}) \tag{1.33}$$

where $u_{\boldsymbol{k}}(\boldsymbol{r})$ is a periodic function of $\boldsymbol{r}$.

The periodic properties in direct space translate simply into periodic properties of the eigenvalues in the reciprocal space. If $\Psi_{\boldsymbol{k}}(\boldsymbol{r})$ is an eigenfunction of H for the vector $\boldsymbol{k}$, it is also an eigenfunction for the vector $\boldsymbol{G} + \boldsymbol{k}$, where $\boldsymbol{G}$ is a vector of the reciprocal lattice. It is thus sufficient to restrict the study of the eigenvalues of H to one period of the reciprocal lattice. We can define the smallest region of interest by taking the Wigner–Seitz cell of the reciprocal space, also known as the first Brillouin zone. Figure 1.5 shows the Brillouin zones of the square and hexagonal lattices.

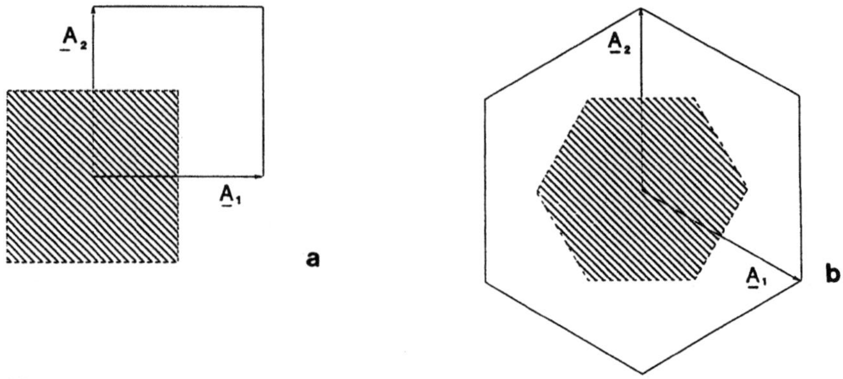

Fig. 1.5. Brillouin zones of two-dimensional lattices: square (a), hexagonal (b)

Finally the surface is usually considered as infinitely extended in both directions of the surface plane. For calculational purpose, one may thus use the periodic (Born–Von Kármán) boundary conditions. The conditions on the eigenfunctions are:

$$\Psi_{\boldsymbol{k}}(\boldsymbol{r}) = \Psi_{\boldsymbol{k}}(\boldsymbol{r} + N_1\boldsymbol{a}_1) = \Psi_{\boldsymbol{k}}(\boldsymbol{r} + N_2\boldsymbol{a}_2)\,, \tag{1.34}$$

where N_1 and N_2 are the number of unit cells in each direction of the plane. Writing $\Psi_{\boldsymbol{k}}(\boldsymbol{r})$ as a Bloch function shows that the $\boldsymbol{k}$ vectors allowed by the boundary conditions are thus quantized and can be written

$$\boldsymbol{k} = \frac{P_1}{N_1}\boldsymbol{A}_1 + \frac{P_2}{N_2}\boldsymbol{A}_2 + \alpha\boldsymbol{n}\,. \tag{1.35}$$

1.2.2 Surface-Induced Features

To study the effect of the surface on the electronic levels, we may start from the molecular model developed above. We have shown in the molecular model that the coupling between orbitals of adjacent atoms leads to a bonding and an anti-bonding level per bond in the bulk. For nearest neighbor interactions, the surface may be introduced simply by setting the hopping integral (1.7) equal to zero for all bonds crossing the surface plane. For instance, on the (111) surface of a tetrahedrally coordinated homopolar crystal, one orbital per surface atom would not be connected to a neighbor. The electronic levels of the semi-infinite solid are then the same bonding and anti-bonding levels as in the bulk together with an atomic level at the energy of the free atom. This latter state has a degeneracy equal to the number of surface dangling bonds. It is called a surface state. The inclusion of further interactions will be studied in Sect. 2.4. It will be shown that the surface state is broadened into a band of width $N_s\beta^*$ where N_s is the number of first surface neighbors and β^* is an effective interaction energy between dangling bonds of first neighbors on the surface. This effective interaction is usually very small compared to those in the bulk. Thus the surface band is broadened very little in general compared to the bulk band widths.

Another way of viewing the surface stems from a nearly-free-electron analysis. Let us start from the electron in a three-dimensional box. The energy levels are then given by

$$E(\boldsymbol{k}) = \frac{\hbar^2 \boldsymbol{k}^2}{2m} \tag{1.36}$$

and the wavevector is quantized in analogy to (1.35).

The addition of a small periodic potential may be studied in a perturbation framework. We shall use atomic units ($\hbar = e = m = 1$) in this section. Let $|\boldsymbol{k}\rangle$ represent a plane wave of wavevector $\boldsymbol{k}$. It is an eigenvector of the free electron Hamiltonian with an eigenvalue $E^0_{\boldsymbol{k}}$. Then let V be a periodic potential of small amplitude. The order of magnitude of the amplitude should be such that $|V| \ll (G_0/2)^2/2$ where $G_0/2$ is the wavevector at a Brillouin zone edge. V may be expanded in a Fourier series. Then

$$\begin{aligned} \langle \boldsymbol{k}|V|\boldsymbol{k}'\rangle &= \frac{1}{\Omega}\int_\Omega V e^{i(\boldsymbol{k}'-\boldsymbol{k})\cdot \boldsymbol{r}} d_3\boldsymbol{r} = \frac{1}{\Omega}\sum A_{\boldsymbol{k}} \int_\Omega e^{i(\boldsymbol{k}'-\boldsymbol{k}+\boldsymbol{G})\cdot \boldsymbol{r}} d_3\boldsymbol{r} \\ &= \sum_{\boldsymbol{G}} A_{\boldsymbol{G}} \delta(\boldsymbol{k}' - \boldsymbol{k} + \boldsymbol{G}) \, . \end{aligned} \tag{1.37}$$

As long as the non-perturbed states, $E^0_{\boldsymbol{k}}$ and $E^0_{\boldsymbol{k}+\boldsymbol{G}}$ are non-degenerate, one may apply simple second-order perturbation theory to yield

$$E_{\boldsymbol{k}} = E^0_{\boldsymbol{k}} + V_0 + \sum_{\boldsymbol{k}'} \frac{|\langle \boldsymbol{k}|V|\boldsymbol{k}'\rangle|^2}{E_{\boldsymbol{k}} - E_{\boldsymbol{k}'}} \, , \tag{1.38}$$

which shows that over most of the band structure, the perturbation induced by the periodic potential is very weak apart from a constant shift, V_0. The interesting points are the Brillouin zone edges where $E^0_{\boldsymbol{k}}$ and $E^0_{\boldsymbol{k}+\boldsymbol{G}}$ are degenerate. In this case, the perturbed eigenvalues are solutions of the determinant

$$\begin{vmatrix} E^0_{\boldsymbol{k}} + \langle \boldsymbol{k}|V|\boldsymbol{k}\rangle - E & \langle \boldsymbol{k}|V|\boldsymbol{k}+\boldsymbol{G}\rangle \\ \langle \boldsymbol{k}|V|\boldsymbol{k}+\boldsymbol{G}\rangle^* & E^0_{\boldsymbol{k}} + \langle \boldsymbol{k}|V|\boldsymbol{k}\rangle - E \end{vmatrix} = 0 \tag{1.39}$$

and are given by

$$E_{\boldsymbol{k}} = E^0_{\boldsymbol{k}} \pm |\langle \boldsymbol{k}|V|\boldsymbol{k}+\boldsymbol{G}\rangle| \, . \tag{1.40}$$

This shows the opening of gaps at the zone edges due to the potential. Since we expect from tight binding that surface states exist in the gaps, it is then interesting to have a closer look at these regions of energy. This may be done exactly in the one-dimensional case. Let z be the real space coordinate along the one-dimensional axis and k the wavevector in one dimension. We may expand the eigenvectors of the nearly-free-electron Hamiltonian as a Bloch function:

$$|\Psi\rangle = \sum_{G \in \text{ reciprocal lattice}} \alpha_G e^{i(G+k)\cdot z} \, . \tag{1.41}$$

The Schrödinger equation may then be written, using a periodic potential expanded in a Fourier series as in (1.37),

$$\tfrac{1}{2}\sum_{G}\alpha_G(G+k)^2 e^{i(G+k)\cdot z}+\sum_{G,G'}\alpha_G A_{G'}e^{i(G'+G+k)\cdot z}=E\sum_{G}\alpha_G e^{i(G+k)\cdot z}. \quad (1.42)$$

Dividing both sides of (1.42) by e^{ikz}, and projecting onto e^{iGz} one obtains the following linear system in α

$$\alpha_G\left[\frac{(G+k)^2}{2}-E\right]+\sum_{G'}\alpha_{G'}A_{G-G'}=0\,. \quad (1.43)$$

We solve this for the simplest situation where V is a sinusoid, $V=V_0(e^{iG_0z}+e^{-iG_0z})$. To obtain a more symmetrical final expression, we may also fix the origin of wavevectors at the zone edge $G_0/2$. Then $\Psi=\alpha_+e^{i(k+G_0/2)z}+\alpha_-e^{i(k-G_0/2)z}$; α_+ and α_- are solutions of (1.43) only for values of E for which

$$\begin{vmatrix}\frac{1}{2}\left(k+\frac{1}{2}G_0\right)^2-E & V_0\\ V_0 & \frac{1}{2}\left(k-\frac{1}{2}G_0\right)^2-E\end{vmatrix}=0 \quad (1.44)$$

This may be further simplified by choosing the origin of energies at the free-electron-zone-edge value, $(G_0/2)^2/2$ and by restricting k to the vicinity of the zone edge. Then k^2 may be neglected in relation to kG_0. The determinant of (1.44) then reduces to

$$\begin{vmatrix}\frac{1}{2}kG_0-\varepsilon & V_0\\ V_0 & -\frac{1}{2}kG_0-\varepsilon\end{vmatrix}=0$$

whose solutions are $\varepsilon_\pm=\pm\sqrt{V_0^2+1/4k^2G_0^2}$. Let us now write k as a function of ε:

$$k^2=4\frac{(\varepsilon^2-V_0^2)}{G_0^2}\,. \quad (1.45)$$

It is then apparent that in the zone edge gaps ($\varepsilon<V_0$), there exist solutions to the Schrödinger equation with imaginary wavevector (see Fig. 1.6). These solutions are inacceptable in the bulk since they would diverge. However, in presence of a surface, they are acceptable since they may be bounded by the surface. These

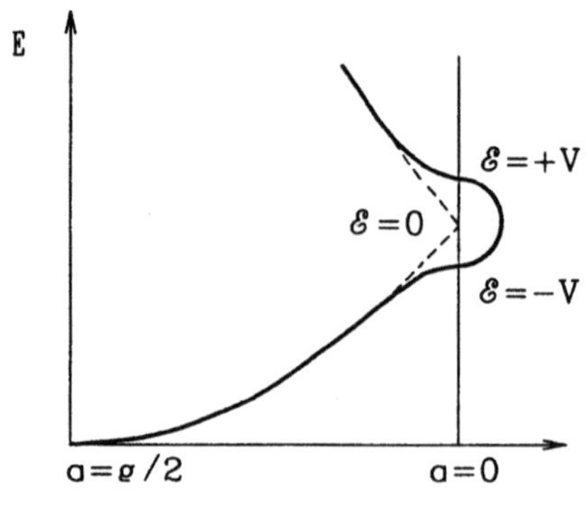

Fig. 1.6. Dispersion relation of a nearly free electron system. The dotted line represents the folding of free-electron bands due to the periodicity, the full line shows the gap due to the lifting of the degeneracy at the zone edge. The half circle at $y>G_0/2$ corresponds to the analytic continuation of the eigenvalues for complex wavevectors

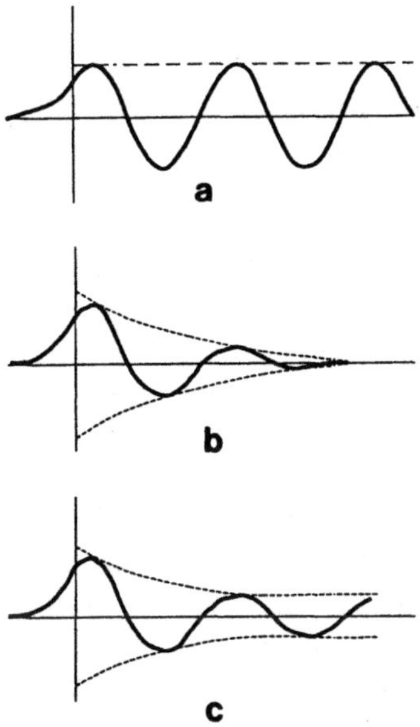

Fig. 1.7a–c. Bulk and surface states. Three different situation; (a) stationary bulk state normally ending on the surface, (b) true surface state decreasing exponentially towards the bulk, (c) surface resonance

states decrease exponentially into the bulk of a one-dimensional crystal. They are surface states.

The three-dimensional case cannot be treated so easily in an analytical way. However, it is not difficult to predict the main features of the three-dimensional problem. We have seen that surface states may only exist in a forbidden gap of the bulk band structure. Such a gap may exist in one direction in reciprocal space but be in an allowed band in other directions. Two main cases are then possible:

- The vanishing state is in a true gap (forbidden energies in all k space directions). The surface state is then a true surface state and may not connect to the bulk.
- There exists one direction in k-space where the energies corresponding to the gap are allowed bulk states. One then observes a surface resonant state which may connect to a bulk state. The surface density of states however is higher than in the bulk.

These situations are depicted in Fig. 1.7.

In this section we have introduced the notion of a surface state. Such states are very important for determining the physics of surfaces and interfaces, especially in the case of semiconductors. Indeed, surface states, although of infinitely small density compared to bulk states, fall in the forbidden gap. The filling of such states may significantly affect the electrostatics of the semiconductor by depleting the near surface region of its majority carriers. This is essential in determining the transport properties of heterojunction structures. We shall make repeated use of the concept of surface states in the remainder of this book.

1.3 Geometrical Structure of "Real" Surfaces

How does the surface periodicity relate to the bulk periodicity? Traditionally, surfaces structures have been studied mostly by means of diffraction techniques. Thus, the reciprocal space properties of the surface (or more precisely of the semi-infinite crystal) have been the most apparent. Two cases could be clearly distinguished: either the semi-infinite crystal showed no new features compared to the bulk diffraction pattern, or novel features did appear. In the first case the surface is said to be relaxed, but not reconstructed. In the latter case it is also reconstructed. Theoretically, these two cases do not call for different concepts.

The reconstruction of a surface can lead to a two-dimensional primitive cell that is an integer multiple of the projection of the bulk unit cell onto the surface plane. In some cases the surface periodicity can be an irrational multiple of the projected bulk periodicity. We shall make use of standard conventions to describe reconstructions. The projection of the bulk unit cell onto the surface defines two generating vectors $\boldsymbol{a}_1$ and $\boldsymbol{a}_2$ such that $(\boldsymbol{a}_1, \boldsymbol{a}_2, \boldsymbol{n})$, where n is the surface normal, forms an orthonormal basis. If the unit cell of a reconstructed surface is n_1 times larger along $\boldsymbol{a}_1$, and n_2 times larger along a_2, we speak of an $(n_1 \times n_2)$ reconstruction. The surface unit cell may also be centered, in which case the structure is denoted by $c\,(n_1 \times n_2)$. A further possibility is a unit cell that is rotated with respect to the vectors $(\boldsymbol{a}_1, \boldsymbol{a}_2)$ by an angle θ. This is described by $(n_1 \times n_2)R\theta$. Some examples of these classes of reconstruction will be encountered in the next section.

1.3.1 Examples of Relaxed and Reconstructed Surfaces of Semiconductors

In the following, we give some examples of relaxed and reconstructed surfaces. The examples are mostly of semiconductor surfaces, where the covalency of bonds leads to well-defined reconstructions. We shall come back in a later chapter to the physics behind these structural changes.

a) The Relaxed GaAs(110) Surface

The best known relaxed surface is probably the (110) surface of GaAs, on which very detailed studies have been made over the last fifteen years. We show in Fig. 1.8 the unrelaxed (a, b, c) and relaxed (d) (110) surfaces of GaAs. As is evident from a simple ball and rod model, the relaxation can be achieved through an angle relaxation with very little change in bond lengths. There is a good consensus from many experimental determinations that the bond angle lies in the range 25–29°. The second and third layer distortions are less well established, however. Most probably the second layer As atoms move inwards by 0.06 Å whereas the second layer Ga rise by 0.06 Å [1.5].

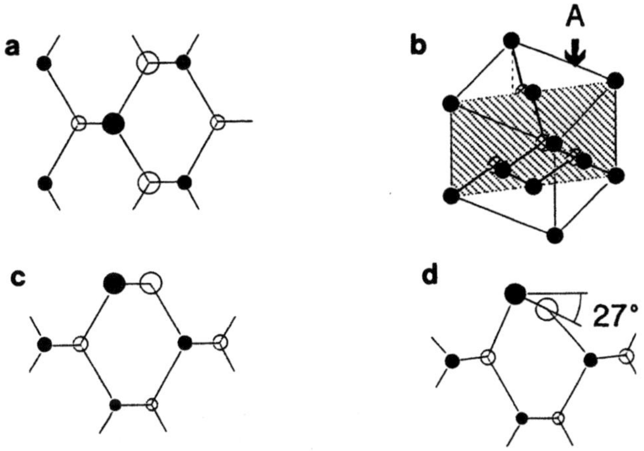

Fig. 1.8a–d. Relaxation of the GaAs surface. (a) and (c) show the top and side views of the unrelaxed surface, (b) the intersection of the (110) plane with the GaAs bulk unit cell and (d) the side view of the relaxed (110) surface

b) The (100) Surface of Si

In low energy electron diffraction experiments the Si(100) surface shows a 2 × 1 reconstruction. Theoretically, it has been studied in detail including subsurface strains of up to five layers beneath the surface [1.7]. A simpler and maybe more reliable view is the asymmetric dimer model of *Chadi*. The ideal (100) diamond or zinc-blende surface is composed of parallel rows of atoms with two broken bonds. The atoms on two adjacent rows are quite close to each other. This can be seen in Fig. 1.9. Two adjacent broken bonds can combine to give a dimer type of level at the cost of a deformation energy needed to draw the atoms towards each other (Fig. 1.10). This corresponds indeed to a minimum in the total surface energy as computed by *Chadi*. It is furthermore possible to deform the dimer to make an asymmetric dimer as shown in Fig. 1.11. This lowers the energy of the surface even more.

This model was also found to be a good description of the 4 × 2 and $c4 \times 4$ reconstructions of the (100) GaAs surface [1.8]. The doubling of the periodicity from the 2 × 1 was suggested in the initial papers to be due to a phasing of the asymmetric dimers. More recently, tunneling electron microscope experiments showed that the doubling of the periodicity is related to the absence of one of the dimer rows [1.9]. This was predicted from simple electrostatic arguments long ago [1.10] but confirmed experimentally only very recently.

The last semiconductor surface that we want to detail is Si(111). This surface is known to reconstruct in different ways depending on its preparation. We shall describe in turn the 2 × 1 and the 7 × 7 reconstruction. Also, in presence of an adsorbate such as Al, the surface can show a $\sqrt{3} \times \sqrt{3}$ reconstruction.

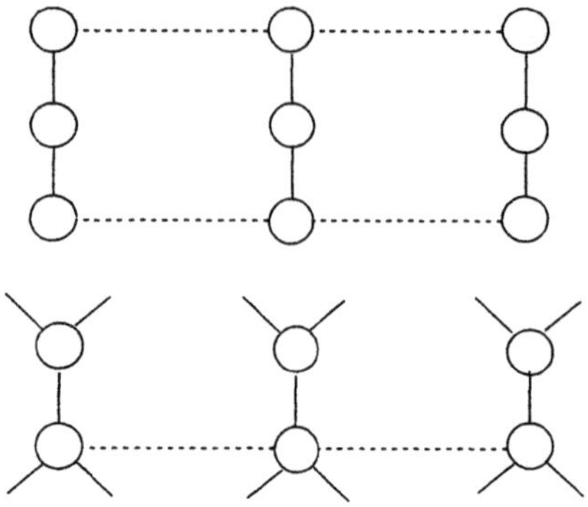

Fig. 1.9. Si(100) ideal surface. Top (*upper part*) and side (*lower part*) views of the ideal (100) diamond structure

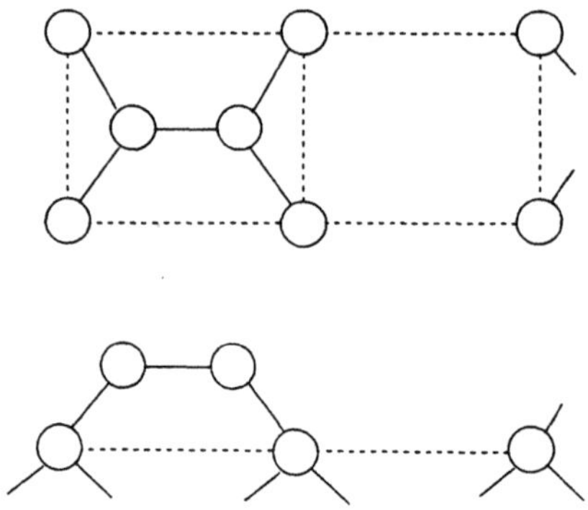

Fig. 1.10. Si(100) symmetric dimer. Top (*upper part*) and side (*lower part*) view of the symmetric dimer geometry

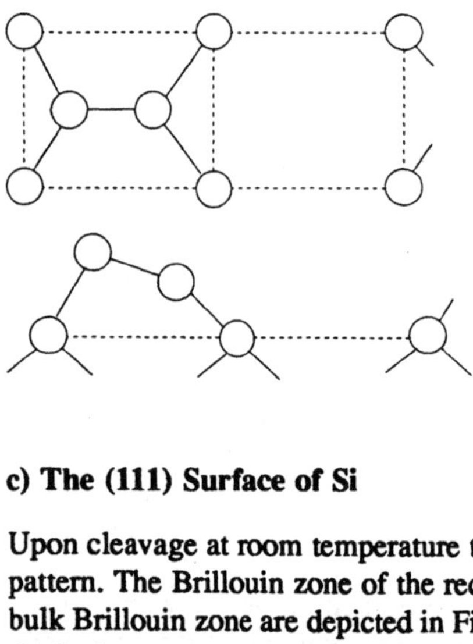

Fig. 1.11. Si(100) 2×1 reconstruction. Top (*upper part*) and side (*lower part*) view of the asymmetric dimer geometry of the 2×1 reconstruction after [1.6]

c) The (111) Surface of Si

Upon cleavage at room temperature the (111) face of Si reconstructs with a 2×1 pattern. The Brillouin zone of the reconstructed surface and the projection of the bulk Brillouin zone are depicted in Fig. 1.12. This surface has been studied extensively by angle-resolved photoemission which shows some very strong surface features. The first model that was proposed assumed a buckled surface where the atoms of the top layer were alternately raised and lowered [1.11]. This, however, was incompatible with dynamic calculations of low energy electron diffraction intensities, with the large core level shifts (3 eV) observed in photoemission and with the energy band dispersion observed in angle-resolved photoemission.

The π-bonded chain was proposed by *Pandey* [1.13]. It is shown in Fig. 1.13 together with the buckling model. The excellent agreement between the calculated and measured band dispersion for the π-bonded chain model is shown in Fig. 1.14. The π-bonded chain is the only model investigated that shows the strong surface state dispersion seen on the figure. All other models require this dispersion to be interpreted as two different surface states.

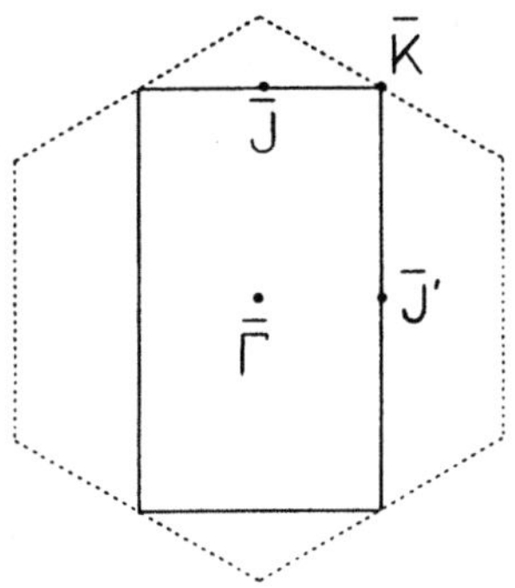

Fig. 1.12. Surface Brillouin zone of Si(111) 2 × 1. Superposition of the 1 × 1 (*dotted line*) and 2 × 1 (*full line*) surface Brillouin zones for Si(111)

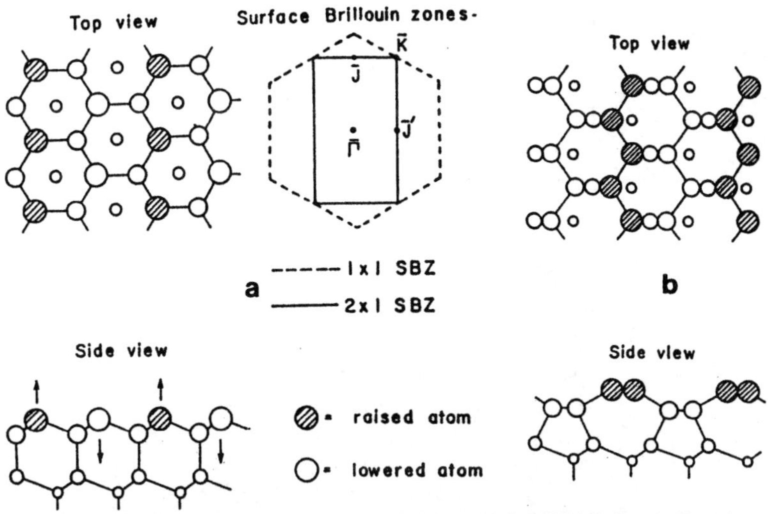

Fig. 1.13. The buckling and the π-bonded chain model for Si(111) 2 × 1. Top view and side view of the buckling model (*left*) and the π-bonded chain (*right*) [1.12]

The 7×7 reconstruction diffraction pattern has been known since 1969 [1.14], but it is only recently [1.15] that a consensus seems to have evolved about the structure. This reconstruction is obtained experimentally from the 2 × 1 by annealing the sample at 380°C. The primitive cell contains 98 atoms. Thus a determination of the surface structure by dynamic calculations of the diffraction intensities for various possible geometries is not feasible. Results from a variety of ultra high vacuum transmission electron microscopy techniques eventually converged to yield the proposed structure. The cell consists of two triangular units that are similar apart from a stacking fault in one of them. In Fig. 1.15 the stacking fault is in the left triangle. The sequence of planes in the right triangle is normal ("bCcA" up to the last two planes, "aB" for the last two planes). For the left triangle, the stacking fault appears between the two final planes (same sequence up to the last two planes, but with "aC" for the last planes). The connection between the two faulted and unfaulted triangles is through the "a"

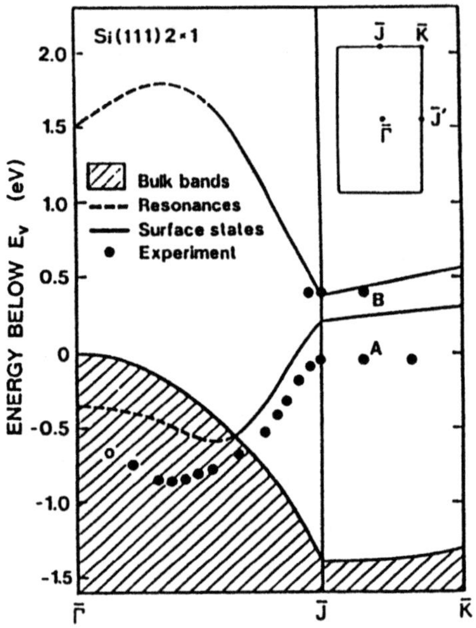

Fig. 1.14. Band dispersion for the π-bonded chain model. Comparison between experimental and theoretical results for the surface-state dispersion on Si(111) 2 × 1. Theory is in full lines, data points are discrete symbols. From [1.12]

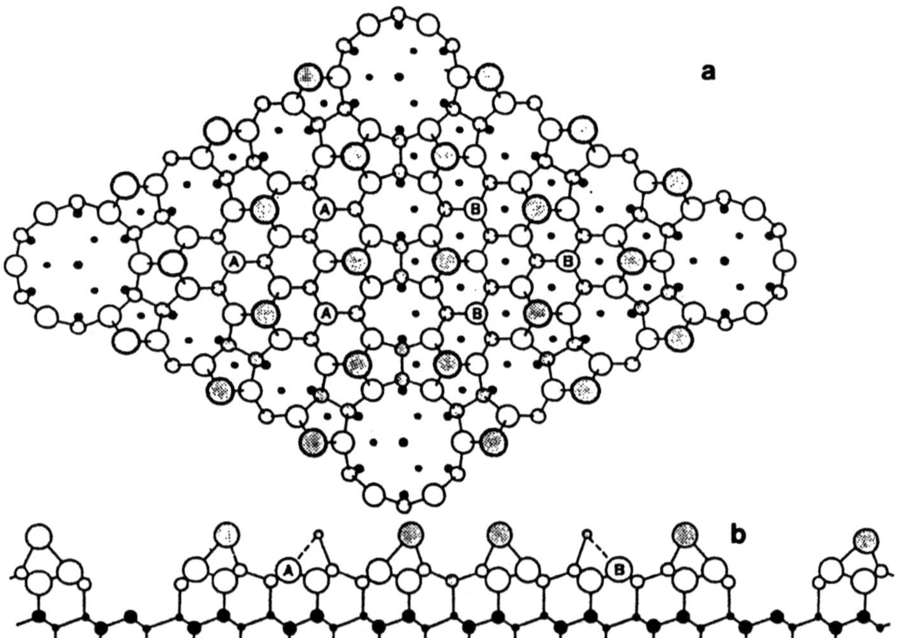

Fig. 1.15a,b. The dimer-adatom-stacking fault model of the 7 × 7 Si(111). Top and side views of the DAS model [1.14]. (a) In the top view, the size of the atoms indicates their proximity to the surface. The bold atoms are the first layer of undistorted bulk. The structure contains: 1. dimers (slanting hashes), 2. adatoms (12 largest circles), 3. stacking faults which can be seen on the left triangle where the first layer of bulk atoms is not visible. (b) The side view is a projection on the $(10\bar{1})$ lattice plane. Atoms along the diagonal of the 7 × 7 unit cell are the largest circles. The stacking fault is apparent in this view too

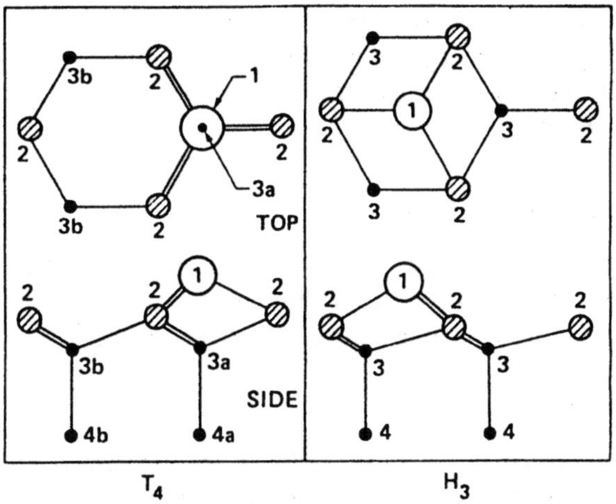

Fig. 1.16. Alternative models for the $\sqrt{3} \times \sqrt{3}$ adsorbate induced reconstruction of Si(111). On the right, the so-called $H3$ model where the metal adatom is placed next to a normal diamond lattice position. On the left, in the model $T4$, the metal atom is above the second layer Si atom. Top (*above*) and side (*below*) views are shown. The figure contains enough primitive cells to see the $\sqrt{3}$ relation with the substrate

plane of atoms. To compensate the absence of two connecting "B" type atoms, the "a" type atoms on the sides of the cell dimerize. There are thus 9 dimers in the cell. Above the last plane are 12 adatoms. Finally, the corners of the cell may or may not have a vacancy. The figure shows a vacancy.

To conclude our brief survey of the Si(111) face, which is rich in different reconstructions, we would like to show one example of adsorbate-induced reconstruction. When covered with a metal, the Si(111) surface exhibits a number of different reconstructions. For example, the $(\sqrt{3} \times \sqrt{3})$ Ag, Al, Ga and In, the $(\sqrt{7} \times \sqrt{7})$Al or the $(\sqrt{19} \times \sqrt{19})$Ni, but this is not an exhaustive list. The $(\sqrt{3} \times \sqrt{3})$Al was one of the first observed [1.16]. Some recent first-principles calculations [1.17] and photoemission measurements [1.18, 19] have been used to decide between one of two geometries for the Al, Ga and In covered surfaces. In Fig. 1.16 we show the alternative positions used in these studies. The reconstruction is mainly due to the adatom geometry, to which one may add some relaxation of the Si network under the surface. Both reconstructions proposed in this figure saturate the dangling bond. The so-called $H3$ reconstruction has an adatom placed as in the usual diamond cell, whereas the $T4$ structure has the adatom above the second layer Si. We shall discuss the reconstruction for the case of an Al adatom.

Total energy calculations for each geometry in an unrelaxed lattice show the $H3$ configuration to be more stable than the $T4$. However, significant repulsion exists between the adsorbate atom and the Si atom of the second layer in the $T4$. This Si atom is the nearest neighbor of the adatom in the unrelaxed case. The adatom-to-first-layer Si distance is 4.72 a.u. whereas it is 4.63 a.u. to the second layer Si. Thus must one let the near-surface atoms relax to their equilibrium positions. In this process, the Si atoms of the first layer come nearer to the adatoms whereas the second and third layer Si undergo a large downward displacement. The effect of this relaxation is such that the $T4$ structure then becomes more stable than the $H3$ by 0.3 eV/adatom.

1.3.2 Relaxations and Reconstructions of Metals

Experimentally, metals exhibit a large variety of reconstructions and relaxations. We summarize some results on perpendicular displacement of layers (inward relaxation and periodic relaxation) as well as incommensurate reconstructions.

a) Inward Relaxation of Metals

Many metals show an inward relaxation on open surfaces as determined by LEED. These are the fcc(110), bcc(001) and (111). On other surfaces little or no relaxation occurs. The inward relaxation is often of the order of −9% for the (110) surface of fcc metals: Ag(110) [1.20], Al(110) [1.21], Cu(110) [1.22], Ni(110) [1.23]. It can be as large as −15% for Fe(111) [1.24].

Pair potentials (such as Lennard–Jones potentials) predict an expansion. For such potentials the bulk interatomic distance is smaller than the radius of the minimum pair interaction energy. This is due to the fact that further neighbors fall in the attractive part of the potential. The presence of the surface reduces the number of neighbors and thus drives the atoms outwards [1.25]. This gives a good description of rare gas solids but is not well suited for metals where large electronic rearrangements can take place near the surface.

For sp-bonded crystals, this electronic rearrangement may be viewed in simple terms [1.26]. The electrons near the surface have a high surface tension and tend to smooth the initial bulk Wigner–Seitz cell (Fig. 1.17). This reduces the surface area. It also displaces the electrical center of gravity of the negative charges in one Wigner–Seitz cell towards the bulk. The positive ion follows this displacement as the result of electrostatic forces. This rather crude argument is

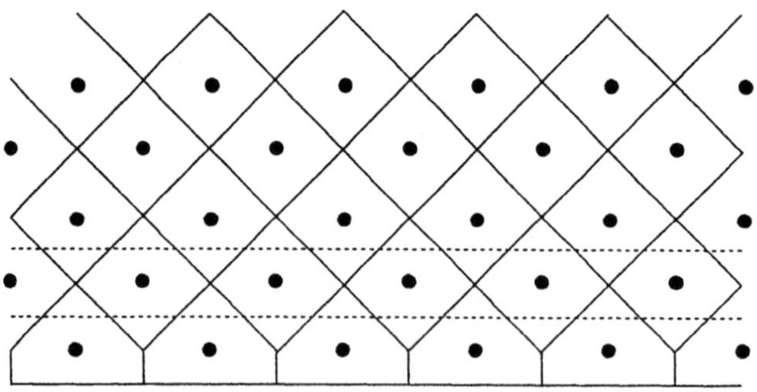

Fig. 1.17. Smoothing of the electron density near the surface; from [1.26]. Case of two-dimensional square lattice where the electron density is assumed to terminate uniformly at the surface instead of filling a square unit cell

in good qualitative agreement with experience. It predicts larger contractions for the open surfaces as is actually observed. The calculated values [16% for the (110) surface of fcc metals] are somewhat in excess of the experimental ones, however.

b) Multilayer Relaxation

LEED measurements on a number of metals have shown a damped oscillatory behavior for the interlayer spacing. Examples are cited in Table 1.1. The most usual case which is found for (100) and (110) surfaces exhibits a (− + −+) behavior but the sequence (− − +−) is observed near the bcc Fe(111) surface. The damping of the oscillation is small for an open surface such as the fcc(110) but large for a dense surface such as the bcc(100).

Table 1.1. Multilayer relaxation: Interlayer spacing variations for Al(110) [1.21], Cu(110) [1.22, 27], Ni(110) [1.23, 28], Al(311) [1.29], and Fe(211) [1.30].

	*d*1	*d*2	*d*3
Al(110)	− 9%	+5%	−2%
Cu(110)	− 9%	+2%	
Ni(110)	− 9%	+2%	
Al(311)	−12%	−4%	+9%
Fe(211)	−10%	+5%	−1%

This relaxation occurs even for charge densities without Friedel oscillations. First principles calculations yield good agreement with experimental values but fail to predict trends since they must be carried out separately for each individual case. A recent publication gives some simple models to account for this effect [1.31, 32].

The simplest model is the point-ion model, which is an extension of the *Smoluchowski* [1.26] model. Instead of limiting the relaxation to a perpendicular movement of the surface layer one can extend it to a larger number of Wigner cells. The electrostatic energy is split into an intralayer term and an interlayer term [1.33]. The intralayer tries to draw the nucleus towards the center of mass of its electronic energy charge. The interlayer interaction resembles a Madelung energy. This model supposes a uniform electronic charge which is certainly not the case. It can only converge to physically acceptable results by multiplying the intra-atomic term by a factor larger than 1 [1.33]. This is done for one surface and is assumed to be transferable to other surfaces. The agreement for fcc Al and bcc Fe is good for the perpendicular displacements. The sequences (− + −+) for Al and (− − +−) for Fe are correctly predicted.

More elaborate models account for the oscillating relaxation and give some insight into the origin of the relaxation. We shall come back to this point in Sect. 3.3.3 using the minimization of the total energy, and in Sect. 8.3.1 in terms of an asymptotic phonon at frequencies tending towards zero.

c) Reconstruction of Transition Metals

We give a slightly more detailed description of transition metal reconstructions since these will be explicitly studied in Chap. 3.

Low Temperature Reconstruction of the W(001) Surface. The W(001) surface exhibits a smooth phase transition from a high temperature 1×1 to a low temperature $\sqrt{2} \times \sqrt{2}$ structure when cooled from 300 K to below 200 K. This transition is reversible. The LT reconstruction has a two-fold symmetry instead of the four-fold bulk symmetry. Investigations by LEED [1.34, 35] are compatible with domains of different orientation. In these domains, the surface W atoms are displaced in the [110] direction (or the $[1\bar{1}0]$ direction in the other orientation) to produce chains as shown in Fig. 1.18. The lateral displacement is 0.2 Å and the surface layer interspacing is contracted by 6%.

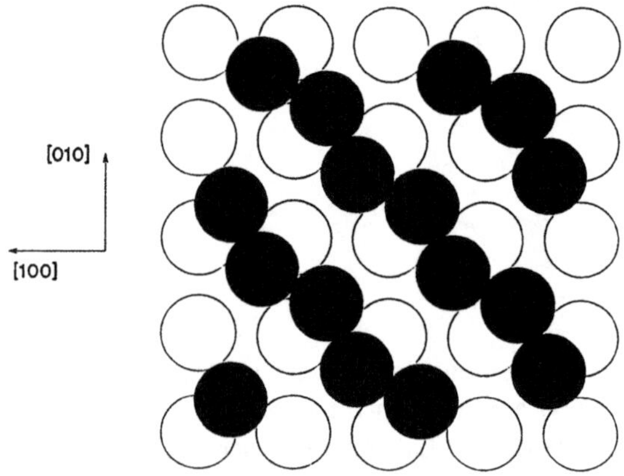

Fig. 1.18. Low temperature $\sqrt{2} \times \sqrt{2}$ structure of the W(001) surface; from [1.34]. Filled circles: surface atoms; open circles: atoms of the first sub-surface layer

Low Temperature Reconstruction of Mo(001). The reconstruction of Mo is very similar to that of W as might be expected from their comparable electronic structure (half-filled d-band). It also has a smooth transition from a 1×1 to a reconstructed surface when cooled. However the LEED spots due to reconstruction are split into four. This stems from the coexistence of domains in which a cosine modulation of the lateral displacement occurs with a wavevector (8/9,8/9) insteand of (1,1). It is likely that the Mo reconstruction resembles that of W but with a long wavelength modulation. A detailed study of the LEED intensities suggests that the modulation wavevector and the lateral displacement are parallel [1.36].

Exercises

1.1 *Parameters of the molecular model:* Determine the numerical values of these parameters for the III-V and II-VI compounds in the following way:

a) For α_A^0 and α_B^0 take the sp^3 averages of the free atom levels as discussed in [1.37].
b) Express β in terms of the interactions β_{ss}, $\beta_{s\sigma}$, $\beta_{\sigma\sigma}$ between s and p nearest neighbor atomic orbitals defined in Sect. 2.1.2. Give the result in terms of the interatomic distance by using Harrison's rules (4.9).
c) How can one estimate J_A and J_B of (1.13) from the free atom properties?
d) Show that there is near cancellation of $J_A + J_B - \frac{2\alpha e^2}{R}$ for all compounds.
e) Assuming complete cancellation of this quantity, determine Phillips' ionicity $f = C/Eg$ and compare it with Phillips' empirical value [1.3].

1.2 *Band structure of covalent systems*: Draw the dispersion curves $E(\boldsymbol{k})$ obtained in the simple model of covalent systems described in Sect. 1.1.3, with the values $\Delta = -1.75\,\text{eV}$ and $\beta = -4.1\,\text{eV}$ appropriate for silicon. Compare these curves with the results of more refined calculations as given in all standard text books and draw general conclusions about the accuracy of the description.

1.3 *The 2D surface lattice*: Consider the 2D lattice formed by the atoms belonging to the surface planes of the (111) (100) and (110) faces of zinc-blende compounds. Give the basis vectors and the unit cell of these lattices. Determine the corresponding reciprocal lattice.

1.4 *The molecular model of Se and Te*: Selenium and tellurium have chain structures with an interbond angle θ that lies between the normal tetrahedral value of 109° and the simple cubic value of 90°. Show how it is possible to build a meaningful molecular model based on non-hybridized s and p atomic orbitals, when the difference $E_p - E_s$ between the corresponding atomic levels is large (for this it is necessary to make a judicious choice of the direction of the three p orbitals on each atom).
Numerical application: Take $E_s = -20.3\,\text{eV}$, $E_p = -9.5\,\text{eV}$ for Se and $E_s = -17.1\,\text{eV}$, $E_p = -8.6\,\text{eV}$ for Te with nearest neighbor interactions given by Harrison's formula (4.9), the interatomic distances being 2.32 Å for Se and 2.86 Å for Te.

Se and Te having 6 electrons on their valence shell, what is the filling of these molecular levels?

1.5 *The molecular model of SiO_2*: In most phases, SiO_2 is characterized by tetrahedral coordination of the Si atoms and a twofold coordination of the O atoms (the SiO-Si angle θ being $\sim$ 140°). Describe a molecular model based on sp^3 hybrids on Si and using the fact that the s atomic state of oxygen is much lower in energy than all other relevant atomic states.

Application: Take $E_s = -13.5\,\text{eV}$, $E_p = -6.5\,\text{eV}$ for Si and $E_s = -29.1\,\text{eV}$, $E_p = -14.1\,\text{eV}$ for O, the interatomic parameters being given by Harrison's rules (4.9), with an interatomic spacing of 1.54 Å. Show that two types of bonds occur, differing in the strength of their covalent character.

1.6 *Surface Friedel oscillations*: Consider a one-dimensional free electron gas in a square potential well such that

$$V(x) = 0 \quad \text{for} \quad 0 \le x \le L$$

$$V(x) = \frac{1}{2}k_0^2 \quad \text{for} \quad x \le 0\,, \quad x \ge L$$

This schematically represents the potential for conduction electrons in a metal.

- Show that, for electron energies $1/2\,k^2 \le 1/2\,k_0^2$ the wave function ψ_k is given by $A\sin(kx+\eta)$ within the well, and by $B\,\mathrm{e}^{qx}$ and $\pm B\,\mathrm{e}^{-q(x-L)}$ outside the well. Express q, $\eta(k)$, A and B in terms of k, k_0 and $L/2$ and show that $0 \le \eta \le \pi/2$ (use must be made of the normalization condition).
- Write down the condition determining the allowed values of k and express the density of states $n(k)$ in terms of $d\eta(k)/dk$.
- Assuming that L is large enough write the electron density as an integral over k up to k_{F} (where $1/2k_{\mathrm{F}}^2$ is the Fermi energy at 0 K).
- Show that, for $k_{\mathrm{F}} \ll k_0$ one can get fairly simple analytical results for $\varrho(x)$ near $x = 0$.
- Discuss the physical significance of such results.

1.7 *Surface relaxation with pair potentials*: We want to show here that the use of pair potentials is likely to produce a surface expansion. For this we consider the simplest case of a semi-infinite system of parallel and equidistant planes. We assume that its total energy can be written as a sum of pair potentials between nearest and second nearest neighbor planes.

Write the equilibrium condition for the bulk infinite system in terms of the derivatives of the pair potential at the nearest and next nearest neighbor distances. Use this to draw conclusions about their relative sign.

Assuming all lattice planes of the semi-infinite system to be fixed except the surface one, express the energy of this plane as a function of the distance Δ to the first subsurface plane.

Differentiate this energy with respect to Δ and show that one automatically gets an expansion of the system.

1.8 *Density of states for a linear chain in tight binding*: From the definition (1.28) and the expression for the energy dispersion relation for a tight binding s band linear chain find the density of states as a function of energy.

2. General Methods for Calculating the Electronic Structure of Surfaces

There has been much progress recently in the theoretical understanding of the electronic properties of surfaces and interfaces. Such progress is at least partly due to the development of more or less sophisticated methods of calculation. Among these, two of them dominate, differing in their basic philosophy: the local density plus pseudopotential method and the empirical tight binding approximation. The first one is a first principles theory; the second one aims at simulating defect or surface properties with the help of simple rules.

In the first part of this chapter we describe the general principles of each method, giving some comments about their accuracy and the improvements that can be expected in the near future. We then discuss different techniques that can be used in quantitative calculations of surface properties. We describe various ways of calculating surface Green functions in the tight binding case. Finally we give some simple illustrations: a semi-infinite chain and the (100) face of the simple cubic lattice in tight binding and the jellium surface in local density theory.

2.1 The Two Basic Theoretical Methods

We present here the two most common theoretical methods used in the calculation of surface properties: the local density plus pseudopotential method with its recent improvements and the empirical tight binding theory.

2.1.1 The Local Density Approximation and Its Improvements

This is an extension of the Thomas–Fermi approximation based on the *Hohenberg* and *Kohn* Theorem [2.1] which shows that the ground state properties of an electron system are entirely determined by the knowledge of its electron density $\varrho(\boldsymbol{r})$. The total energy of the interacting electron system can then be written:

$$E[\varrho] = T[\varrho] + \frac{1}{2}\int \frac{\varrho(\boldsymbol{r})\varrho(\boldsymbol{r}')}{|\boldsymbol{r}-\boldsymbol{r}'|} d\boldsymbol{r}\, d\boldsymbol{r}' + \int V_{\text{ext}}(\boldsymbol{r})\varrho(\boldsymbol{r})d\boldsymbol{r} + E_{\text{xc}}[\varrho] \quad (2.1)$$

where T represents the kinetic energy, the second term gives the electrostatic interelectronic repulsion, V_{ext} is the potential due to the nuclei and E_{xc} is the exchange and correlation energy. A variational solution of the problem [2.2] allows one to derive a set of one particle Schrödinger equations of the form:

$$\left[t + V_{\text{ext}} + \int \frac{\varrho(\boldsymbol{r}')d\boldsymbol{r}'}{|\boldsymbol{r}-\boldsymbol{r}'|} + V_{\text{xc}}(\boldsymbol{r})\right] \psi_k(\boldsymbol{r}) = \varepsilon_k \psi_k(\boldsymbol{r}) \tag{2.2}$$

with

$$t = \text{the one electron kinetic energy,}$$

$$\varrho(\boldsymbol{r}) = \sum_{\text{occupied } k} |\psi_k(\boldsymbol{r})|^2 , \tag{2.3}$$

$$V_{\text{xc}}(\boldsymbol{r}) = \frac{\delta E_{\text{xc}}[\varrho(\boldsymbol{r})]}{\delta \varrho(\boldsymbol{r})}$$

The practical solution of (2.2, 3) is usually performed by replacing the atomic potentials by pseudopotentials, avoiding the explicit consideration of atomic core states. Originally these pseudopotentials were treated empirically but now there exist methods which allow one to determine them quantitatively from the free atom properties [2.3]. The knowledge of these pseudopotentials plus the local density treatment allow a complete determination of the solutions Ψ_k of equation (2.2). The wave functions ψ_k can be calculated either by an expansion in plane waves or in orbitals localized on the atoms (usually taken as Gaussian functions).

Equations (2.2) and (2.3) have to be solved in a self-consistent way. Up to now this formulation is exact, the problem being that the quantity E_{xc} and thus V_{xc} is not known in general. The local density approximation then consists in the assumption that, locally, the relation between E_{xc} and $\varrho(\boldsymbol{r})$ is the same as for the free electron gas of identical density, which is known quite accurately. This approximation turns out to give quite satisfactory results as regards the prediction of structural properties of molecules and solids. For instance in solids (either with *sp* bonds or *d* bonds as in transition metals) the cohesive energy, the interatomic distance and the elastic properties are predicted with a precision in general better than 5%. This remains true for diatomic molecules except for the binding energy which is overestimated by about 0.5 to 1 eV (see [2.4, 5] for recent reviews on the subject).

It is tempting to use the differences between the eigenvalues of empty and occupied states given by (2.2) as particle excitation energies. This is not justified in general as evidenced by the predicted values for the energy gap $\varepsilon_G = \varepsilon_C - \varepsilon_V$ in semiconductors and insulators, taken as the difference in the energies ε_C of the first empty state and ε_V of the last filled state. The local density value of ε_G is always found to be substantially smaller than the experimental one [2.6]. It is equal to 0.6 eV for silicon instead of 1.2 eV and even vanishes for germanium. The origin of the errors cannot be traced to the use of an overly simplified exchange and correlation potential such as LDA. This has been most clearly shown by the almost identical results obtained using an improved non-local exchange-correlation potential [2.7]. The exchange-correlation term thus cannot be reduced to a simple local potential.

To correct for this deficiency in defect calculations one simple approach has been to use a "scissor" operator [2.8] which corrects for the bandgap error by a rigid shift of the conduction band states. It was shown later [2.9, 10] that this procedure was closer to reality than expected since the one electron exchange correlation potential of the local density formalism must experience a discontinuity across the gap. This is correctly handled by the scissor operator for bulk semiconductors but the applicability to defect levels is still questionable. In particular, in the case of extended defects such as surfaces it is doubtful that the defect states are correctly obtained. The advantage of this correction is mainly that it does not add any computational requirements compared to conventional LDA calculations.

The more sophisticated way of improving the density functional theories is to evaluate the electron self-energy operator $\Sigma(\boldsymbol{r}, \boldsymbol{r}', E)$ [2.10–12]. Σ contains the effects of exchange and correlation. It is non-local, energy dependent and non-Hermitian. Its non-Hermiticity means that the eigenvalues of the new Schrödinger equation:

$$(T + V_{\text{ext}} + V_H)\psi_{nk}(\boldsymbol{r}) + \int d\boldsymbol{r}\, \Sigma(\boldsymbol{r}, \boldsymbol{r}', E_{nk})\psi_{nk}(\boldsymbol{r}') = E_{nk}\psi_{nk}(\boldsymbol{r}) \tag{2.4}$$

will in general be complex. The imaginary part of Σ gives the lifetime of the quasiparticle.

The self-energy operator can be estimated using the GW approximation [2.13]. The self-energy is expanded in a perturbation series of the screened Coulomb interaction, W. The first term of the expansion (the GW term) only is retained. Both the Green's function G and the screened Coulomb interaction must then be calculated. G is evaluated using the LDA wave functions but with a self-consistent quasiparticle spectrum. The computation of W requires the frequency dependent inverse dielectric matrix. Different schemes have been devised for this evaluation [2.11, 14]. This is the most demanding step in computation time.

The GW approximation gives excitation energies in C, Si and Ge accurate to a few percent, and has been applied to superlattices and surfaces as we shall see in Chap. 6.

2.1.2 Tight-Binding Empirical Methods

In this section we discuss the approximations involved in the tight binding (TB) approach that was introduced in Sect. 1.1. We show how the minimal basis set may be more rigorously introduced; we discuss the neglect of the overlap integral and the two-center approximation that were implicitly used in Sect. 1.1.

a) Basis of the LCAO and Tight-Binding Approximations

An illustration of the physical basis of LCAO techniques is depicted in Fig. 2.1 with the formation of energy bands in diamond starting from the atomic limit. For large interatomic distances R, perturbation theory shows that the correct zeroth

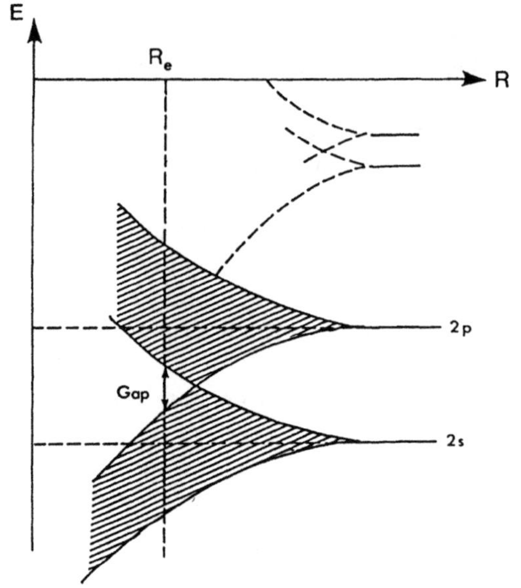

Fig. 2.1. Broadening of atomic levels (in this case $2s$ and $2p$) into bands as the interatomic distance decreases

order basis set corresponds to the free atom state. LCAO theory assumes that such a basis set remains useful (in the variational sense) near the equilibrium interatomic distance R_e. The wave function ψ is then written as:

$$\psi = \sum_{i,\alpha} c_{i\alpha} \Phi_{i\alpha} \tag{2.5}$$

where $\Phi_{i,\alpha}$ is the αth free atom orbital of atom i, at position $\boldsymbol{R}_i$. As each complete set of such orbitals belonging to any given atom forms a basis of Hilbert space, the whole set of $\Phi_{i\alpha}$ is complete (actually overcomplete) and expansion (2.5) can yield the exact wave function of the whole system.

In practice one has to truncate the sum over α in this expansion. In many simplified calculations it has been assumed that the valence states of the system can be described in terms of a "minimal basis set" which only includes free atom states up to the outer shell of the free atom (e.g. $2s$ and $2p$ in diamond). It is that description which provides the most appealing physical picture, allowing one to understand clearly the formation of bands from the atomic limit as is apparent in Fig. 2.1. The "minimal basis set" approximation is also used in most semi-empirical calculations.

When the sum over α in (2.5) is limited to a finite number, the energy levels E of the whole system are given by the secular equation

$$\det |H - ES| = 0 \tag{2.6}$$

where H is the Hamiltonian matrix in the atomic basis and S the overlap matrix of elements

$$S_{i\alpha,j\beta} = \langle \Phi_{i\alpha} | \Phi_{j\beta} \rangle \; . \tag{2.7}$$

The tight binding approximation is defined as the use of a minimal basis set formed from free atom wave functions plus a neglect of interatomic overlaps. The secular equation thus becomes:

$$\det |H - EI| = 0 \tag{2.8}$$

where I is the unit matrix. The solution of the problem then requires the knowledge of the Hamiltonian matrix elements. In the empirical tight-binding approximation these are obtained from a fit to the bulk band structure. For this one always truncates the Hamiltonian matrix in real space, i.e. one only includes interatomic terms up to first, second or at most third nearest neighbors. Also, in most cases one makes use of a two-center approximation as discussed by *Slater* and *Koster* [2.15]. In such a case all Hamiltonian matrix elements $\langle \Phi_{i\alpha} | H | \Phi_{j\beta} \rangle$ can be reduced to a limited number of independent terms which we can call $H_{\alpha\beta}(i,j)$ for the pair of atoms i,j and the orbitals α, β. For an "s,p" basis valid for group IV, III-V and II-VI semiconductors symmetry considerations applied to the two-center approximation give only the following independent terms

$$H_{\alpha,\beta}(i,j) = H_{ss}(i,j), \quad H_{s\sigma}(i,j), \quad H_{\sigma s}(i,j),$$
$$H_{\sigma,\sigma}(i,j), \quad H_{\pi\pi}(i,j), \quad H_{\sigma,\pi}(i,j) \tag{2.9}$$

where $H_{\sigma\pi}$ is strictly zero in a two-center approximation. s stands for the s orbital, σ for the p orbital along axis i,j with positive lobe in the direction of the neighboring atom and π is a p orbital perpendicular to the axis i,j. With these conventions all matrix elements are in general negative.

Similar considerations apply to transition metals with s,p and d orbitals. Simple rules obtained for the $h_{\alpha\beta}(i,j)$ in a nearest neighbor approximation are given in [2.16].

b) Validity of the Tight-Binding Approximation

An interesting discussion of the validity of the use of a minimal basis set has been given by *Louie* [2.17]. Starting from the minimal basis set $|\Phi_{i\alpha}\rangle$ one can increase the size of the basis set by adding other atomic states $|\chi_{i\mu}\rangle$, called the peripheral states, which must lead to an improvement in the description of the energy levels and wave functions. However, as we have seen before, this will rapidly lead to problems related to overcompleteness, i.e. the overlaps between different atomic states will become more and more important. To overcome this difficulty, *Louie* proposes three steps which have the virtue of justifying the use of a minimal basis set. These are the following:

1) Orthogonalize symmetrically the states $|\Phi_{i\alpha}\rangle$ belonging to the minimal basis set between themselves. This leads to an orthonormal set $|\overline{\Phi}_{i\alpha}\rangle$.
2) The peripheral states $\chi_{i\mu}\rangle$ overlap strongly with the $|\overline{\Phi}_{i\alpha}\rangle$. It is then necessary to orthogonalize them to these $|\overline{\Phi}_{i\alpha}\rangle$, which yields new states $|\overline{\chi}_{i\mu}\rangle$ defined as

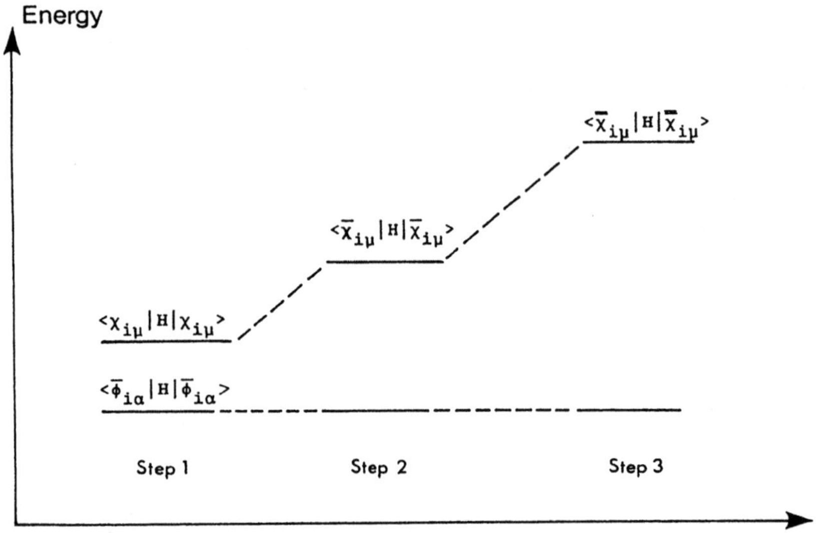

Fig. 2.2. Behavior of the peripheral state energy of Si in the orthogonalization scheme of Louie. The increase in energy as compared to the minimal basis set leads to a decreased coupling with the minimal set [2.17]

$$|\bar{\chi}_{i\mu}\rangle = |\chi_{i\mu}\rangle - \sum_{j\alpha} |\bar{\Phi}_{j\alpha}\rangle\langle\bar{\Phi}_{j\alpha}|\chi_{i\mu}\rangle \tag{2.10}$$

3) The new states $|\bar{\chi}_{i\mu}\rangle$ are then orthogonalized between themselves leading to a new set of states $|\bar{\bar{\chi}}_{i\mu}\rangle$.

Louie has shown that, at least for silicon, the average energies of these atomic states behave as shown in Fig. 2.2. After step 3, the peripheral states $|\bar{\bar{\chi}}_{i\mu}\rangle$ are much higher in energy and their coupling to the minimal set is reduced. They only have a small (although non negligible) influence, justifying the use of the minimal set as the essential step in the calculation.

The quantitative value of LCAO techniques for covalent systems such as diamond and silicon has been demonstrated by *Chaney*, *Lin* and *Lafon* [2.18]. They have shown that the minimal basis set gives good results for the valence bands, and slightly less (but still meaningful) results for the lower conduction bands. Such conclusions have been confirmed by *Kane* [2.19], *Chadi* [2.20] and *Louie* [2.17] but working with pseudopotentials instead of true atomic potentials.

The great interest of minimal basis set LCAO calculations is that they provide the direct connection between the valence states of the system and the free atom states. This becomes still more apparent with the TB approximation which we have discussed and which allows one to obtain extremely simple and physically sound descriptions of many systems. A good illustration of this has been its application to the band structure of group IV semiconductors, either crystalline [2.21] or amorphous [2.22], the extension to defect tetrahedral compounds such as IV-VI compounds for instance [2.23, 24], and its impressive success in transition metals (Chap. 3).

Viewed as an approximate version of LCAO theory, tight binding must lead to incorrect energy levels (for instance the band structure will in all cases be narrower than the corresponding LCAO one). However one can obtain a simple secular equation like (2.8) in various ways, of which the most popular is to write (2.6) under the alternative form

$$\det|S^{-1/2}H\,S^{-1/2}-EI|=0\,. \tag{2.11}$$

This corresponds to orthogonalizing the atomic functions by Löwdin's symmetrical procedure [2.25] which keeps their atomic-like character and symmetry. Thus, in (2.11), the effective Hamiltonian to be diagonalized is not the original one H but $S^{-1/2}H\ S^{-1/2}$ which can be very different in the case of strong overlaps as in diamond and silicon.

We have seen that the tight binding parameters are determined from a fit to known bulk band structures. However care must be taken when applying such parameters to defect or surface problems. For instance one should not forget in principle the change in $S^{-1/2}H\ S^{-1/2}$ due to the change in S induced by the presence of the defect or surface. This is particularly true in semiconductors where S is important, less for the "d" band of transition metals where S is much smaller. Also the two-center matrix elements depend in principle on the structure. Another modification of the matrix elements can be due for instance to charge transfer occurring near a surface. This requires a self-consistent treatment which will be developed later for specific cases.

c) Total Energies in Tight Binding

Up to now we have discussed how it is possible to get one particle eigenstates, obtained by diagonalization of the tight binding matrix. However it is also possible to derive an empirical technique which allows a determination of total energies. For this let us consider the simpler case of a nearest-neighbor approximation. One then assumes that all independent nearest neighbors $H_{\alpha\beta}(i,j)$ vary in the same way as functions of the interatomic distance R_{ij}, for instance as

$$H_{\alpha\beta}(R_{ij})=H^0_{\alpha\beta}\exp(-qR_{ij}) \tag{2.12}$$

or also like some inverse power of R_{ij}, see [2.16]. Then by summing over the energies of filled states one can obtain the band structure energy E_{BS} as a function of the atomic positions.

This band structure energy has an attractive character. To determine the crystal stability it is necessary to add a repulsive part corresponding to terms which have been neglected or counted twice in E_{BS}. In a non self-consistent scheme such terms correspond to the repulsion of neutral atoms [2.26]. They are short-ranged and we can simulate them simply by Born–Mayer pair potentials

$$V(R_{ij})=V^0\exp(-pR_{ij})\,. \tag{2.13}$$

The parameters p, q and V^0 are obtained empirically from a fit to the observed cohesive energy, lattice parameter and compressibility. They can then be used to calculate phonon dispersion curves and relaxation or reconstruction energies near defects or surfaces. This technique has been applied with much success in transition metals [2.27] as well as in *sp*-bonded semiconductors [2.26]. Its application to surfaces will be discussed later.

2.2 Direct Calculations of Surface Properties

We now describe three computational techniques which have been used to calculate the surface electronic structure of metals and semiconductors. We do not adopt chronological order but instead introduce first the technique whose principle is closer in spirit to the bulk calculations.

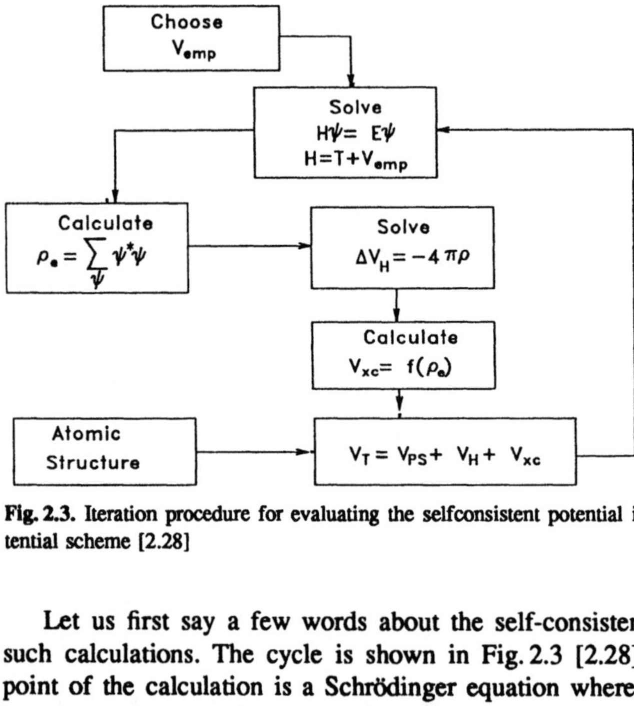

Fig. 2.3. Iteration procedure for evaluating the selfconsistent potential in a local density pseudopotential scheme [2.28]

Let us first say a few words about the self-consistent cycle to be used in such calculations. The cycle is shown in Fig. 2.3 [2.28]. Usually the starting point of the calculation is a Schrödinger equation where the potential is taken to be the sum of empirical pseudopotentials V_{emp}. The solution of this equation yields zeroth-order wave functions and charge densities. These, in turn, allow the computation of a first order potential which is then reinjected into the Schrödinger equation.

2.2.1 The Periodic Slab Geometry

This was first used by *Chelikowsky* et al. [2.29]. The technique is extremely simple. One studies a slab of the material, which is repeated periodically, the slabs being separated by a finite interval of vacuum. The main advantage of this geometry is that it has translational periodicity in the three directions of space. The solution thus proceeds exactly as for a bulk crystalline material. The only difference is the large size of the unit cell in the z-direction perpendicular to the surface. The dimensions of the slab and of the vacuum intervals have to be chosen in such a way that there is no interaction between any pair of surfaces. This usually requires a slab thickness corresponding to at least ten atomic planes parallel to the surface, while the vacuum interval must correspond to a few Thomas–Fermi wavelengths.

Once these requirements are fulfilled, the calculation proceeds exactly as for a perfect crystal. The main simplification comes from the fact that the eigenstates $\Phi_i(\boldsymbol{r})$ can be written under the form of Bloch states:

$$\psi_{\boldsymbol{k}}(\boldsymbol{r}) = u_{\boldsymbol{k}}(\boldsymbol{r})\mathrm{e}^{\mathrm{i}\boldsymbol{k}\cdot\boldsymbol{r}} \qquad (2.14)$$

where $u_{\boldsymbol{k}}(\boldsymbol{r})$ is periodic in the three directions of space. In actual calculations one can use different basis sets, as for the bulk, i.e. plane waves, mixed basis sets or localized orbitals.

2.2.2 The Single Slab Geometry

This represents an alternative method for studying solid surfaces. The first self-consistent calculation using this geometry was done by *Alldredge* and *Kleinman* [2.30]. The Schrödinger equation is solved for a slab of material embedded in an infinite potential well of width larger than the slab. The infinite potential walls are placed a few lattice spacings away from the surface for the same reason as before. The system is periodic along the surface and the wave functions $\psi_{\boldsymbol{k}_\parallel}(\boldsymbol{r})$ can be written as Bloch functions under the form:

$$\Psi_{\boldsymbol{k}_\parallel}(\boldsymbol{r}) = u_{\boldsymbol{k}_\parallel}(\boldsymbol{r})\mathrm{e}^{\mathrm{i}\boldsymbol{k}_\parallel\cdot\boldsymbol{r}_\parallel} \qquad (2.15)$$

where $\boldsymbol{r}_\parallel$ is the component of $\boldsymbol{r}$ parallel to the surface plane. The part $\boldsymbol{u}_{\boldsymbol{k}_\parallel}(\boldsymbol{r})$ must be periodic along the surface and can be expanded in a Fourier series, i.e. in plane waves of the form $\exp(\mathrm{i}\boldsymbol{G}_\parallel\cdot\boldsymbol{r}_\parallel)\exp(\mathrm{i}\boldsymbol{k}_\parallel\cdot\boldsymbol{r}_\parallel)$. The coefficients of these plane waves are functions of z which must vanish on the potential walls, taken to be at $z = 0$ and $z = L$. A complete basis set for the expansion of $\Psi_{\boldsymbol{k}_\parallel}(\boldsymbol{r})$ is defined by the functions:

$$\Phi_{\boldsymbol{k}_\parallel,\boldsymbol{G}_\parallel,k_n}(\boldsymbol{r}) = \exp(\mathrm{i}(\boldsymbol{k}_\parallel + \boldsymbol{G}_\parallel)\cdot\boldsymbol{r}_\parallel)\sin(k_n z) \qquad (2.16)$$

where

$$k_n = n\frac{\pi}{L}\,, \quad n = 1,2,\dots\,. \qquad (2.17)$$

The calculation then proceeds along the same lines as for the periodic slab geometry. It has been applied to the case of lithium leading to a work function of 3.71 eV. This value is greater than the experimental one by 1 eV, the discrepancy being due to the choice of the pseudopotential which did not give a very good band structure for bulk Li.

2.2.3 The Method of Apprelbaum and Hamann

This method is the first self-consistent calculation for surfaces [2.32]. The formulation uses the periodicity along the surface, expanding the potential $V(\boldsymbol{r})$ in a Fourier series and writing the wave function $\Psi_{\boldsymbol{k}_\parallel}(\boldsymbol{r})$ as a Bloch function. This gives:

$$\begin{aligned} V(\boldsymbol{r}) &= \sum_{\boldsymbol{G}_\parallel} \exp(\mathrm{i}\boldsymbol{G}_\parallel \cdot \boldsymbol{x}_\parallel) V_{\boldsymbol{G}_\parallel}(z) \\ \Psi_{\boldsymbol{k}_\parallel}(\boldsymbol{r}) &= \sum_{\boldsymbol{G}_\parallel} \exp[\mathrm{i}(\boldsymbol{k}_\parallel + \boldsymbol{G}_\parallel) \cdot \boldsymbol{x}_\parallel] u_{\boldsymbol{G}_\parallel}(z) \,. \end{aligned} \tag{2.18}$$

Here $\boldsymbol{G}_\parallel$ are reciprocal lattice vectors of the two-dimensional lattice parallel to the surface. Again the periodic part u of the Bloch function has been expanded in a Fourier series of coefficients $u_{\boldsymbol{G}_\parallel}(z)$. Expressions (2.18) can be injected into the Schrödinger equation giving a set of coupled equations between the unknowns $u_{\boldsymbol{G}_\parallel}(z)$:

$$\left\{ -\frac{1}{2}\frac{d^2}{dz^2} + \frac{1}{2}(\boldsymbol{k}_\parallel + \boldsymbol{G}_\parallel)^2 - E \right\} u_{\boldsymbol{G}_\parallel}(z) + \sum_{\boldsymbol{K}_\parallel} V_{\boldsymbol{G}_\parallel - \boldsymbol{K}_\parallel} u_{\boldsymbol{K}_\parallel}(z) = 0 \,. \tag{2.19}$$

For fixed values of $\boldsymbol{k}_\parallel$ and E this equation can be integrated numerically between two planes z_1 and z_2 as shown in Fig. 2.4. The coordinate z_1 is chosen to be deep enough within the material that for $z < z_1$ one recovers the bulk solutions

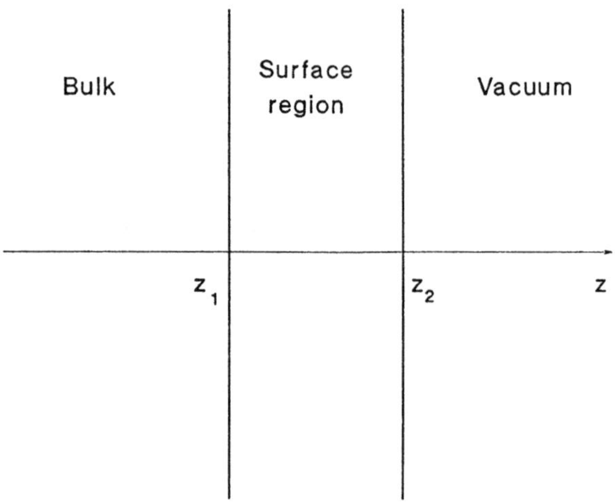

Fig. 2.4. Limits of integration of (2.19) in the surface region. For $z < z_1$ the bulk solutions are obtained. For $z > z_2$ the wave functions are exponentially decreasing functions. $z_1 < z < z_2$ is the extent of the surface perturbation

and for $z > z_2$ one has exponentially decaying functions of z. The numerical integration allows the calculation of quantities such as $u_{G_\parallel}(z)$ and $(d/dz)u_{G_\parallel}(z)$ at $z = z_2$ if the corresponding quantities at $z = z_1$ are assumed to be known. This relation can be formally written, in matrix form, as:

$$u_2 = T_{21} u_1 \tag{2.20}$$

where u_1 and u_2 are column vectors built from the quantities $u_{G_\parallel}(z)$ and $(d/dz)u_{G_\parallel}(z)$ at $z = z_1$ and $z = z_2$ respectively. The matrix T_{21} is given by numerical integration of Schrödinger's equation between these two planes. As the logarithmic derivative of the wave function must be continuous on these planes one can rewrite equation (2.20) in the form:

$$u_v = T_{21} u_b \tag{2.21}$$

where now u_b and u_v correspond to bulk and vacuum wave functions whose form is known.

When E falls within a bulk allowed energy band, then u_b is taken to represent an incident propagating wave of unit amplitude and equation (2.21) allows one to calculate u_b and u_v completely. The situation is completely different when E falls within a forbidden energy gap. There are no propagating bulk waves at this energy and the solution, if it exists, decays in both directions away from the surface. In that case equation (2.21) can only be satisfied for special values of the energy which are the bound surface states.

Appelbaum and *Hamann* first applied this technique to the case of Na(100) [2.32]. They found a work function 0.3 eV lower than the jellium value, consistent with the first-order value of *Lang* and *Kohn*. Other applications have been made to semiconductor surfaces.

2.3 The Green's Function Method

We describe here a very powerful technique for calculating the change in physical properties due to local perturbations. In the local density scheme its application has mainly been restricted to point defects but there have been a few applications to surface problems [2.33]. This is not the case in tight-binding where it has often been used in surface or interface calculations.

2.3.1 Definitions and Useful Relations

We consider a system described by its Hamiltonian H and define its resolvent $G(E)$ as:

$$G(E) = \lim_{\eta \to 0+} (E - H + \mathrm{i}\eta)^{-1} \tag{2.22}$$

where E is the energy. If we make use of the eigenstates $|k\rangle$ of H of energy E_k

we can write G in the form:

$$G(E) = \sum_k \frac{|k\rangle\langle k|}{E - E_k + \mathrm{i}\eta} \tag{2.23}$$

where it is understood that one should take $\eta \to 0^+$. The Green's function $G(\boldsymbol{r}, \boldsymbol{r}', E)$ is defined as the matrix element of G between two position vectors $|\boldsymbol{r}\rangle$ and $|\boldsymbol{r}'\rangle$:

$$G(\boldsymbol{r}, \boldsymbol{r}', E) = \sum_k \frac{\langle \boldsymbol{r}|k\rangle\langle k|\boldsymbol{r}'\rangle}{E - E_k + \mathrm{i}\eta} \tag{2.24}$$

$$= \sum_k \frac{\Psi_k(\boldsymbol{r})\Psi_k^*(\boldsymbol{r}')}{E - E_k + \mathrm{i}\eta} \tag{2.25}$$

where $\Psi_k(\boldsymbol{r})$ is the eigenfunction $\langle \boldsymbol{r}|k\rangle$. It is interesting to define a local density of states $n(\boldsymbol{r}, E)$ at point $\boldsymbol{r}$ given by:

$$n(\boldsymbol{r}, E) = \sum_k |\Psi_k(\boldsymbol{r})|^2 \delta(E - E_k)\,, \tag{2.26}$$

which can be expressed as

$$n(\boldsymbol{r}, E) = -\frac{1}{\pi} \mathrm{Im}\{G(\boldsymbol{r}, \boldsymbol{r}, E)\} \tag{2.27}$$

by using the fact that $\mathrm{Im}\{E - E_k + \mathrm{i}\eta\}^{-1}$ is equal to $-\frac{1}{\pi}\delta(E - E_k)$, Im denoting the imaginary part. From (2.26) it is easy to show that the integral of $n(\boldsymbol{r}, E)$ over $\boldsymbol{r}$ is equal to the total density of states $n(E)$. In other words we have:

$$n(E) = -\frac{1}{\pi} \mathrm{Im}\{\mathrm{Tr}\, G(E)\} \tag{2.28}$$

since Tr G can be written as the integral over $\boldsymbol{r}$ of $G(\boldsymbol{r}, \boldsymbol{r})$. Using the basis of the eigenstates $|k\rangle$ we can also write:

$$n(E) = -\frac{1}{\pi} \mathrm{Im} \left\{ \sum_k \frac{1}{E - E_k + \mathrm{i}\eta} \right\} \tag{2.29}$$

which can be transformed to:

$$n(E) = -\frac{1}{\pi} \frac{d}{dE} \mathrm{Im}\{\ln \det G\}\,. \tag{2.30}$$

Another very important quantity is the electron density $n(\boldsymbol{r})$ which is the integral of $n(\boldsymbol{r}, E)$ given by (2.27) over the energy of occupied states. We easily get:

$$n(\boldsymbol{r}) = -\frac{1}{\pi} \int_{\mathrm{occ.E}} \mathrm{Im}\{G(\boldsymbol{r}, \boldsymbol{r}, E)\} dE \tag{2.31}$$

where one should take into account the spin index for the summation over k defining G.

We see that the knowledge of G allows the determination of $n(\boldsymbol{r})$ which, in turn completely defines H in the local density approximation. However, the whole calculation must again be performed self-consistently.

2.3.2 Green's Functions for Localized Perturbations

We now consider a perfect crystal of Hamiltonian H_0 and resolvent operator G_0. Application of a perturbation V leads to a Hamiltonian $H = H_0 + V$ and resolvent G. The Schrödinger equation of the perturbed system is:

$$(E - H_0)\Psi = V\Psi \; . \tag{2.32}$$

The solution Ψ of this differential equation is the sum of the general solution Ψ_0 of the equation without the second member, i.e.:

$$(E - H_0)\Psi_0 = 0 \tag{2.33}$$

and a particular solution which can be written $G_0 V \Psi$. We get:

$$\Psi = \Psi_0 + G_0 V \Psi \; . \tag{2.34}$$

This is valid for energies E within the allowed energy bands of the perfect system. However, if this is not the case, i.e. if E falls within a forbidden gap then we have:

$$\Psi = G_0 V \Psi \; . \tag{2.35}$$

Solutions of this equation correspond to localized states. Expressing Ψ in a given basis set we get a linear homogeneous system of equations for the coefficients of the expansion. This has non vanishing solutions only if:

$$\det(I - G_0 V) = 0 \tag{2.36}$$

which gives the allowed energy values.

As discussed in [2.34] equation (2.36) is very interesting when the perturbation V can be written in the form of a matrix of small size (for instance when it is localized in real space). In this case the relevant part of $\det|I - G_0 V| = 0$ is of the same size as V itself so that solutions of (2.36) can be easily obtained. The same is true when one calculates the change $\delta n(E)$ in density of states induced by the perturbation. From (2.30) one finds:

$$\delta n(E) = \frac{1}{\pi}\frac{d}{dE}\operatorname{Im}\left\{\ln\det\left(G G_0^{-1}\right)\right\} \; . \tag{2.37}$$

From the fact that G^{-1} is equal to $G_0^{-1} - V$ we obtain Dyson's equation:

$$G = G_0 + G_0 V G \tag{2.38}$$

$$= (I - G_0 V)^{-1} G_0 \tag{2.39}$$

so that $\delta n(E)$ becomes:

$$\delta n(E) = -\frac{1}{\pi}\frac{d}{dE}\mathrm{Im}\,\{\ln\det(I - G_0 V)\} \tag{2.40}$$

which, again, only requires the calculation of a determinant having the same size as the matrix V itself. Such simplifications have allowed the application of Green's function techniques to point defects in semiconductors within the local density theory [2.35, 36]. Even self-consistency can be achieved since $\delta n(\boldsymbol{r})$ is short ranged (for neutral defects at least) so that it can be computed. The situation for surfaces is different since the perturbation is localized perpendicular to the surface but delocalized parallel to it. Green's function theory can then be applied by working partly in real space (in the perpendicular direction), partly in reciprocal space (in the parallel direction). However, up to now, local density calculations have been made preferentially using the other techniques described in Sect. 2.2. This is why we concentrate in the following on the tight-binding approximations for which there exist a variety of different possibilities.

2.3.3 Tight-Binding: Recursion Method and Moments of Partial Densities of States

We now consider a system of orthonormal local states $|\alpha\rangle$ (for instance an atomic state in the tight-binding limit) and define, in analogy with (2.26), the partial density of states:

$$n_\alpha(E) = \sum_k |\langle\alpha|k\rangle|^2 \delta(E - E_k)\ . \tag{2.41}$$

The pth moment $\mu_{p,\alpha}$ of $n_\alpha(E)$ is defined as:

$$\mu_{p,\alpha} = \int n_\alpha(E) E^p dE \tag{2.42}$$

and can be transformed to:

$$\mu_{p,\alpha} = \sum_k \langle\alpha|k\rangle\langle k|\alpha\rangle E_k^p \tag{2.43}$$

or equivalently to:

$$\mu_{p,\alpha} = \langle\alpha|H^p|\alpha\rangle\ . \tag{2.44}$$

A knowledge of all the moments allows one in principle to reconstruct the exact $n_\alpha(E)$. In particular, the first order moment ($p = 1$) is the center of gravity of the density of states and the second order moment is related to its width. For this purpose expression (2.44) is particularly useful since it can also be rewritten, assuming that $(|\alpha_i\rangle)$ is a complete basis of the Hilbert space:

$$\mu_{p,\alpha} = \sum_{\alpha_1\alpha_2\ldots\alpha_{p-1}} \langle\alpha|H|\alpha_1\rangle\langle\alpha_1|H|\alpha_2\rangle\ldots\langle\alpha_{p-1}|H|\alpha\rangle\ . \tag{2.45}$$

The calcuation of $\mu_{p,\alpha}$ is equivalent to the determination of all possible walks of p jumps from one local orbital to another one, starting and ending on $|\alpha\rangle$, and corresponding to non vanishing matrix elements. In simple cases (i.e. nearest-neighbors tight-binding s-band) this can be evaluated from purely geometrical considerations [2.37]. In most realistic situations the moments are computed numerically.

For systems with no forbidden gap and for which fine structure in the densities of states has a negligible effect, the method of moments can lead to extremely powerful approximations. One of the best and most widely used is to approximate $n_\alpha(E)$ by the following Gaussian curve:

$$n_\alpha(E) = (2\pi\mu_{2,\alpha})^{-1/2} \exp - \frac{(E-\mu_{1,\alpha})^2}{2\mu_{2,\alpha}} \tag{2.46}$$

adjusted to the exact first two moments. Many successful applications of this to transition metals will be discussed in the next chapter, ranging from self-consistency to total energy calculations.

In cases where structure in $n_\alpha(E)$ is important it is interesting to use the recursion method. For this one first notices that, in analogy to (2.27), we can write:

$$n_\alpha(E) = -\frac{1}{\pi} \operatorname{Im}\{G_{\alpha\alpha}(E)\} \tag{2.47}$$

where $G_{\alpha\alpha}(E)$ stands for $\langle\alpha|G(E)|\alpha\rangle$ and can be expanded as a continued fraction:

$$G_{\alpha\alpha}(E) = \cfrac{1}{E - a_1 - \cfrac{b_1}{E - a_2 - \cfrac{b_2}{E - a_3 \dots}}} \; . \tag{2.48}$$

The coefficients a_n and b_n can be generated in a systematic way which corresponds to a tridiagonalization of the Hamiltonian matrix [2.38, 39]. For this one starts from a vector $|1\}$ equal to $|\alpha\rangle$. The state $|2\}$ is defined as:

$$|2\} = H|1\} - a_1|1\} \tag{2.49}$$

where a_1 is chosen to orthogonalize $|2\}$ to $|1\}$. One can proceed further and define for $n > 1$:

$$|n+1\} = H|n\} - a_n|n\} - b_{n-1}|n-1\} \; . \tag{2.50}$$

With this prescription the matrix becomes tridiagonal (i.e. $\{n|H|m\}$ vanishes for $|n-m| > 1$) and the Green function automatically takes the form (2.48) with:

$$a_n = \frac{\{n|H|n\}}{\{n|n\}} \tag{2.51}$$

$$(b_{n-1})^{1/2} = \{n-1|H|n\} = \{n|H|n-1\} \tag{2.52}$$

The coefficients a_n, b_n of the continued fraction are related to the moments $\mu_{p,\alpha}$. In practice only a limited number of these coefficients can be calculated, up to

a_N and b_N for instance. One method is to stop the continued fraction at this level, leading to the equivalent expression:

$$G_{\alpha\alpha}(E) = \sum_{i=1}^{N} \frac{\omega_i}{E - E_i + \mathrm{i}\eta} \tag{2.53}$$

for which the density of states $n_\alpha(E)$ is the sum of N delta functions. Smoothing procedures can then be used to get a continuous curve. Good approximations usually require $N > 10$ to 15.

Another termination is to say that a_n and b_n take a constant value for $n > N$. This corresponds to writing (2.48) as:

$$G_{\alpha\alpha} = \cfrac{1}{E - a_1 - \cfrac{b_1}{E \ldots \cfrac{b_{N-1}}{E - a_N - b_N G}}} \tag{2.54}$$

with G equal to:

$$G = \frac{1}{E - a_N - b_N G} \tag{2.55}$$

whose solution is:

$$G = \frac{E - a_N \pm \sqrt{(E - a_N)^2 - 4b_N}}{2b_N} . \tag{2.56}$$

Such a termination is only valid if the system behaves truly like a linear chain for large n. More elaborate terminations have been worked out [2.40, 41] and have been shown to be efficient even for strongly structured densities of states. In particular for the case of semiconductors, the correct value of the gap as well as the steepness of the DOS at the edges of the gap is a good test of the termination.

2.3.4 Surface Dispersion Relation

We now outline how it is possible to get detailed information about surface-induced modifications in the band structure. In most cases studied the system is periodic in directions parallel to the surface. The wave functions can then be taken as Bloch states specified by their wavevector $\boldsymbol{k}_\parallel$ to the surface.

Except for those using slab geometries, all tight-binding calculations of surface properties are based on Green's functions. Different versions are the following:

a) Dividing the Bulk Crystal into Two Parts

Here one uses Dyson's equation (2.38) where H^0 and G^0 are the perfect crystal Hamiltonian and resolvent operators. One defines H and G to correspond to the two half crystals of Fig. 2.5. The perturbative potential V then consists in breaking the matrix elements connecting the two half crystals plus eventually making some changes in the parameters near the surface. Dyson's equation is

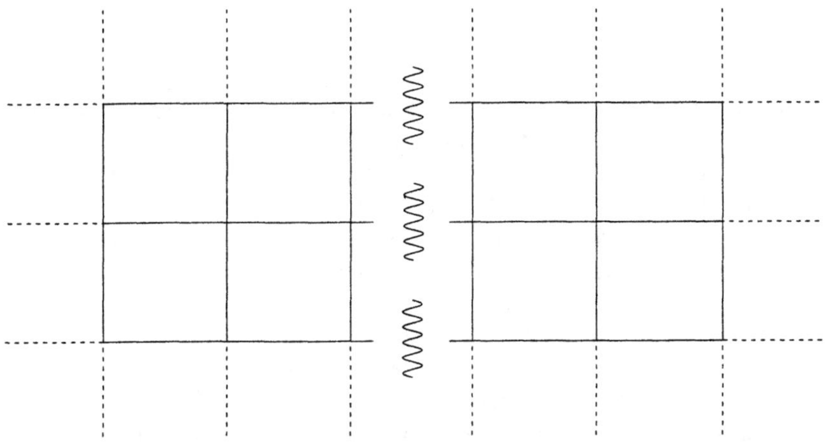

Fig. 2.5. Broken bonds at the surface. A surface may be represented as the perturbation introduced by cutting one plane of bonds in a perfect crystal

a matrix relation using the atomic orbitals as a basis, which we now label as $\Phi_{l,j\alpha}$, where l is the plane index, j the index of the unit cell in planes parallel to the surface, and α the index of the orbital in the unit cell. The problem can be factorized at $\boldsymbol{k}_\parallel$ by introducing the Bloch sums:

$$\psi_{l,\alpha}(\boldsymbol{k}_\parallel) = \frac{1}{\sqrt{N}} \sum_j \mathrm{e}^{\mathrm{i}\boldsymbol{k}_\parallel \cdot \boldsymbol{R}_j} \varphi_{l,j\alpha} \, . \tag{2.57}$$

The central quantity in these treatments is the $\det(I - G_0 V)$. As the matrix $I - G_0 V$ can be factorized into different blocks corresponding to each value of $\boldsymbol{k}_\parallel$, one can write this determinant as:

$$\det(I - G_0 V) = \prod_{\boldsymbol{k}_\parallel} \det\left(I - G_0(\boldsymbol{k}_\parallel) V(\boldsymbol{k}_\parallel)\right) \tag{2.58}$$

where $G_0(\boldsymbol{k}_\parallel)$ and $V(\boldsymbol{k}_\parallel)$ represent the submatrices of G_0 and V in the basis of the $\psi_{l\alpha}(\boldsymbol{k}_\parallel)$. The localized surface states are given by the zeros of $\det(I - G_0 V)$ outside the bulk band. Clearly there is one such equation per value of $\boldsymbol{k}_\parallel$, a fact which allows one to calculate the dispersion curves for the surface bands. Another factor is the total change in density of states $\delta n(E)$ given by (2.40):

$$\delta n(E) = -\frac{1}{\pi} \frac{d}{dE} \mathrm{Im}\{\log \det(I - G_0 V)\} \tag{2.59}$$

which can be rewritten, from (2.58), as:

$$\delta n(E) = \sum_{\boldsymbol{k}_\parallel} \delta n(E, \boldsymbol{k}_\parallel) \, . \tag{2.60}$$

There have been only a few attempts to calculate the surface states along these lines [2.28, 36] the method being technically more difficult to implement.

b) Direct Application of Dyson's Equation to the Semi-Infinite System

To describe this method we first consider a simplified situation where there are interactions only between nearest neighbor planes. We also assume that there is only one s orbital per unit cell in the planes parallel to the surface. We can thus build only one Bloch function $\Psi_l(\boldsymbol{k}_\parallel)$ per plane of index l, defined by (2.57) (we have dropped the index α since it takes only one value). At fixed $\boldsymbol{k}_\parallel$ the semi-infinite system of Fig. 2.6a is equivalent to the semi-infinite linear chain of Fig. 2.6b.

We now label G_{ll}, the matrix elements $\langle\Psi_l(\boldsymbol{k}_\parallel)|G|\Psi_l(\boldsymbol{k}_\parallel)\rangle$ of G between these Bloch states, the resolvent G corresponding to the Hamiltonian H of the semi-infinite system under consideration. We define H^0 and G^0 as the operators corresponding to the same system where the interactions between planes $l = 0$ and $l = 1$ have been suppressed. We can relate G and G^0 through Dyson's equation (2.38) where V is the interaction between the first two planes. In terms of matrix elements this gives:

$$G_{00} = G^0_{00} + G^0_{00} V_{01} G_{10} \tag{2.61}$$

$$G_{10} = G^0_{11} V_{10} G_{00} \tag{2.62}$$

which can be rewritten in terms of G_{00} alone:

$$G_{00} = (1 - G^0_{00} V_{01} G^0_{11} V_{10})^{-1} G^0_{00} \,. \tag{2.63}$$

From Fig. 2.6 it is clear that, for H^0, the plane $l = 0$ is disconnected, so that we have:

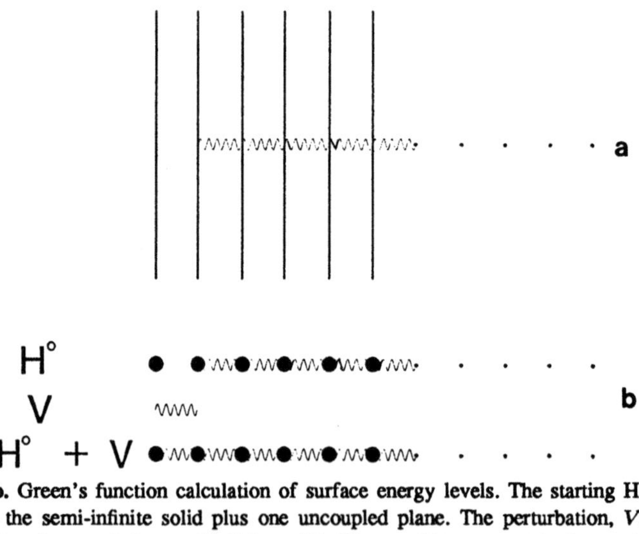

Fig. 2.6a,b. Green's function calculation of surface energy levels. The starting Hamiltonian corresponds to the semi-infinite solid plus one uncoupled plane. The perturbation, V, is the coupling between the plane and the semi-infinite solid. The resulting system is the semi-infinite solid. A formal equivalence with the semi-infinite linear chain is depicted at the bottom of the figure

$$G^0_{00} = (E - H^0_{00})^{-1} \tag{2.64}$$

where H^0_{00} is a function of $\boldsymbol{k}_{\parallel}$. Furthermore, the two semi-infinite chains of Fig. 2.6 are equivalent so that we can write:

$$G^0_{11} = G_{00} \,. \tag{2.65}$$

Inserting (2.64) and (2.65) into (2.63) we obtain:

$$G_{00} = \frac{1}{E - H^0_{00} - |V_{01}|^2 G_{00}} \,. \tag{2.66}$$

This represents a quadratic equation for the unknown G_{00} whose solution turns out to be given by:

$$\begin{aligned} G_{00} &= \frac{E - H^0_{00}}{2|V_{01}|^2}\left(1 - \sqrt{1 - \frac{4|V_{01}|^2}{|E - H^0_{00}|^2}}\right) \quad \text{for} \quad |E - H^0_{00}| > 2|V_{01}| \\ G_{00} &= \frac{E - H^0_{00}}{2|V_{01}|^2} - \frac{i}{|V_{01}|}\sqrt{1 - \left|\frac{E - H^0_{00}}{2V_{01}}\right|^2} \quad \text{for} \quad |E - H^0_{00}| < 2|V_{01}| \,. \end{aligned} \tag{2.67}$$

From this we can get the partial density of states $n_{00}(\boldsymbol{k}_{\parallel})$ as:

$$n_{00}(\boldsymbol{k}_{\parallel}) = -\frac{1}{\pi}\,\mathrm{Im}\{G_{00}\} \,. \tag{2.68}$$

This method can be extended to more general situations (i.e. several orbitals per unit cell and more distant interactions between planes). For this one associates neighboring planes in groups in such a way that each group of planes only interacts with its nearest neighbors. Then equations (2.61) and (2.62) remain valid but in matrix form, the size of each matrix being the product of the number of planes in each group times the number of orbitals per unit cell. Again we can make use of the identity (2.65) allowing us to write (2.63) as:

$$G_{00} = \left[E - H^0_{00} - V_{01} G_{00} V_{10}\right]^{-1} \,. \tag{2.69}$$

However, the solution of this equation is no longer analytic and has to be performed iteratively. The main difficulty is that, at a given $\boldsymbol{k}_{\parallel}$, G_{00} exhibits divergences characteristic of one-dimensional systems. The calculation of G_{00} near these divergences requires a large number of iterations ($\sim$ 50) which makes the method of restricted interest. It can be advantageously replaced by another iterative procedure which we now describe.

c) Iterative Slab Procedure – Decimation

To describe this technique we consider the two identical slabs of Fig. 2.7 which contain planes. We use assumptions similar to those in 2.3.4b except that now H^0 and G^0 correspond to the non interacting slabs while G and H correspond to the

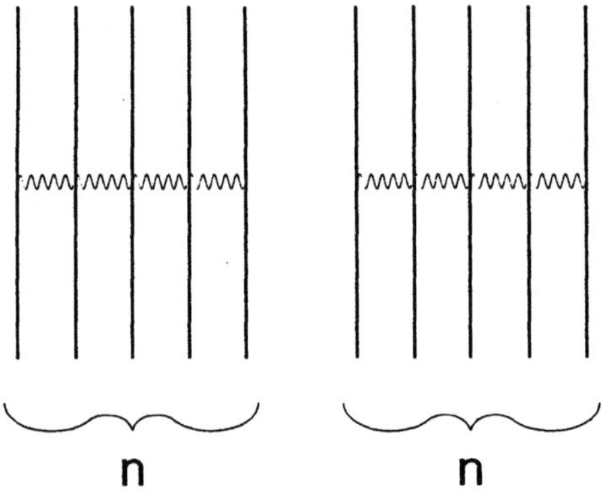

Fig. 2.7. Slabs used in the iterative Green's function method. The eigenvalues and eigenfunctions of the surface and bulk atoms may be obtained iteratively by connecting two crystals of n planes

interacting situation. Again we use Dyson's equation where V is the perturbation connecting the two slabs, i.e. it has only one matrix element:

$$V_{n,n+1} = H_{n,n+1} \ . \tag{2.70}$$

Direct application of Dyson's equation allows us to write:

$$\begin{aligned} G_{11} &= G^0_{11} + G^0_{1n} H_{n,n+1} G_{n+1,1} \\ G_{n+1,1} &= G^0_{n+1,n+1} H_{n+1,n} G_{n,1} \\ G_{n,1} &= G^0_{n,1} + G^0_{n,n} H_{n,n+1} G_{n+1,1} \\ G_{2n,2n} &= G^0_{2n,2n} + G^0_{2n,n+1} H_{n+1,n} G_{n,1} \\ G_{2n,1} &= G^0_{2n,n+1} H_{n+1,n} G_{n,1} \ . \end{aligned} \tag{2.71}$$

One can solve for $G_{n,1}$ in these equations and obtain:

$$G_{n,1} = \left[I - G^0_{n,n} H_{n,n+1} G^0_{n+1,n+1} H_{n+1,n}\right]^{-1} G^0_{n,1} \ . \tag{2.72}$$

As the two slabs are identical we make use of the identities:

$$G^0_{n+1,n+1} = G^0_{11} \qquad G^0_{2n,n+1} = G^0_{n,1} \qquad G^0_{2n,2n} = G^0_{n,n} \tag{2.73}$$

and rewrite the relations as:

$$\begin{aligned} G_{n,1} &= \left[I - G^0_{n,n} H_{n,n+1} G^0_{1,1} H_{n+1,n}\right]^{-1} G^0_{n,1} \\ G_{n+1,1} &= G^0_{1,1} H_{n+1,n} G_{n,1} \\ G_{1,1} &= G^0_{1,1} + G^0_{1,n} H_{n,n+1} G_{n+1,1} \\ G_{2n,1} &= G^0_{n,1} H_{n+1,n} G_{n,1} \\ G_{2n,2n} &= G^0_{n,n} + G^0_{n,1} H_{n+1,n} G_{n,1} \ . \end{aligned} \tag{2.74}$$

This last set of equations allows one to determine $G_{1,1}$, $G_{2n,2n}$ and $G_{2n,1}$ for the slab with $2n$ planes in terms of the equivalent Green functions $G^0_{1,1}$, $G^0_{n,n}$ and $G^0_{n,1}$ for the slab of half thickness, i.e. with n planes. One can thus start an iterative procedure with one plane $n = 1$ in which case one has only one matrix element of G^0, i.e.:

$$G_{11}^0 = \frac{1}{E - H_{1,1}^0} \ . \tag{2.75}$$

From this the set of equations (2.71) with $n = 1$ allows one to determine G_{11}, G_{22} and G_{21} for the slab of two planes. These three Green's functions then serve as G^0 for $n = 2$, from which one determines G_{11}, G_{44} and G_{41} for $n = 2$, and so on .... Thus each step of this iterative procedure is equivalent to considering a slab of doubled thickness. When n is large enough for $G_{2n,1}$ to become negligible $G_{1,1}$ and $G_{2n,2n}$ provide the correct surface Green functions. This procedure has many advantages: it is fast and requires a small number of iterations (~ 5) corresponding to one matrix inversion and 14 matrix products; contrary to continued fraction expansions it naturally leads to well defined gaps. As in subsection b) it can be extended to any general situation by considering groups of planes with nearest neighbor interactions.

The original presentation of the method also called the decimation technique [2.43] was made using the formalism of transfer matrices. We have preferred here to write equations directly in terms of Green functions, obtaining a simpler formulation. We shall give later some numerical results obtained with this method.

2.4 Simple Examples of Tight-Binding Surface Bands

In order to have some feeling for the modifications of the band structure due to surfaces, we now treat in detail two very simple illustrations: the semi-infinite linear chain and the (100) surface of a simple cubic lattice. The interest of treating the semi-infinite chain is due to the fact that surface problems at fixed $\boldsymbol{k}_{\parallel}$ are formally equivalent to a semi-infinite linear system. On the other hand the (100) surface of the simple cubic lattice is a nice simple illustration of the fact that surface states can be localized at fixed $\boldsymbol{k}_{\parallel}$ and at the same time fall in the continuum of bulk states.

2.4.1 The Semi-Infinite Linear Chain

a) Boundary States

Let us consider the semi-infinite chain of Fig. 2.8. We use the tight-binding approximation, with one s-orbital per atom. As usual we assume these to form an orthonormal set. We also take the Hamiltonian matrix to be of the form:

$$\langle \Phi_l | H | \Phi_l \rangle = 0 \quad \text{for} \quad l \neq 0 \tag{2.76}$$

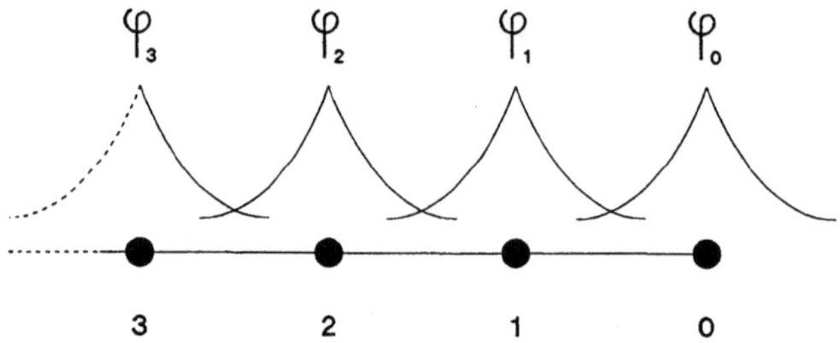

Fig. 2.8. Semi-infinite linear chain. s-type wave functions

$$\begin{aligned}\langle \Phi_0 | H | \Phi_0 \rangle &= U \\ \langle \Phi_l | H | \Phi_{l+1} \rangle &= -\beta\end{aligned} \tag{2.77}$$

where Φ_l is the s orbital centered on atom l. This means that we use a nearest neighbor approximation and take the bulk diagonal element as the origin of energies. In (2.76,77) we have allowed the terminal atom to have a different diagonal term U. As we shall see later this simulates the effect of a possible dipole layer at the surface, induced either by the free surface itself or by the simulation of a chemisorbed layer.

Let us look at the possibility of the existence of localized states. For this we write the wave function Ψ in the form:

$$\Psi = \sum_{l=0}^{\infty} a_l \Phi_l \,. \tag{2.78}$$

To solve the problem, we write the equations:

$$\langle \Phi_l | (E - H) | \Psi \rangle = 0 \tag{2.79}$$

for any value of l. This is equivalent to the following set of equations:

$$\begin{aligned}(E - U) a_0 &= -\beta a_1 \\ E a_1 &= -\beta (a_0 + a_2) \\ E a_k &= -\beta (a_{k-1} + a_{k+1}) \,.\end{aligned} \tag{2.80}$$

If we look for the possibility of a localized state with a wave function having an exponential decay, it is obvious that we have to try a solution with a_l of the form:

$$a_l = a_0 K^l \,. \tag{2.81}$$

The set of equations (2.79) reduces to:

$$\begin{aligned}E &= U - \beta K \,, \\ E &= -\beta \left(K + \frac{1}{K} \right) \,,\end{aligned} \tag{2.82}$$

whose solutions are:

$$K = -\frac{\beta}{U} ,$$
$$E_L = U + \frac{\beta^2}{U} . \tag{2.83}$$

Such solutions are only valid if $|K|$ is smaller than unity, since one must have an exponential decrease at infinity. This means that one must have:

$$\left|\frac{U}{\beta}\right| > 1 \tag{2.84}$$

which leads to:

$$|E| > 2|\beta| . \tag{2.85}$$

It is well known that the continuum of bulk states extends between -2β and $+2\beta$ so that this solution corresponds to a true bound state lying outside the band. For $U > 0$ (repulsive potential) it is above the band, the reverse being true for attractive potentials. One can also see from (2.82) and (2.83) that the state becomes more localized near the boundary as the absolute value of the energy $|E|$ increases. This is a quite general behavior.

b) Local Density of States at the Boundary

We shall now obtain further insight into this problem by calculating the local density of states on the surface atom. For this we first make a direct application of Dyson's equation as in Sect. 2.3.4, the only difference being that our basis functions Ψ_l are atomic s orbitals instead of Bloch sums $\Psi_l(\boldsymbol{k}_\parallel)$ at fixed $\boldsymbol{k}_\parallel$. Let us for the moment consider the situation where the intra-atomic term on the surface atom vanishes. In this case we can apply (2.67) with $H^0_{00} = 0$ and $V_{01} = -\beta$. Taking the local density of states $n_{00}(E)$ as $-\frac{1}{\pi}\mathrm{Im}\, G_{00}(E)$ we obtain:

$$n_{00}(E) = \frac{1}{2\pi\beta^2}\sqrt{4\beta^2 - E^2} . \tag{2.86}$$

This is compared to the local density of states on a bulk atom $n_{\mathrm{B}}(E)$ equal to $1/\pi\sqrt{4\beta^2 - E^2})$ in Fig. 2.9. One clearly sees that the main effect corresponds to an overall narrowing of the density of states for the surface atoms which, as we shall see later, is related to a reduction of the second moment near the surface.

Let us now look at what happens when we introduce the perturbation U on the surface atom. The corresponding Hamiltonian becomes H', of resolvent G', which can be obtained from G making use of Dyson's equation (2.38). This leads to:

$$G'_{00} = G_{00} + G_{00}UG'_{00} = \frac{G_{00}}{1 - UG_{00}} . \tag{2.87}$$

The poles of G'_{00} correspond to the bound states, which can only occur for

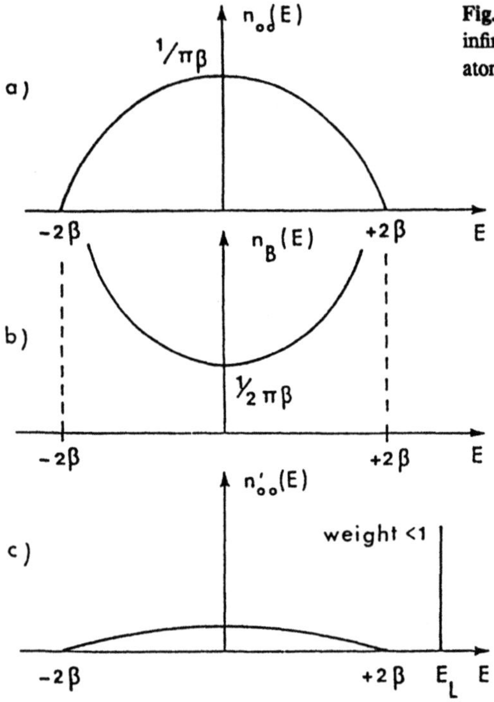

Fig. 2.9a–c. Local densities of states for a semi-infinite chain. (a) Surface atom $U = 0$. (b) Bulk atom. (c) Surface atom $U > \beta$

$|E| > 2\beta$, i.e. outside the band where G_{00} is purely real. From the expression (2.87) we find the same solution as given by (2.83). However the advantage of the present formulation is that it allows the calculation of the new local density of states on the surface atoms. An example is given in Fig. 2.9c, corresponding to $U > \beta$, i.e. to the case of a bound state occurring at an energy E_L above the bulk band. The local density of states $n'_{00}(E)$ is positive and normalized to unity. The weight of the delta function at E_L is smaller than unity and tends towards this value when U becomes infinite.

Another situation of some interest is illustrated in Fig. 2.10 and corresponds to a simplified model of chemisorption. The local density of states on the chemisorbed atom can be determined by the method used above, and some typical situations are illustrated in Fig. 2.11. Figure 2.11a shows the case where $\beta_c \ll \beta$, $U = 0$, i.e. the adsorbed atom is only weakly coupled. The local density of states is then reminiscent of the free atom limit and corresponds to a narrow resonance. The case $\beta_c = \beta$, $U = 0$ (Fig. 2.11b) corresponds to the semi-infinite

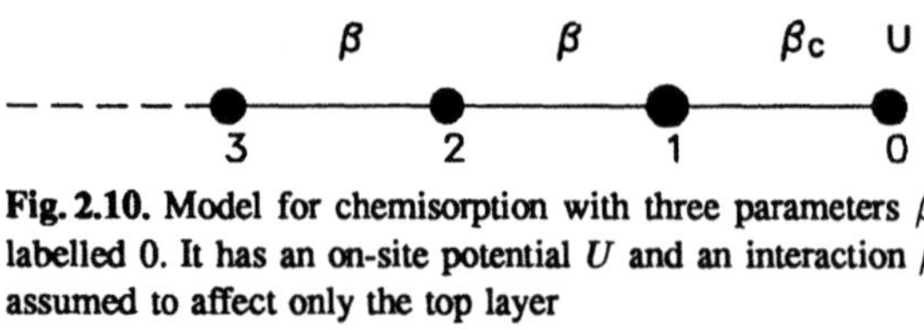

Fig. 2.10. Model for chemisorption with three parameters β, U and β_c. The surface layer atom is labelled 0. It has an on-site potential U and an interaction β_c with the second layer. The surface is assumed to affect only the top layer

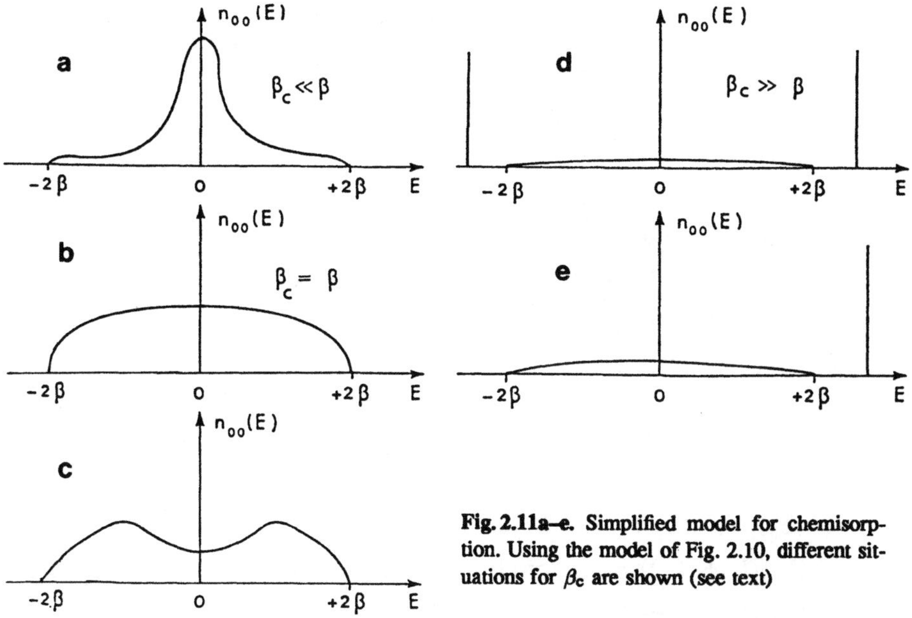

Fig. 2.11a–e. Simplified model for chemisorption. Using the model of Fig. 2.10, different situations for β_c are shown (see text)

chain of Fig. 2.8 and the broadening is of order 2β. Another limit corresponds to $\beta_c \gg \beta$, $U = 0$ (Fig. 2.11d) where the adsorbed atom forms a strong covalent bond with its neighbor, practically giving a bonding state at $-\beta_c$ and an anti-bonding one at $+\beta_c$. An intermediate case $\beta_c > \beta$ is illustrated in Fig. 2.11c, where the bonding and anti-bonding states are beginning to build up. Finally we recall the situation $0 < \beta_c \lesssim \beta$, $U > \beta$ for which a localized state appears above the band (Fig. 2.11e).

Figure 2.11a–e depicts most of the possible situations for chemisorbed atoms and shows the main trends which persist in more realistic descriptions of the electronic properties.

2.4.2 The (100) Face of the Simple Cubic Lattice

We now consider the tight-binding s-band model for the (100) surface of a simple cubic lattice (Fig. 2.12). The model that we adopt is exactly the same as for the linear chain, with an interaction $-\beta$ between any pair of nearest neighbors. Again each diagonal matrix element of the Hamiltonian is taken to be zero except for the surface atoms where it is equal to U. We number the atomic planes by the index l($l = 0$ to ∞) and the atoms within one plane by the index j.

As before, we make use of Bloch functions $\Psi_l(\boldsymbol{k}_\parallel)$ defined by (2.57) where we drop index α since there is only one orbital per unit cell. The only non vanishing matrix elements of the Hamiltonian H are given by:

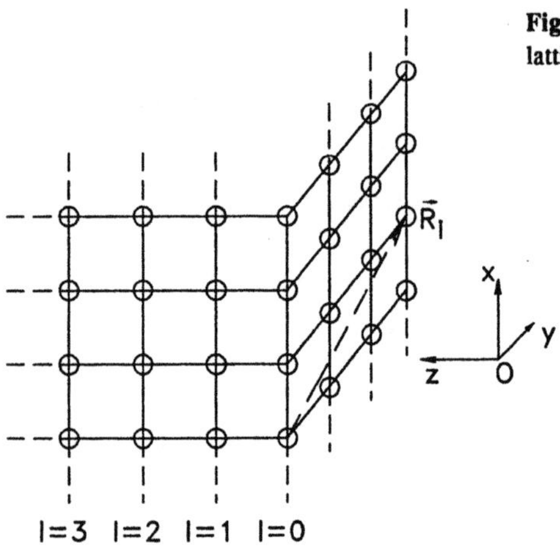

Fig. 2.12. The (100) surface of the simple cubic lattice. The plane is the surface

$$
\begin{aligned}
&\langle\psi_0(\boldsymbol{k}_\parallel)|H|\psi_0(\boldsymbol{k}_\parallel)\rangle = U - 2\beta(\cos k_x a + \cos k_y a) \\
&\langle\psi_l(\boldsymbol{k}_\parallel)|H|\psi_l(\boldsymbol{k}_\parallel)\rangle = -2\beta(\cos k_x a + \cos k_y a) \quad \text{for} \quad l \neq 0 \\
&\langle\psi_l(\boldsymbol{k}_\parallel)|H|\psi_{l+1}(\boldsymbol{k}_\parallel)\rangle = -\beta
\end{aligned}
\tag{2.88}
$$

where k_x and k_y are the components of $\boldsymbol{k}_\parallel$ along the basis vectors of the square lattice representing the (100) planes parallel to the surface. It is clear that $-2\beta(\cos k_x a + \cos k_y a)$ is the dispersion relation for such planes in the bulk.

From the parameters defined by relations (2.88) we see that our system is rigorously identical, at fixed $\boldsymbol{k}_\parallel$, to the semi-infinite chain discussed above with the condition that all diagonal terms are shifted by the quantity $-2\beta(\cos k_x a + \cos k_y a)$. All the results obtained in this case can immediately be transposed. The first of them corresponds to the possibility of obtaining bound surface states. Again one must have the condition $|U| > \beta$, independent of $\boldsymbol{k}_\parallel$, in which case the bound state energy $E_L(\boldsymbol{k}_\parallel)$ is given by:

$$E_L(\boldsymbol{k}_\parallel) = -2\beta(\cos k_x a + \cos k_y a) + U + \frac{\beta^2}{U} , \tag{2.89}$$

as was obtained in (2.83). However it is important to refer this two-dimensional dispersion relation to the bulk band. It is well known that the bulk dispersion relation is given by:

$$E(\boldsymbol{k}) = -2\beta(\cos k_x a + \cos k_y a + \cos k_z a) , \tag{2.90}$$

whose limits $E_{\mathrm{M}}(\boldsymbol{k}_\parallel)$ and $E_{\mathrm{m}}(\boldsymbol{k}_\parallel)$, at fixed $\boldsymbol{k}_\parallel$ are given by:

$$E_{\mathrm{M}}(\boldsymbol{k}_\parallel) = +2\beta - 2\beta(\cos k_x a + \cos k_y a) , \tag{2.91}$$

$$E_{\mathrm{m}}(\boldsymbol{k}_\parallel) = -2\beta - 2\beta(\cos k_x a + \cos k_y a) . \tag{2.92}$$

The same results can be obtained from the Green's function calculation of

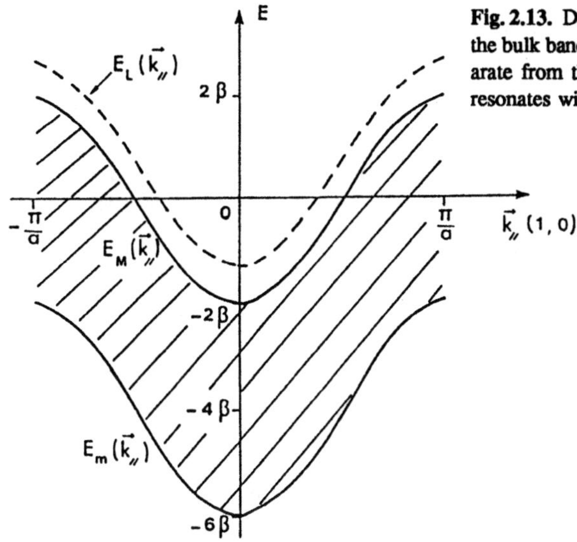

Fig. 2.13. Dispersion relation for $E_L(\boldsymbol{k}_{\parallel})$ and the bulk bands. Although the surface state is separate from the continuum, one may see how it resonates with the bulk states

Sect. 2.4.1 since the continuum of bulk states corresponds to an energy range for which $|E + 2\beta(\cos k_x a + \cos k_y a)|$ lies between 0 and 2β. Then the local density of states in the surface plane is the same as for the semi-infinite chain, again translated by the dispersion relation of the square lattice.

An important and general situation corresponds to the case pictured in Fig. 2.13. There it is seen that, although the surface states are true localized states at fixed $\boldsymbol{k}_{\parallel}$, only part of the surface band lies at energies higher than the top of the bulk band. This shows that localized surface states can exist in the bulk continuum provided that they fall in a forbidden gap, for their value of $\boldsymbol{k}_{\parallel}$. Such gaps often occur for a continuum which results from a superposition of different bands. As will be seen later this can also occur for gaps which only exist for some values of $\boldsymbol{k}_{\parallel}$ and can be considered as "windows" in the bulk bands when these are plotted versus $\boldsymbol{k}_{\parallel}$.

2.4.3 Moment Analysis

We can readily discuss the main features of the above results by determining the moments of the local density of states for surface atoms relative to bulk atoms. For the "s" band cases discussed above, assuming no shift in diagonal matrix elements of H, the second order moment of any atom i can be written from (2.45) in the form:

$$\mu_{2i} = Z_i \beta^2 \tag{2.93}$$

where Z_i is the coordination number (number of nearest neighbors) of this atom. As we have seen before (2.46) the width W_i of the corresponding local density of states ($\sqrt{\mu_{2i}}$) is proportional to $\sqrt{Z_i}$, so that the ratio of the widths for surface

(W_s) and bulk (W_b) atoms is given by:

$$\frac{W_s}{W_b} = \sqrt{\frac{Z_s}{Z_b}} . \tag{2.94}$$

As Z_s is always smaller than Z_b the local density of states at the surface is narrower than in the bulk as is confirmed by the analytical models treated above. This is a general conclusion which has important consequences, especially for surface magnetism and reconstruction of transition metal surfaces as will be discussed later.

2.5 Surface Properties in the Jellium Model

We describe briefly the pioneering calculation on surfaces performed by *Lang* and *Kohn* [2.44]. This was the first application of the local density approximation to a surface problem. It used an idealized description of *sp*-bonded metals, i.e. the jellium model. In this model the ionic charges are described by a continuous background of positive charge. In the bulk this is exactly compensated by a constant electron density. The potential is constant; its solutions are plane waves and the model is automatically coherent. This is no longer true for the surface problem in which one deals with a semi-infinite background of positive charge. The electron density is still constant inside the metal but has a tendency to spread out over the surface as is shown in Fig. 2.14.

The problem is still very simple since all quantities such as the electron and nuclear density, the potential etc. ... are constant in planes parallel to the surface. The Schrödinger equation only depends upon the coordinate z perpendicular to the surface plane. If we take the semi-infinite system to extend from $z = -\infty$ to $z = 0$, then the Hamiltonian can be written:

$$H = T + V_{es}(z) + V_{xc}(z) , \tag{2.95}$$

where $V_{es}(z)$ is the electrostatic potential energy due to the sum of the electronic density $\varrho_e(Z)$ and the positive background charge ϱ_0 which extends from $z = -\infty$ to $z = 0$. $V_{es}(z)$ can be obtained as the solution of Poisson's equation:

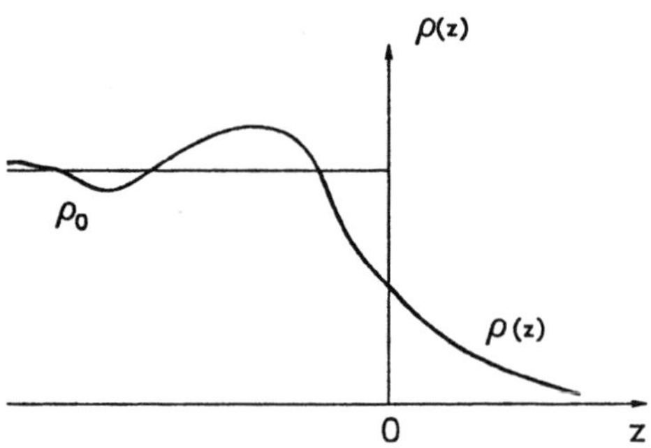

Fig. 2.14. Electron density $\varrho_e(z)$ near the jellium surface. Near to the surface the density oscillates around the bulk value and decays exponentially into the vacuum

$$\frac{d^2}{dz^2} V_{es}(z) = -4\pi[\varrho_e(z) - \varrho_0\theta(z)] \ , \tag{2.96}$$

where $\theta(z)$ is unity for $z < 0$ and zero for $z > 0$. Again $\varrho_e(z)$ is deduced from the wave functions $\psi_i(\boldsymbol{r})$ which take the simple form:

$$\psi_i(\boldsymbol{r}) = \Phi_i(z) \cdot e^{i\boldsymbol{k}_\parallel \cdot \boldsymbol{r}} \ , \tag{2.97}$$

where $e^{i\boldsymbol{k}_\parallel \cdot \boldsymbol{r}}$ is a plane wave with wavevector $\boldsymbol{k}_\parallel$ parallel to the surface plane.

The jellium model is characterized by a single parameter r_s which is the radius of the sphere occupied by one electron, given by:

$$\frac{4}{3}\pi r_s^3 = \frac{1}{\varrho_0} \ . \tag{2.98}$$

Lang and *Kohn* have obtained a self-consistent numerical solution of this problem over a large range of values for r_s($2\,\text{a.u} < r_s < 6\,\text{a.u}$) covering all *sp*-bonded solids. They have used the local-density form of $V_{xc}(z)$ derived from homogeneous electron gas, i.e. the Wigner formula:

$$V_{xc} = -\varrho_e^{1/3}\left[0.984 + \frac{0.944 + 8.77\,\varrho_e^{1/3}}{(1 + 12.57\,\varrho_e^{1/3})^2}\right] \ . \tag{2.99}$$

A typical form of the results for $\varrho_e(z)$, $V_T(z) = V_{es}(z) + V_{xc}(z)$, is given in Fig. 2.15. The first important quantity which can be obtained from such a calculation is the work function. The plot of the results as a function of r_s is given in Fig. 2.16 where it is compared to some experimental values. The agreement is excellent,

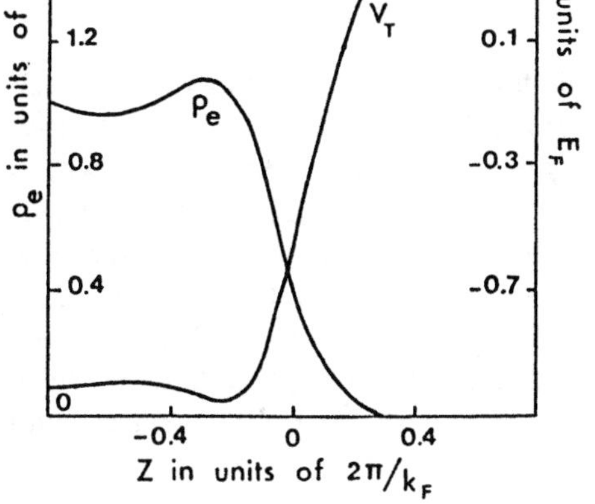

Fig. 2.15. Jellium results for ϱ_e, V_{es} and the total potential V_T; $r_s = 4$

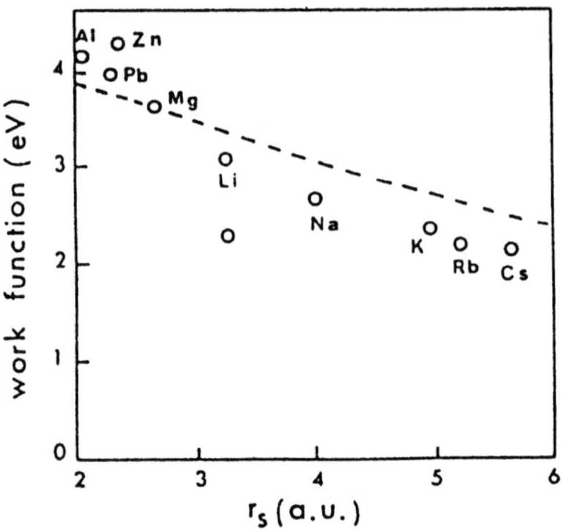

Fig. 2.16. Work function for various metals in the jellium model [2.44]. (o – experimental value, - - - - jellium model)

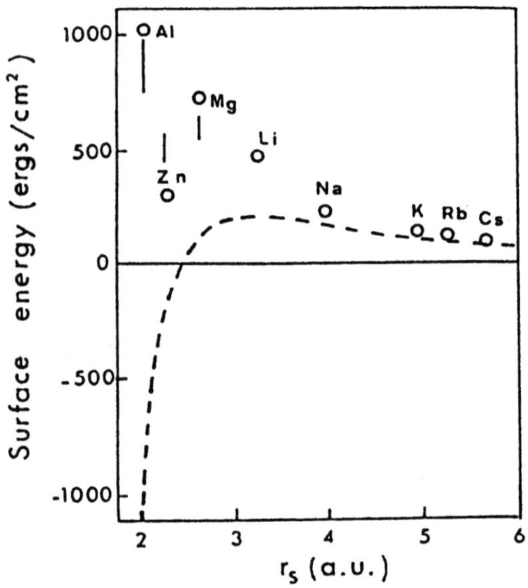

Fig. 2.17. Surface energy. (o – experimental, - - - - jellium model without correction). Vertical bars are theoretical results for bcc (110) and fcc (111) surfaces with first-order pseudopotential correction

showing that the jellium model is valuable for that physical quantity. This is not so true for the surface energy, plotted in Fig. 2.17. For small r_s, i.e. high density systems, the surface energy becomes negative. *Lang* and *Kohn* have shown, to first order, that replacing the average lattice ($\varrho_0\theta(z)$) by a lattice of discrete ions represented by atomic pseudopotentials corrects for most of this discrepancy. However, in view of the large corrections involved, the validity of such a first-order approximation is not guaranteed, and more quantitative calculations are needed. Such a calculation, based on local atomic pseudopotentials, has been reported recently [2.45]. It essentially confirms the jellium results, the maximum correction being of the order of 0.2 eV.

Exercises

2.1 Consider a homonuclear diatomic molecule (e.g. H_2) treated in the tight binding approximation with one s orbital on each atom. Write its total energy using the assumptions of Sect. 2.1.2c. Find the stability condition and express the second-order derivative of the energy in terms of its cohesive energy. Extend this derivation to the molecular model of covalent systems like silicon.

2.2 Determine the first three moments of the density of states for an s band in: the simple cubic, fcc, bcc and diamond lattices.

2.3 Calculate one diagonal matrix element of the Green's function for a tight binding s band on a simple cubic lattice with nearest neighbor interaction β. Use the recursion method and determine the first two steps (i.e. 4 parameters of the tridiagonalized matrix).

2.4 Separate $G(E)$ of (2.23) into its imaginary part (written in terms of delta functions) and its real part which involves a principal part. Show that the latter can be written as an integral transform (Hilbert transform) of the former.

2.5 Starting from the density of states determined in exercise 1.8 calculate the diagonal Green's function G_{00} on an arbitrary site of the infinite linear chain (tight-binding s band). Do this by expressing the real part of G_{00} as a Hilbert transform (exercise 2.4) and calculating it for $|E| > 2\beta$ and $|E| < 2\beta$.
Deduce from this the diagonal term of G on the end atom of the semi-infinite chain by cutting one bond of the infinite chain and using Dyson's equation.

2.6 Consider one trivalent silicon atom connected to a silicon Bethe lattice (i.e. a system where each atom has the normal tetrahedral coordination with normal bond length and bond angles, but where there is no closed loop of bonds as in crystalline silicon). Describe its electronic structure by the tight-binding model of exercise 1.2 with the two parameters Δ and β defined in terms of sp^3 hybrid orbitals. Calculate G_{00} the diagonal matrix element of the Green's function on the dangling sp^3 hybrid of the trivalent Si atom. For this use Dyson's equation relating the Green's function matrix G to the unperturbed one g corresponding to the problem where the trivalent atom is decoupled from the rest of the system. In this way relate G_{00} to some g_{ii} and discuss why these quantities must be equal. Show that this gives an equation for G_{00}, solve it and derive the corresponding density of states. Show that it has a localized state in the gap and discuss the origin and localization of this state.

3. Transition Metal Surfaces

Transition metals represent very interesting materials in which one can fairly readily analyze trends in their properties along the transition series. Although there have been recent calculations of their bulk and surface properties in the local-density approximation, our discussion here will be mainly based on the tight-binding method. This is because the latter technique, even at its simplest level of sophistication, has been very successful in describing chemical trends associated with the filling of the *d*-band. To begin with we thus recall the main features of the band structure of these materials and how it is possible to derive a simple model for their cohesive properties, elastic constants and even phonon dispersion curves.

3.1 Bulk Properties

We recall here a picture originally proposed by *J. Friedel* [3.1] for analyzing the bulk properties of transition metals. It is based on the fact that the radial extension of the *d*-orbitals is rather small (~ 0.5 Å) compared to the interatomic distance. This leads to a narrow *d* band whose width lies in the range 5 to 10 eV. Such a situation is ideally suited for a tight binding description. This is not true of the *s*- and *p*-orbitals which are much more extended, leading to a broad *sp* band which overlaps the narrow *d*-band as pictured in Fig. 3.1.

In this picture the chemical trends along the transition series can be understood by the filling of the *d*-band (there are 5 *d*-states allowing up to 10 electrons

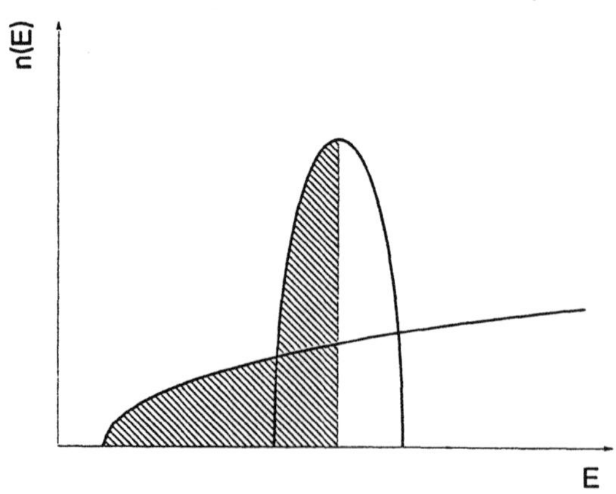

Fig. 3.1. *s* and *d* densities of states in transition metals. The peak corresponds to the *d* part of the density of states

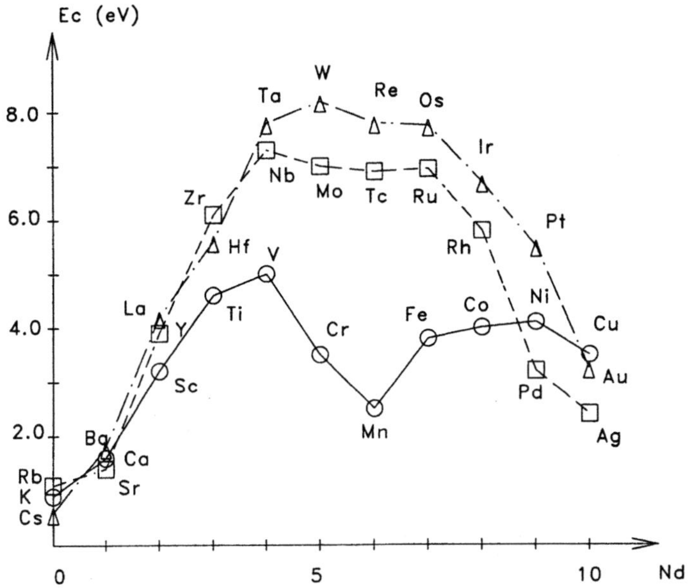

Fig. 3.2. Cohesive energy E_c for the three series of transition metals [3.2]

per atom) while the number of s-electrons remains practically constant such that the atomic configuration in the solid is close to $d^{n-1}\,s^1$. This description is consistent with the observation of a large paramagnetic susceptibility and a large electronic specific heat, which are indications of a large density of states at the Fermi level.

A very interesting feature of transition metals is the behavior of their cohesive properties along the transition series (Fig. 3.2). The overall parabolic shape and the large maximum value ($\sim$ 8 eV) correspond again to the filling of a narrow band with a high density of states. To show that this is the case let us use the original argument of *Friedel* and approximate the exact density of states by a rectangular one of width W (Fig. 3.3).

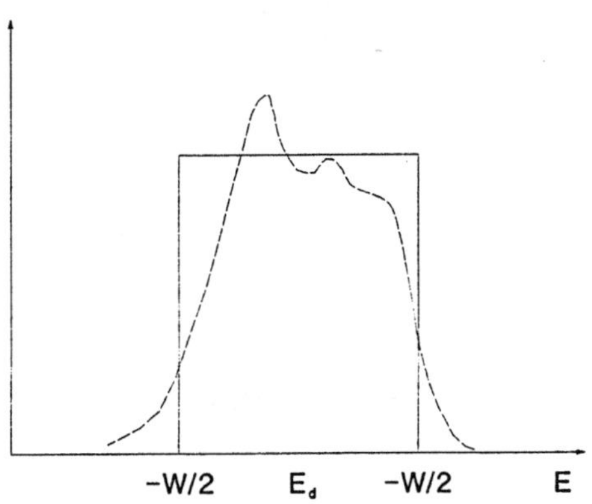

Fig. 3.3. Approximation of the d densities of states by a rectangle compared to a typical profile of the true density of states (*dotted line*)

As in any simple tight-binding treatment the center of gravity of the band remains the atomic level, with the rectangular density of states centered on the atomic d-level E_d. Its height must be $5/W$ in order to accommodate the five d-states per atom. Taking E_d as the energy origin, the band structure energy E_{bs} (i.e. the sum of one-electron energies), which is defined as

$$E_{bs} = 2\int^{E_F} n(E)E\,dE \;, \tag{3.1}$$

can be expressed as a function of the number N_d of d-electrons per atom, given by

$$N_d = 2\int^{E_F} n(E)dE \;. \tag{3.2}$$

For the rectangular shape one readily finds

$$E_{bs} = -\frac{W}{20}N_d(10 - N_d) \;. \tag{3.3}$$

If we consider that N_d varies from 0 to 10 along the transition series, $-E_{bs}$ already has the general shape of the experimental curve of Fig. 3.2, with its maximum at the center of the series. Of course E_{bs} is not the only contribution to the total energy per atom since one should correct it for electron-electron interactions (which have been counted twice) and for nuclear repulsions. It has been argued [3.3] that all corrective terms could be expressed as sums of pair potentials, corresponding to the electrostatic repulsion of neutral atoms. As mentioned in Sect. 2.1.2 this results in a repulsive contribution to the total energy, of the form

$$c_0 \exp(-pR_{ij}) \tag{3.4}$$

for a pair of atoms i, j at a distance R_{ij} from one another. Usually this sum is truncated to nearest neighbors.

Before obtaining an estimate of the cohesive energy we need a more precise evaluation of the rectangular bandwidth W in terms of the actual tight binding interatomic interactions. The rectangular shape is an approximation to the exact partial atomic density of states (as defined in Sect. 2.3.3) and the best value of W is obtained by equating the second order moment of the two curves (Sect. 2.3.3). For the rectangular shape the second-order moment whose general definition is

$$\mu_2 = \int n(E)E^2 dE \tag{3.5}$$

turns out to be given by

$$\mu_2 = 5\frac{W^2}{12} \;, \tag{3.6}$$

while the exact value of μ_2 can be deduced from (2.44) leading, for atom i, to

$$\mu_{2i} = \sum_{\alpha}\sum_{j\beta} |\langle i,\alpha|H|j,\beta\rangle|^2 \;, \tag{3.7}$$

where α represent the five d-orbitals of atom i, and β those of atom j. If one assumes that the d-d matrix elements in (3.7) are truncated to nearest neighbors and that they scale with interatomic distance R_{ij} as $\exp(-qR_{ij})$, then one finds

$$\mu_{2i} = 5\beta_0^2 \sum_j \exp(-2qR_{ij}) \, , \tag{3.8}$$

where β_0 is a parameter characteristic of the average strength of d-d interactions. In the perfect crystal μ_{2i} is independent of i and using (3.6) and (3.8) we obtain

$$W = \beta_0\sqrt{12Z}\exp(-qR) \tag{3.9}$$

where R is the nearest neighbor distance and Z is the coordination number. In the same way the repulsive contribution reduces to $Z/2C_0\exp(-qR)$ so that the total energy per atom becomes

$$E_{\mathrm{T}} = -\frac{\beta_0\sqrt{12Z}}{20}N_d(10-N_d)\exp(-qR) + \frac{Z}{2}C_0\exp(-pR) \, . \tag{3.10}$$

This expression depends on N_d mainly through the product $N_d(10-N_d)$ while β_0, c_0, q and p have a much weaker dependence [3.3] and can be assumed constant in a zero order treatment. Minimizing (3.10) with respect to R one easily sees that the equilibrium interatomic distance R_e obeys the relation

$$R_e = R_e(N_d = 5) - \frac{1}{(p-q)}\ln\frac{N_d(10-N_d)}{25} \, , \tag{3.11}$$

whose general behavior is in agreement with experiment [3.3]. For the cohesive energy E_c [equal to $-E_{\mathrm{T}}(R_e)$] one can write

$$E_c = E_c(N_d = 5)\frac{N_d(10-N_d)}{25} \, , \tag{3.12}$$

whose parabolic shape corresponds excellently with the experimental results of Fig. 3.2. It is also possible to deduce the compressibility K from the same model. Using

$$K = \frac{R_e^2}{9}\left(\frac{d^2E_{\mathrm{T}}}{dR^2}\right)_{R_e} \, , \tag{3.13}$$

one easily arrives at the following relation

$$K = \frac{pqR_e^2}{9}E_c \, . \tag{3.14}$$

Good average values of p and q are such that

$$\begin{aligned} pR_e &\sim 9 \\ qR_e &\sim 3 \end{aligned} \tag{3.15}$$

so that one concludes that K must have the same shape as the cohesive energy E_c as a function of N_d, the filling of the d band. These values of p and q can be obtained from direct calculations [3.4]. The q value is slightly smaller than the one given by Harrison's empirical law [3.5] in which the d-d interactions scale like R_e^{-5} whose logarithmic derivative $(-5/R_e)$ is equal to $-q$. However *Harrison*'s expression also contains a corrective factor that varies from element to element so that direct comparison is not straightforward.

The agreement between the theoretical laws (3.11–13) and experiment shows that the variation of N_d is indeed the leading factor in describing bulk properties of transition metals. However the predicted curves are symmetrical with respect to $N_d = 5$, contrary to Fig. 3.2. A possible way of improving on this point is to introduce a constant shift $-\alpha$ of all diagonal terms. This introduces an extra term $-N_d\alpha$ in the band structure energy. If furthermore α is taken to have the same dependence as $W[\exp(-qR)]$, the previous theory can be extended fairly easily and provides the required asymmetric behavior of E_c and K. The numerical values deduced from a best fit to experiment are given in Table 3.1.

Table 3.1. Width W and shift α of the d band in the three transition series

Transition series	1	2	3
W [eV]	7.5	9	10.5
α [eV]	0.37	0.45	0.52

Of course this treatment only includes d-electrons and neglects the s and p contributions as well as s-d mixing. It is thus more appropriate near the center of the series ($N_d \sim 5$) where E_c is much larger than for typical sp metals for which $E_c \sim 1\,\text{eV}$. A further proof of the essential validity of this description is its application to the prediction of phonon dispersion curves, which has given fairly good results for the cases investigated [3.6]. However it must be noticed that the model used here can sometimes lead to incorrect predictions for some particular physical quantities. For instance this can occur for the elastic constant C' in fcc systems: for $N_d \simeq 5$ the second moment model leads to $C' < 0$, while a more complete treatment including moments up to the fifth provides a correct value; for $N_d \simeq 9$ the reverse is true [3.7]. This last discrepancy is probably related to the fact that the model is less valid for almost filled or almost empty bands.

3.2 Work Function, Dipole Layer and Core-Level Shifts

An important quantity characterizing crystal surfaces is the work function $e\Phi$ which, as shown in Fig. 3.4, is the minimum ionization energy of an electron from the Fermi level to the vacuum level. As is apparent in Fig. 3.4 it is the sum of a pure bulk quantity $|\mu|$ (where μ is the chemical potential) and a pure surface contribution χ, the dipole layer. As we shall see the most important step

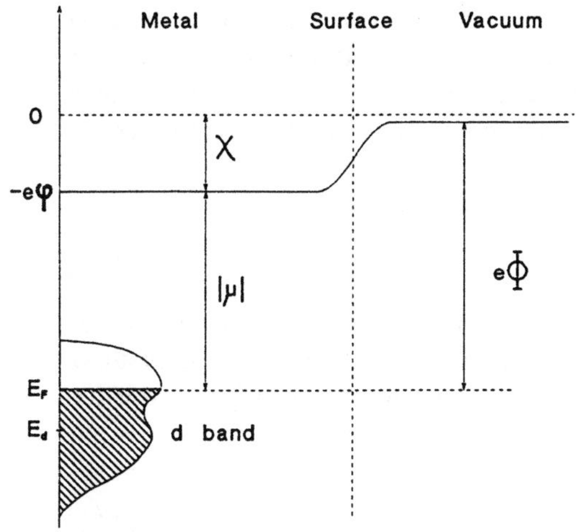

Fig. 3.4. Surface of a transition metal: ϕ – work function; χ – dipole layer; μ – chemical potential

in calculating $e\Phi$ lies in a proper determination of μ. However the contribution of the surface dipole layer is not negligible as evidenced by the substantial core level shifts that have been measured at the surface [3.8]. In the following we discuss these effects separately.

3.2.1 Determination of the Chemical Potential

The easiest way to obtain the chemical potential in these systems is to consider the d-electrons and determine the absolute position of the Fermi level from the filling of the d-band. In a tight binding scheme this requires a knowledge of the d-level in the solid from which one can obtain the Fermi level position. The average d level in the solid is obtained from the following expression

$$E_d = \tfrac{1}{5} \sum_i \langle \varphi^i_{d0} | H | \varphi^i_{d0} \rangle \ , \qquad (3.16)$$

where φ^i_{d0} is the ith d-wave function of the central site atom. This is equivalent to writing

$$E_d = \langle \varphi_{d0} | H | \varphi_{d0} \rangle \ , \qquad (3.17)$$

where φ_{d0} is the common radial part of the d-orbitals. The one electron Hamiltonian H can be written

$$H = T + V_{\rm H} + V_{\rm xc} \ , \qquad (3.18)$$

where T is the kinetic energy, $V_{\rm H}$ the Hartree potential and $V_{\rm xc}$ the exchange–correlation potential.

The Hartree potential can be obtained from simple arguments. First the s electron charge density can be separated into atomic parts $\varrho_{s,i}$ contained in the Wigner and Seitz sphere centered on atom i. As to the d electron density we shall, in view of the small overlap of neighboring d orbitals, approximate it by a superposition of free atom densities. From this we can easily write V_H as a superposition of atomic contributions $(V_H)_i$. It is clear that, for $i \neq 0$, $(V_H)_i$ gives a negligible contribution to (3.18), due to the small extension of φ_{d0}, which allows the nuclear charge of the neighbors to be almost perfectly screened by their electronic charge distribution.

From these arguments, for the central site we have

$$H \sim T + (V_H)_0 + V_{xc} . \tag{3.19}$$

In contrast to the situation in nearly-free-electron systems, one has no precise knowledge of V_{xc} in transition metals, so must use some type of approximation. We shall retain here the reasonable assumption of an atomic exchange–correlation hole.

Within this context one can write

$$E_d = E_{d0} + \langle \varphi_{d0} | \delta V_s | \varphi_{d0} \rangle \tag{3.20}$$

where E_{d0} is the free-atom value and δV_s is the modification of the Hartree potential due to the change in the s electron density between the free atom and the crystal (the d electron density is identical to the free atom one).

To estimate the second term in (3.20) we have taken into account the fact that φ_{d0} is fairly concentrated on the atom so that

$$\langle \varphi_{d0} | \delta V_s | \varphi_{d0} \rangle \simeq \delta V_s(0) \tag{3.21}$$

δV_s being the potential at the atomic site.

To proceed further one must have information about the s-wave functions in the free atom and in the crystal. For the free atom we use Slater-type orbitals

$$\varphi_{n,s} = N_{n,s} r^{n-1} e^{-\beta r} \tag{3.22}$$

for which optimal values of α have been tabulated for most atoms by *Clementi* [3.9]. In the crystal, we use the *Wigner–Seitz* approximation with

$$\varphi'_{n,s} = N'_{n,s} r^{n-1}(1 + \gamma r) . \tag{3.23}$$

The choice of (3.23) has been guided by the following considerations:

i) the wave function must behave like the free-atom function (3.22) near the origin, i.e. like r^{n-1} in our case, this fact offering the advantage of a cancellation between the errors in $\delta V_s(0)$ introduced by the neglect of oscillations of both wave functions inside the atomic core;
ii) the crystal wave function must have some flexibility, introduced here by the term γr^n, in order to satisfy the Wigner–Seitz boundary conditions.

$$\left(\frac{d\varphi'_{ns}}{dr}\right)_{r=\Gamma} = 0 \tag{3.24}$$

where Γ is the radius of the Wigner–Seitz sphere.

This leads to

$$\gamma = -\frac{n-1}{n\Gamma} \,. \tag{3.25}$$

The shift $\delta V_s(0)$ can then be easily evaluated and written as

$$\delta V_s(0) = \frac{1+\theta(n)}{\Gamma} - \frac{\beta}{n} \,, \tag{3.26}$$

where $\theta(n)$ is a smooth function of n which takes the values 0.19 for $n = 4$; 0.16 for $n = 5$; and 0.13 for $n = 6$. The shifts (3.26) have been evaluated for the three transition metal series and are given in Table 3.2, assuming a $d^{p+1}s$ configuration for all atoms.

Table 3.2. Shifts $\delta V(0)$ vs the number of electrons in the d band for the three transition series. The values of Γ_s have been taken from [3.11], and the values of β are from [3.9] after a slight correction [3.12]

	N_d	1	2	3	4	5	6	7	8	9	10
	first series		3.3	3.8	4.1	4.7	4.35	4.35	4.35	4.35	3.5
$\delta V_s(0)$	$n = 4$										
[eV]	$n = 5$		2.7	3.8	3.8	4.1	4.35	4.22	4.08	3.54	
	$n = 6$			3.3	3.5	3.8	4.1	4.1			

Now to obtain E_d one must know the free-atom level E_{d0} which is equal to the opposite of the ionization potential from the $d^{p+1}s$ configuration to the $d^p s$ of the positive ion. This can be deduced either from experimental data [3.10], or from the results of self-consistent Hartree–Fock computations [3.11]. We have chosen the latter, but instead of tabulating the values for all transition metals, we have chosen to draw an average curve of E_{d0} for the three transition series, as a function of N_d, the number of d electrons per atom. The results are nicely represented by the straight line of Fig. 3.5, described by

$$E_{d0} = -3.5 - N_d \text{ in eV} \,. \tag{3.27}$$

To this one must add the shift of (3.26) due to s-electrons, which remains almost constant throughout the series and is of the order of ~ 4 eV. Thus the level E_d in the solid is given by

$$E_d \sim 0.5 - N_d \text{ in eV} \,. \tag{3.28}$$

To find the absolute position of the Fermi level E_F we must know its relative position ε_F with respect to E_d, which can be deduced from the knowledge of the density of states. This can be done numerically but here we prefer to make use again of the simplified rectangular density of states. This gives

$$\varepsilon_F = \frac{N_d - 5}{10} W \,. \tag{3.29}$$

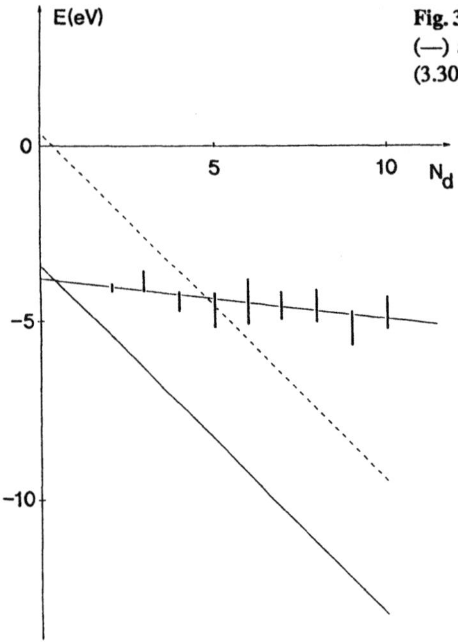

Fig. 3.5. Straight lines corresponding to equations (3.27) (—) and (3.28) (- - -) with error bars on E_F as given by (3.30) and experiment

To obtain an average prediction for the three transition series we take the value $W = 9\,\mathrm{eV}$ appropriate for the second series (Table 3.1). Combining (3.28) and (3.29) we obtain the following expression

$$E_F \simeq -4 - 0.1 N_d \text{ in eV} . \tag{3.30}$$

Figure 3.5 shows that this expression for E_F is in very good agreement with experiment [3.14]. Again this shows the predictive power of the simple representation of the d-band for transition metals. Improvements in the details of the density of states will only bring minor modifications as shown by comparison with the results of [3.15].

3.2.2 The *d*-Electron Dipole Layer

We analyze in this section the d-electron distribution near the surface, ignoring for the moment the s-contribution and its possible influence. This will allow us to have a deeper insight into the d electron screening properties near the surface. We first present the full tight binding self-consistent treatment and then describe a very simple approximation to it based on a local charge neutrality argument. We then show that the dipole layer contribution to the work function is likely to be small, confirming the picture of Sect. 3.2.1. We treat the d-electrons in a tight-binding formalism, the basis consisting of five d-orbitals per atom. We neglect interatomic overlap integrals, only taking into account the hopping integrals and intra-atomic matrix elements of the Hamiltonian. The physical arguments presented here are essentially equivalent to those of [3.16].

The advantage of using the tight-binding model is that we get a very simple and efficient description in terms of atomic charges. In the perfect crystal each atom has the same number of d electrons and is neutral. The creation of a surface will change this situation so that it becomes necessary to perform a self-consistent calculation to determine the new charge distribution. This difficult problem can be simplified by considering that only the diagonal part of the Hamiltonian matrix is affected by the corresponding Coulomb potential. Furthermore we shall take equal intra-atomic diagonal terms for all d orbitals belonging to the same atom. Calling U_i the change in the intra-atomic term on atom i, we use the following expansion:

$$U_i = \sum_j \gamma_{ij} \delta N_j \ , \tag{3.31}$$

where δN_j is the net excess of electrons on the jth atom, and the γ_{ij} are close to the inverse interatomic distances, except for γ_{ii}, which is the intra-atomic Coulomb repulsion term. One notices that (3.31) is analogous to the CNDO (complete neglect of differential overlap) approximation [3.17].

On the other hand, one can calculate the atomic populations in terms of the perturbation due to the surface. In our model this is obtained by cutting the crystal into two non interacting parts. This leads to a suppression of some interatomic coupling terms. To such a perturbation one must then add the intra-atomic matrix elements U_i so that one can write:

$$\delta N_j = f_j(\{U_k\}) \ . \tag{3.32}$$

N_j is then, in principle, a function of the whole set of U_k, the f_j being obtained by a Green's function technique or a method of moments (Sect. 2.3).

Once the f_j are known it is in principle possible to solve the combined set of equations (3.31) and (3.32). Their solution can be simplified by noticing that for a free surface problem the N_i and U_i take the same value for all the atoms of a given plane parallel to the surface. One can then simply use (3.31) and (3.32) where the indices now correspond to the planes instead of the atoms. The only difference is that the γ_{ij} are given by the sum of all the Coulomb terms relating one atom of the ith plane to all atoms of the jth plane. This involves sums over planes of terms such as $1/R_{ij}$ which can be best obtained by Ewald summation techniques [3.18].

To simplify the evaluation of the atomic excess charges δN_i we proceed as at the beginning of this chapter. We replace the local density of states $n_i(E)$ of an atom i by its rectangular approximation (Fig. 3.3) but where the d level is shifted by an amount U_i and the width W_i depends on the position of the atom on the surface. In this way we can easily integrate $n_i(E)$ to obtain the number of d electrons per atom

$$N_i = 5 + 10\frac{\varepsilon_F - U_i}{W_i} \ , \tag{3.33}$$

where ε_F is the relative position of the metal Fermi level with respect to the bulk d-level. Following the same approach as in (3.7–9) the width W_i can be expressed as

$$W_i = \sqrt{12Z_i}\,\beta\,, \tag{3.34}$$

where β stands for $\beta_0 \exp(-qR)$ in (3.9). We then find a very simple expression for the change δN_i in electron population with respect to the bulk situation

$$\delta N_i = \frac{10}{\sqrt{12}\beta}\left(\frac{\varepsilon_F - U_i}{\sqrt{Z_i}} - \frac{\varepsilon_F}{\sqrt{Z_b}}\right)\,, \tag{3.35}$$

where Z_b is the bulk coordination number. Equation (3.35) gives a local relationship between δN_i and U_i. Equations (3.31) coupled with (3.35) represent the tight-binding analog of the Thomas–Fermi approximation [3.19]. Thus they cannot produce the Friedel oscillations [3.20] but these do not play an important role for most transition metals in view of the large electron density characteristic of these systems. Here, however, instead of solving these equations completely we shall use a very simple model which proves to be quite accurate. In view of the high density of states at the Fermi level for most transition metals and of the large values taken by the Coulomb interactions the δN_i will be small enough that a very good approximation to the exact self-consistent U_i is obtained by writing $\delta N_i = 0$ in (3.35), i.e. by imposing a condition of local charge neutrality. This leads directly to

$$U_i = \varepsilon_F\left(1 - \sqrt{\frac{Z_i}{Z_b}}\right)\,, \tag{3.36}$$

which can be stated explicitly in terms of N_d, if use is made of (3.29), thus leading to

$$U_i = \left(1 - \sqrt{\frac{Z_i}{Z_b}}\right)\left(\frac{N_d - 5}{10}\right)W\,. \tag{3.37}$$

Within our simplified approach this gives a completely general law for the intra-atomic shift in terms of the coordination number, N_d and W. As the d-states have a small spatial extension this shift of the valence levels is close to the core level shift and, in the following, we shall use (3.37) to discuss corresponding experimental data. Near the surface U_i is also related to the height of the dipole layer as shown in Fig. 3.6. From this figure the height of the dipole layer is simply equal to the value given by (3.37) for the surface atoms. The maximum effect occurs for the minimum Z_i/Z_b, i.e. 1/2. Taking $W \sim 9\,\mathrm{eV}$ as before we get

$$U_0 \sim 0.27(N_d - 5)\ \mathrm{in\,eV}\,. \tag{3.38}$$

The dipole layer is found positive for $N_d > 5$, negative otherwise. Its contribution

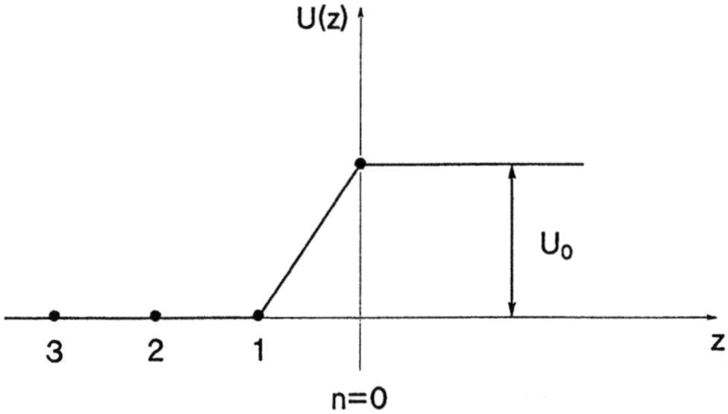

Fig. 3.6. Intra-atomic matrix element and dipole layer when the dipole layer is confined to the first two planes

is negligible at the center of the series but not for almost filled or empty bands where it can amount to more than 1 eV. However we shall see that relaxation effects will reduce this value so that the dipole layer will be found to play a minor role in the evaluation of the work function.

Although these conclusions are based on an extremely simple description they are completely confirmed by the results of numerical calculations [3.21] as we shall see in the next section.

3.2.3 Core-Level Shifts

The determination of the surface core-level shifts relative to the bulk is a powerful source of information as regards the electronic properties in the vicinity of the surface. As an extensive review of this problem has been published relatively recently [3.22] we shall only comment here on systematic trends that can be obtained directly from the simple models described previously.

For this we start from the definition of the binding energy E_B of a core level which can be expressed quite generally as the difference

$$E_B = E_{tot}(n_c - 1) - E_{tot}(n_c) \tag{3.39}$$

of the total energies of the system with $n_c - 1$ and with n_c electrons on a given atomic core state. The calculation of this quantity is relatively complex due in particular to electronic relaxation effects. However it can be made much simpler by using Slater's transition state argument [3.23] which consists of expanding both quantities on the right hand side of (3.39) with respect to the intermediate configuration $(n_c - 1/2)$. This leads to

$$E_B = -\left(\frac{\partial E_{tot}}{\partial n_c}\right)_{n_c - 1/2} + O_3 \, , \tag{3.40}$$

where the correction O_3 is of order 3 in the expansion. In usual self-consistent treatments, such as Hartree–Fock or local density, the derivative of E_{tot} with re-

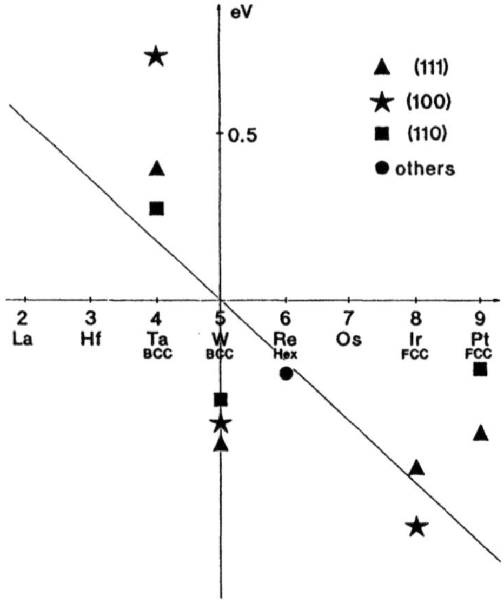

Fig. 3.7. Surface core level shift as a function of N_d for different surface planes (taken from [3.22])

spect to the occupation number of a given state is just its one-electron eigenvalue ε_c. This means that one can rewrite E_B as

$$E_B = -\varepsilon_c \left(n_c - \tfrac{1}{2}\right) , \tag{3.41}$$

an expression that incorporates electronic relaxation effects.

This reasoning applies to any atom and can be used to evaluate the surface core-level shift Δ^S as the negative of the quantity U_i given by (3.37), i.e.,

$$\Delta^S = \left(\sqrt{\frac{Z_s}{Z_B}} - 1\right)\left(\frac{N_d - 5}{10}\right) W , \tag{3.43}$$

where Z_s is the coordination number of a surface atom. To have an estimate of this function we have taken an average value for $\sqrt{Z_s/Z_B} - 1 \sim 1/6$ and have considered the $5d$-series, taking for W the average value of Table 3.1, i.e. 10.5 eV. This gives the straight line of Fig. 3.7 in which are also given the experimental values (taken from [3.22]). It is seen that this straight line gives a fair general order of magnitude and also the average tendency but does not predict correctly individual results and especially the sign reversal between Ta and W. This can be ascribed to the fact that we have used a symmetrical density of states as in any second order moment theory. In fact the computed densities of states have a non negligible third moment and it is the corresponding asymmetry that causes the sign reversal of Δ^s between Ta and W. This is clearly shown in Fig. 3.8 where the surface potential deduced from a neutrality condition but with computed densities of states [3.21] is plotted. The results of this figure now give a quantitative agreement with the experimental values of Fig. 3.7.

A more detailed discussion of the individual cases can be found in [3.22].

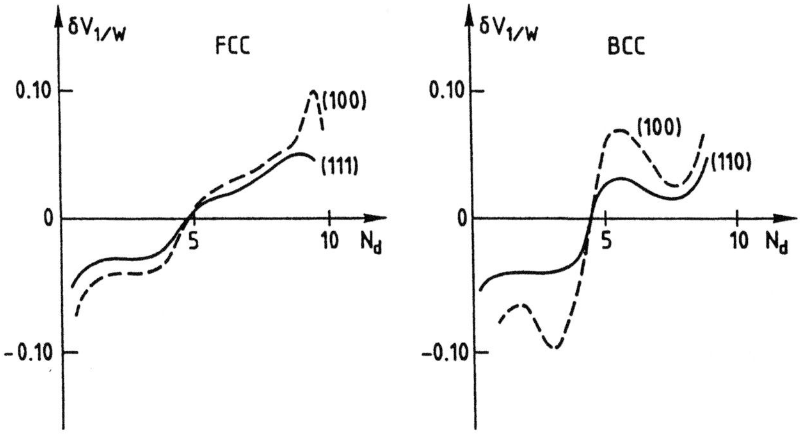

Fig. 3.8. Surface potential computed from the neutrality condition, and the calculated densities of states for (100) and (111) planes of fcc crystals and (100) and (110) planes of bcc crystals [3.21]

3.3 Calculation of the Total Energy

In this section we derive a general expression for the change in total energy with lattice geometry. This will allow us to determine the surface tension and to have an estimate of relaxation energies. We shall conclude from this that atomic relaxation tends to reduce the dipole layer with respect to the value determined previously for the perfect crystal geometry.

3.3.1 Influence of the Lattice Geometry

It is obvious that the lattice geometry is modified in the neighborhood of a surface. As discussed in Chap. 1, two effects can arise: relaxation and reconstruction. We now consider these aspects and try to calculate the change in total energy associated with the new lattice configuration. To achieve this simply we still make use of the successful device consisting in approximating each atomic density of states $n_i(E)$ by a rectangle of height $5/W_i$ extending from $-W_i/2 + U_i$ to $W_i/2 + U_i$. In this way the contribution of atom i to the band structure energy can be written as in (3.3)

$$E_{\text{bs},i} = -W_i \frac{N_i(10 - N_i)}{20} . \tag{3.44}$$

This expression depends implicitly on U_i through N_i. However we have seen that a very good approximation for U_i is the value that gives $N_i = N_d$, i.e. the bulk value. This allows us to express the change $\delta E_{\text{bs},i}$ with respect to the bulk value E_{bs} given by (3.3) as

$$\delta E_{\text{bs},i} = \left(\frac{W_i}{W} - 1\right) E_{\text{bs}} . \tag{3.45}$$

As discussed in Sect. 3.1 the ratio W_i/W can be calculated by a second moment analysis. Using (3.6) and (3.8) this leads to

$$\delta E_{\mathrm{BS},i} = \left\{ \sqrt{\frac{1}{Z_{\mathrm{b}}} \sum_j \exp(-2q\delta R_{ij})} - 1 \right\} E_{\mathrm{bs}} \,, \tag{3.46}$$

where the sum over j corresponds to the nearest neighbors of atom i, δR_{ij} being the change in interatomic distance with respect to the bulk crystal. The same analysis applies to the repulsive contribution. For this we attribute to atom i half the sum of the repulsive pair potentials connecting it to its nearest neighbors. Defining $\delta E_{R,i}$ as the change in this quantity with respect to the perfect crystal value E_R we can write from (3.4)

$$\delta E_{Ri} = \left[\frac{1}{Z_{\mathrm{b}}} \sum_j \exp(-p\delta R_{ij}) - 1 \right] E_R \,. \tag{3.47}$$

From the bulk equilibrium relation $pE_R + qE_{\mathrm{bs}} = 0$ obtained by minimizing E_{T} with respect to R in (3.10), one can rewrite E_R and E_{bs} in terms of the cohesive energy E_{c} per atom. This leads to

$$\begin{aligned} E_{\mathrm{Bs}} &= \frac{E_{\mathrm{c}}}{1 - q/p} \\ E_R &= \frac{q}{p} \frac{E_{\mathrm{c}}}{1 - q/p} \,, \end{aligned} \tag{3.48}$$

so that the contribution δE_i of atom i to the change in total energy can be written

$$\begin{aligned} \delta E_i = \frac{E_{\mathrm{c}}}{1 - q/p} \Bigg[1 - & \sqrt{\frac{1}{Z_{\mathrm{b}}} \sum_j \exp(-2q\delta R_{ij})} \\ & + \frac{q}{p} \left(\frac{1}{Z_{\mathrm{b}}} \sum_j \exp(-p\delta R_{ij}) - 1 \right) \Bigg] \,. \end{aligned} \tag{3.49}$$

This is a completely general expression for the change in total energy in terms of the changes in interatomic distances. Of course it is not as accurate as the result of a first principles calculation would be. However, we shall see that its predictions for the trends are quite correct, whenever they can be checked for consistency.

3.3.2 The Surface Tension

Experimental data have been collected which show the trends in surface tension. As seen in Fig. 3.9 these exhibit the same bell shape as the cohesive energy. From the general expression (3.50) we can calculate the change in energy γ_{s} per surface atom in the absence of any relaxation or reconstruction ($\delta R_{ij} = 0$). When

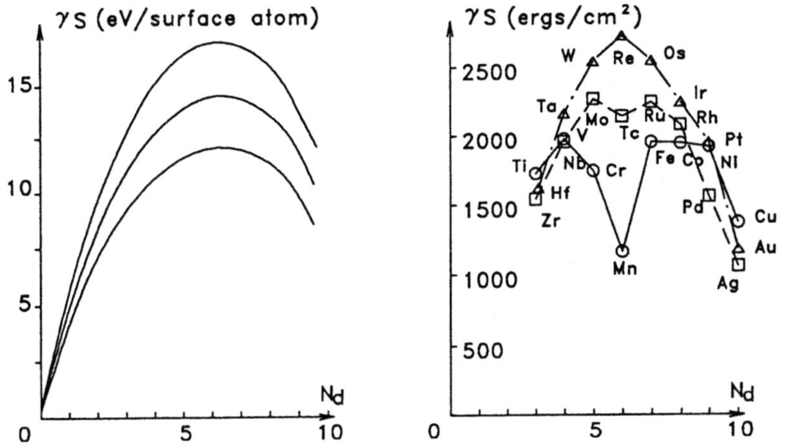

Fig. 3.9. Surface tension for the three series of transition metals: computed (*left*) and experimental (*right*) [3.24]

only surface atoms have broken bonds, we find directly that

$$\gamma_s = \frac{E_c}{1 - q/p}\left[1 - \sqrt{\frac{Z_s}{Z_b}} + \frac{q}{p}\left(\frac{Z_s}{Z_b} - 1\right)\right] \tag{3.50}$$

where Z_s is the coordination number of a surface atom. This proves that, for a given Z_s, γ_s should follow the cohesive energy as a function of N_d, the number of electrons in the d band. This explains directly the observed shape of Fig. 3.9. Expressing Z_s as $Z_b - \delta Z_s$, where δZ_s is the number of broken bonds per surface atom we can make a first-order expansion of γ_s in terms of $\delta Z_s/Z_b$ which even when $\delta Z_s/Z_b = 1/2$ gives an approximation that is valid to within about 10%:

$$\gamma_s \sim \frac{E_c}{1 - q/p}\left(\frac{1}{2} - \frac{q}{p}\right)\frac{\delta Z_s}{Z_b}\,. \tag{3.51}$$

This relation also holds in cases where atoms of the first inner plane have broken bonds. In this case δZ_s has to be the total number of broken bonds per surface atom. With the average value $q/p \sim 1/3$ and considering the (100) surface of a fcc lattice ($\delta Z_s/Z_b = 1/3$) we get, for the maximum value (corresponding to $E_c \sim 8\,\mathrm{eV}$), $\gamma_{s\,\mathrm{MAX}} \sim 0.7\,\mathrm{eV}$ per atom. This is in good agreement with experiment (Fig. 3.9) where $2000\,\mathrm{erg\,cm^{-2}}$ gives 0.7 eV/atom for an interatomic distance of 2.5 Å. It should be noted that a simple broken-bond model would give $\gamma_s \sim E_c\delta Z_s/Z_b$, i.e. a value 4 times larger.

The asymmetric shape of $\gamma_s(N_d)$ can be treated as in Sect. 3.1. For this, the simplest view is that the local shift of the d band should be proportional to the coordination number Z_i of atom i. This adds a positive contribution proportional to $(\delta Z_s/Z_b)N_d$ to γ_s in (3.51) which leads to the correct asymmetry and order of magnitude for $N_d \sim 10$ [3.16].

The anisotropy of the surface tension has been studied in detail taking the correct shape of the d band into account [3.24]. However some approximate information can be obtained from (3.51).

For instance one can compare the values for the (110), (100) and (111) faces of fcc lattices from the quantity $\delta Z_s/Z_b$ which takes the values 6/12, 4/12 and 3/12 respectively. This relative ordering agrees with the experimental data for Ni [3.16]. The situation for bcc metals is more confused and we do not discuss it here (for a complete review see [3.24]).

3.3.3 Magnitude of the Atomic Relaxation Near Surfaces

Expression (3.49) can be expanded to second order in δR_{ij}. The zero order term corresponds to the contribution to γ_s that we have included before. The first- and second-order terms give

$$\delta E_i^{(1,2)} = \frac{q}{Z_b} \frac{E_c}{1-q/p} \left[\left(\sqrt{\frac{Z_b}{Z_i}} - 1 \right) \sum_j \delta R_{ij} + \frac{p}{2} \left(1 - \frac{2q}{p} \sqrt{\frac{Z_b}{Z_i}} \right) \right.$$
$$\left. \times \sum_j \delta R_{ij}^2 + \frac{q}{2Z_i} \sqrt{\frac{Z_b}{Z_i}} \left(\sum_j \delta R_{ij} \right)^2 \right] , \quad (3.52)$$

in which we have written Z_i as the coordination number of atom i since this expression is valid at or away from the surface. The change in interatomic distance δR_{ij} has to be expressed to second-order in the displacements $\boldsymbol{u}_i$ and $\boldsymbol{u}_j$ of atoms i and j from their perfect lattice position, giving

$$\delta R_{ij} = (\boldsymbol{u}_i - \boldsymbol{u}_j) \cdot \boldsymbol{n}_{ij} + \frac{1}{2R} \left\{ |\boldsymbol{u}_i - \boldsymbol{u}_j|^2 - \left[(\boldsymbol{u}_i - \boldsymbol{u}_j) \cdot \boldsymbol{n}_{ij} \right]^2 \right\} \quad (3.53)$$

R being the nearest-neighbor separation in a perfect crystal, and $\boldsymbol{n}_{ij}$ the unit vector of the direction $j \rightarrow i$ in the perfect lattice position.

This expression is valid in any situation, for instance near point defects [3.25], but we apply it here to the perpendicular relaxation of some low index surfaces. The simple derivation which follows is valid in all cases where only atoms belonging to the surface plane have broken bonds. This holds for the (100) and (111) surfaces of the fcc lattice and the (100) and (110) surfaces of the bcc lattice. In such cases the situation of Fig. 3.10 applies where each atom of a plane parallel to the surface has nearest neighbors only in the adjacent planes. Calling θ the angle between the direction of these nearest neighbors and the normal to the plane one can write the contribution of the relaxation to the total energy per surface atom using the form

$$\delta E_T = \frac{q}{Z_b} \frac{E_c}{1-q/p} \left(\alpha_0 d_{1,2} + \frac{\beta_0}{2} d_{1,2}^2 + \frac{\beta}{2} \sum_{i=2}^{\infty} d_{i,i+1}^2 + \gamma \sum_{i=1}^{\infty} d_{i,i+1} d_{i+1,i+2} \right) , \quad (3.54)$$

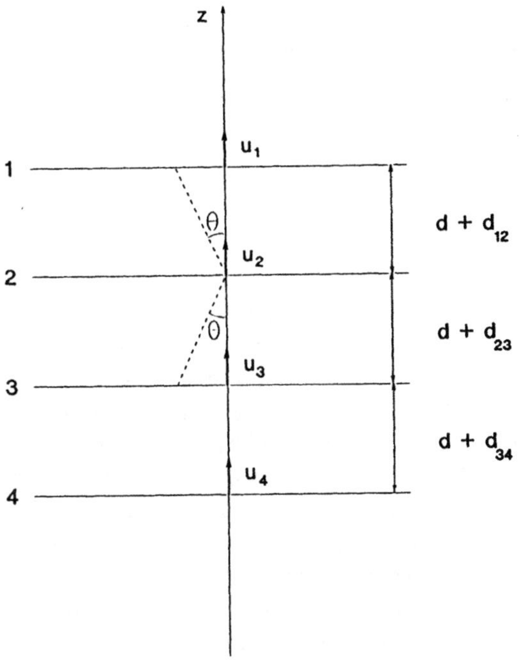

Fig. 3.10. Perpendicular relaxation of low index planes. Only atoms at the surface have broken bonds. Atoms have their nearest neighbors on adjacent planes. The angle between the direction of the nearest neighbors and the normal to the planes is θ

with

$$\begin{aligned}
\alpha_0 &= \delta Z_s \left(\sqrt{\frac{Z_b}{Z_s}} - 1 \right) \cos\theta \,, \\
\beta_0 &= \delta Z_s \left(\sqrt{\frac{Z_b}{Z_s}} - 1 \right) \frac{\sin^2\theta}{R} + p\delta Z_s \cos^2\theta \\
&\quad \times \left[2 - \frac{2q}{p} \left(\sqrt{\frac{Z_b}{Z_s}} + 1 \right) + \frac{q}{p} \left(\sqrt{\frac{Z_b}{Z_s}} \frac{\delta Z_s}{Z_s} + \frac{\delta Z_s}{Z_b} \right) \right] , \\
\beta &= 2p\cos^2\theta\delta Z_s \left(1 - \frac{2q}{p} + \frac{\delta Z_s}{Z_b} \frac{q}{p} \right) , \quad \gamma = \delta Z_s \cos^2\theta \frac{\delta Z_s}{Z_b} \frac{q}{p} \,,
\end{aligned} \tag{3.55}$$

where δZ_s is the number of missing neighbors $Z_b - Z_s$.

Expression (3.54) has exactly the same form as the one analyzed in (8.64) of Sect. 8.3.1 for demonstrating the existence of oscillatory relaxation. From (3.54) and (3.55) it turns out that the changes in interplanar distance $d_{n,n+1}$ obtained by minimizing δE_T are independent of the number of electrons N_d in the d band. This comes from the simple choice of density of states and turns out to be a good approximation. In Table 3.3 we give the results for d_{12}/R and for K, the constant ratio between two consecutive $d_{n,n+1}$.

Table 3.3. Values of the change in first interplanar spacing d_{12} (in units of the nearest neighbor distance R) and of the decay constant K for three low index faces of bcc and fcc lattices

Face	bcc		fcc	
	d_{12}/R	K	d_{12}/R	K
1 0 0	−0.044	−0.17	−0.044	−0.12
1 1 0	−0.032	−0.10		
1 1 1			−0.032	−0.10

Two important results emerge from this analysis. In all cases we predict i) a contraction of the first interlayer spacing, and ii) an oscillatory relaxation, since $K < 0$. The present model was the first to lead to these conclusions [3.26] in disagreement with the predictions of calculations using pair potentials. As mentioned in Chap. 1, such calculations are now confirmed by experimental data as well as numerical calculations [3.27]. Even the orders of magnitude of our predictions agree with experiment. For instance, in the case of Ni, a fit to the phonon dispersion curves gives $pR_e \sim 11$ and $qR_e \sim 2.2$ instead of the average values of equation (3.15). From this one gets the results of Table 3.4 which compare fairly well to the experimental values of [3.28]. Similar agreement is obtained for platinum which exhibits a contraction of the order of 1% [3.29].

Table 3.4. Contraction of the first surface interlayer spacing for Ni

Surface	Coordination number, Z_s	Contraction in % Theory	Contraction in % Experiment
111	9	1.5	1
001	8	2.7	2.5
110	7	6.5	4.5

The advantage of our simple model comes from the fact that we are able to show that the sign of d_{12} (contraction) is imposed by the sign $\alpha_0 (> 0)$ which only depends on the inequality $Z_b > Z_s$, which is always obeyed.

Another interesting figure to discuss is the order of magnitude of the relaxation energy. From the simple arguments derived above its order of magnitude is about 0.01 E_c which is very small and can safely be neglected in calculations of the surface tension.

Finally we discuss the combined effect of the relaxation and reduction of Z_s on the density of states. The contraction of the first interplanar spacing increases the d-d interactions between the atoms belonging to these two planes. This increases the second moment of the density of states for these atoms and tends to compensate for the reduction of this moment due to the breaking of bonds. Taking the logarithmic derivative [see (3.8)] one obtains, for surface atoms

$$\frac{\delta\mu_2}{\mu_2} = -\frac{\delta Z_s}{Z_b}\left(1 + 2qR\frac{d_{12}}{R}\cos\theta\right) \tag{3.56}$$

where the first term is due to the breaking of δZ_s bonds without relaxation, the second one to the contraction. As $qR \sim 3$, taking for d_{12}/R the results of Table 3.3, the second term in parentheses is negative but small compared to unity (it ranges between -0.15 and -0.19). It will act to reduce the magnitude of $\delta\mu_2$ found in the absence of relaxation but this is a relatively small effect.

Of course the model used in this section is too simple to provide truly quantitative answers. The bcc cases might suffer from the limitations discussed previously; higher order moments might give an appreciable contribution, as well as the shift α discussed in Sect. 3.1. Nevertheless, the treatment presented here can be considered as a useful starting point, possibly requiring corrections in some cases.

3.4 The Magnetism of Transition Metal Surfaces

As is known from several studies, the existence of a magnetically ordered state in bulk transition metals requires a large density of states at the Fermi level. In view of the reduced coordination number ($Z_s < Z_b$) the second moment of surface atoms is smaller than in the bulk. The local density of states of surface atoms is thus narrower, leading to an increased maximum value, favorable to the existence of surface magnetism (although the amplitude of the effect is reduced by relaxation). Some years ago the question of the existence of magnetically ordered surfaces for transition metals was somewhat obscured by the existence of dead layers. However recent observations [3.30] have shown that there is indeed an intrinsic surface magnetic ordering.

In this section we follow the approach of *Allan* [3.31] in discussing this problem and analyzing recent results. A first question that arises in this context is the relation between surface magnetism and two-dimensional (2D) magnetism. It is known that although 2D magnetic ordering of an infinite 2D system can in principle exist at zero temperature it is unstable under any thermal excitation. We show this point for 2D phonons since the magnon case is equivalent. It is well known that, in the harmonic approximation, and at high temperature

$$\frac{1}{2}M\omega_k^2u_k^2 \sim \frac{k_BT}{2}\,, \tag{3.57}$$

where u_k is a normal mode of vibration frequency ω_k (M being the mass). Now the mean square displacement $\langle u^2\rangle$ of a given atom is, in a simple situation, the average of u_k^2 which can be written

$$\langle u^2\rangle \sim \frac{kT}{M}\frac{1}{N}\sum_{\boldsymbol{k}}\frac{1}{\omega_k^2}\,, \tag{3.58}$$

where N is the number of modes in the system, with two-dimensional wave vector $\boldsymbol{k}$. If N is infinite the summation over k can be replaced by an integration over the Brillouin zone. However, near $\boldsymbol{k}_\parallel \sim 0$, $\omega_{\boldsymbol{k}_\parallel}^2$ behaves as $\boldsymbol{k}_\parallel^2$

and the 2D integral diverges as $\ln|\boldsymbol{k}_{\parallel}|$. This means that $\langle u^2\rangle$ is infinite and the system is unstable. However truly infinite systems do not exist and for a rectangular sample of dimensions (L_x, L_y) allowed values of $\boldsymbol{k}_{\parallel}$ have components which are multiples of $2\pi/L_x$ and $2\pi/L_y$ with the exception of $\boldsymbol{k}_{\parallel} = 0$ which corresponds to a translation. This means that one must exclude this point from the summation, or equivalently use a low cut-off value k_0 in the integral, such that $\pi k_0^2 = (2\pi)^2/L_xL_y$, corresponding to the surface occupied by one allowed solution in reciprocal space. From this it is trivial that $\langle u^2\rangle$ will now diverge as

$$\langle u^2\rangle = \alpha T \ln N \, . \tag{3.59}$$

Evaluation of α shows that the system becomes unstable (i.e. $\langle u^2\rangle$ is of order of the square lattice parameter) only for temperatures larger than $(10^3/\lg N)$ K, i.e. about 50 K when N takes a typical value of 10^{16}. This conclusion applies to magnetic excitations that behave like phonons so that finite 2D magnetic systems can be stable at low temperatures.

If one now considers the surface case, an important difference is that the system is no longer truly two-dimensional. In the phonon case it has been shown [3.33], that even for an infinite surface the mean square displacements $\langle u^2\rangle$ do diverge as $\ln I/I_s$ where I is the characteristic intra-surface plane force constant while I_s is the parameter describing the coupling with the substrate. Thus even if the ratio I/I_s is large ($\sim$ 100) $\langle u^2\rangle$ remains small and the 2D instability practically disappears.

Let us then now concentrate on the condition for the existence of magnetism near a transition metal surface. To simplify we start from a bulk paramagnetic system and look at the possibility of existence of magnetic moments on each atom. For this we assume that the population $N_{i\uparrow}$ of electrons of up-spin on atom i becomes different from $N_{i\downarrow}$, the corresponding population with down-spin. If this is the case the one-electron Hartree–Fock Hamiltonian $h_\uparrow$ and $h_\downarrow$ will become different due to the exchange terms which are intra-atomic and equal to $-J_iN_{i\uparrow}$ and $-J_iN_{i\downarrow}$ on a given atom i respectively (we allow for a site dependence of the exchange interaction). Each exchange contribution can be written as $-J_i(\bar{N}_i \pm \Delta N_i)$ where $\bar{N}$ is the average of $N_{i\uparrow}$ and $N_{i\downarrow}$ and ΔN is defined as

$$\Delta N_i = \frac{N_{i\uparrow} - N_{i\downarrow}}{2} \tag{3.60}$$

and is a measure of the local magnetic moment. The term $-J_i\bar{N}_i$ is the same for spin up and down electrons and can be absorbed in the spin independent part of the Hamiltonian. Starting from the results of the non-polarized calculation the effect of the spin polarization is to induce on each atomic site a perturbative potential, equal to $-J_i\Delta N_i$ for spin-up electrons and $+J_i\Delta N_i$ for spin-down electrons.

In principle, the ΔN_i have to be determined from a full self-consistent calculation. To illustrate this let us consider for simplicity the case of one orbital per atom and assume that the ΔN_i are fixed at their final self-consistent value. The

Hamiltonian to be solved is thus the unperturbed one, H^0, plus diagonal terms equal to $\mp J_i \Delta N_i$ for electrons with spin up and down respectively. Its eigenstates $\psi_{k\uparrow}$ or $\psi_{k\downarrow}$ can be expanded in the atomic orbital basis with coefficients $C_{ik\uparrow}$ or $C_{ik\downarrow}$. The total energy can be written as

$$E_{\text{tot}} = -\sum_i J_i(\Delta N_i)^2 + \sum_{k_{\text{occ}}}^{i,j}(C^*_{ik\uparrow}C_{jk\uparrow} + C^*_{ik\downarrow}C_{jk\downarrow})(H^0)_{ij} \,. \tag{3.61}$$

In this expression the first term corresponds to the electron–electron interactions $1/2\sum_i(N_{i\downarrow} - N_{i\uparrow})J_i\Delta N_i$ and the second one to the expectation value of the unperturbed Hamiltonian. The values of the unknown $C_{ik\uparrow}$ or $C_{ik\downarrow}$ have to be determined by minimization of E_{tot} with the additional constraints that

$$\tfrac{1}{2}\sum_{k_{\text{occ}}} |C_{ik\uparrow}|^2 - |C_{ik\downarrow}|^2 = \Delta N_i \,, \tag{3.62}$$

where the sum over k is performed over the occupied states, and also that the state be normalized, i.e.

$$\sum_i |C_{ik\uparrow}|^2 = 1 \,, \quad \sum_i |C_{ik\downarrow}|^2 = 1 \,. \tag{3.63}$$

To solve this problem we use the method of Lagrange multipliers and minimize the quantity

$$E_{\text{tot}} - \sum_i \lambda_i \sum_{k_{\text{occ}}}(|C_{ik\uparrow}|^2 - |C_{ik\downarrow}|^2)$$
$$-\sum_{k_{\text{occ}}}\left[\varepsilon_{k\uparrow}\sum_i |C_{ik\uparrow}|^2 + \varepsilon_{k\downarrow}\sum_i |C_{ik\downarrow}|^2\right] \,. \tag{3.64}$$

Differentiating with respect to $C^*_{ik\uparrow}$ and $C^*_{ik\downarrow}$ we get

$$\sum_j (H^0)_{ij}C_{jk\uparrow} - (J_i\Delta N_i + \lambda_i)C_{ik\uparrow} = \varepsilon_{k\uparrow}C_{ik\uparrow} \,, \tag{3.65}$$

$$\sum_j (H^0)_{ij}C_{jk\downarrow} + (J_i\Delta N_i + \lambda_i)C_{ik\downarrow} = \varepsilon_{k\downarrow}C_{ik\downarrow} \,. \tag{3.66}$$

This means that the $C_{jk\uparrow}$ and $C_{jk\downarrow}$ are the solutions of a problem with diagonal perturbation potentials equal to $\mp(J_i\Delta N_i + \lambda_i)$. *Thus $\mp\lambda_i$ plays the role of an external potential chosen in such a way that the spin polarization on site i is equal to ΔN_i*. Multiplying (3.65) and (3.66) by $C^*_{ik\uparrow}$ and $C^*_{ik\downarrow}$ respectively, summing over i and occupied k, and finally inserting into the expression (3.61) for E_{tot} we get

$$E_{\text{tot}} = -\sum_i \left[J_i(\Delta N_i)^2 + 2(J_i\Delta N_i + \lambda_i)\Delta N_i\right] + \sum_{k_{\text{occ}}}(\varepsilon_{k\uparrow} + \varepsilon_{k\downarrow}) \,. \tag{3.67}$$

We are looking here for the possible existence of a magnetic instability. For this we express $\varepsilon_{k\uparrow}$ and $\varepsilon_{k\downarrow}$ using second-order perturbation theory in terms of the perturbation potentials λ_i. In this way we get the second-order change ΔE_{tot} given by

$$\Delta E_{\text{tot}} = -\sum_i J_i \Delta N_i^2 + 2\sum_i (J_i \Delta N_i + \lambda_i)\Delta N_i$$

$$-\sum_{ij} \chi_{ij}(J_i \Delta N_i + \lambda_i)(J_j \Delta N_j + \lambda_j) , \tag{3.68}$$

where the χ_{ij} are the susceptibilities defined by

$$\chi_{ij} = 2\text{Re} \sum_{k_{\text{occ}}, k'} \frac{C_{ik}^* C_{jk'} C_{ik'} C_{jk}^*}{\varepsilon_{k'} - \varepsilon_k} \tag{3.69}$$

in which ε_k are the eigenvalues of the non-magnetic Hamiltonian. It is also easy to show that, to first order, one can write

$$\Delta N_i = \sum_j \chi_{ij}(J_j \Delta N_j + \lambda_j) . \tag{3.70}$$

From this we easily get the expression

$$\Delta E_{\text{tot}} = \sum_i \lambda_i \Delta N_i \tag{3.71}$$

and taking (3.70) into account

$$\Delta E_{\text{tot}} = \sum_{ij} (\chi_{ij}^{-1} - J_i \delta_{ij}) \Delta N_i \Delta N_j . \tag{3.72}$$

To relate this to known situations let us consider a bulk elemental crystal with one atom per unit cell and look at the possible occurrence of magnetic instabilities. For this we introduce the Fourier transform

$$F(\boldsymbol{q}) = \frac{1}{\sqrt{N}} \sum_i \mathrm{e}^{\mathrm{i}\boldsymbol{q}\cdot\boldsymbol{R}_i} F_i \tag{3.73}$$

for all site-dependent quantities (N is the number of atoms and $\boldsymbol{R}_i$ is the lattice vector of atom i). Then the total change in energy becomes ($J_i = J$ now being independent of i)

$$\delta E_{\text{tot}} = \sum_{\boldsymbol{q}} \left(\frac{1}{\chi(\boldsymbol{q})} - J \right) |\Delta N(\boldsymbol{q})|^2 . \tag{3.74}$$

An instability of wave vector $\boldsymbol{q}$ will occur if

$$J\chi(\boldsymbol{q}) \geq 1 \tag{3.75}$$

with

$$\chi(\boldsymbol{q}) = \sum_i e^{i\boldsymbol{q}\cdot\boldsymbol{R}_i} \chi_{0i} \, . \tag{3.76}$$

If we look for ferromagnetism (where all ΔN_i are equal to ΔN) then we must use the condition (3.76) in the limit $\boldsymbol{q} \to 0$, i.e.,

$$J\chi_{\mathrm{F}} > 1 \, , \tag{3.77}$$

$$\chi_{\mathrm{F}} = \sum_i \chi_{0i} \, , \tag{3.78}$$

which is Stoner's condition for the occurrence of ferromagnetism. Although this condition has to be fulfilled it does not allow us to calculate ΔN, i.e., the magnetic moment. For this one has to minimize ΔE_{tot} with respect to ΔN which requires the inclusion of higher order terms in ΔE_{tot}. We can also apply condition (3.75) to antiferromagnetism which corresponds to a $\boldsymbol{q}$ vector such that the quantity $\exp(i\boldsymbol{q} \cdot \boldsymbol{R}_i)$ alternates in sign on successive lattice sites.

The above formalism can be extended to surfaces. Starting from a bulk paramagnetic crystal and looking for ferromagnetism near the surface corresponds to searching for solutions in which all ΔN_i and J_i belonging to a given plane p parallel to the surface are equal to ΔN_p and J_p respectively. In this case the total energy per surface atom, to second order, is

$$\delta\varepsilon_{\mathrm{tot}} = \sum_{p=1}^{\infty} \sum_{p'=1}^{\infty} - \left\{ (\chi^{-1})_{pp'} - J_p \delta_{pp'} \right\} \Delta N_p \Delta N_{p'} \, , \tag{3.79}$$

where the planes away from the surface are numbered from 1 to ∞. The susceptibilities $\chi_{pp'}$ are obvious linear combinations of the $\chi_{kk'}$ ($\chi_{pp'} = \sum_{ij} \chi_{i\in p\, j\in p'}$). The conditions for instabilities are less simple than for a homogeneous system. However (3.79) is a symmetrical quadratic form which can be diagonalized by an orthogonal transformation over the ΔN_p.

We have then to solve an eigenvalue problem that can be written in matrix form

$$(\chi^{-1} - J)|u\rangle = \lambda |u\rangle \, . \tag{3.80}$$

To avoid the difficulty of inverting χ we can rewrite this problem in the form

$$[I - \chi(J + \lambda I)]|u\rangle = 0 \tag{3.81}$$

where I stands for the unit matrix.

For simplicity we consider the following reasonable situation: $\chi_{pp} = \chi_{\mathrm{s}}$ for the surface plane; $\chi_{pp'} = \chi_I$ for all nearest-neighbor planes and $\chi_{pp'} = 0$ for more distant planes; $J_p = J_{\mathrm{s}}$ for the surface plane and $J_p = J$ for all other planes. We write the eigenstate as

$$|u\rangle = \sum_p C_p |p\rangle \tag{3.82}$$

and, by projecting (3.81) onto the basis states, we obtain the following set of equations

$$\begin{aligned}
&\{1-\chi_s(J_s+\lambda)\}C_1-\chi_I(J+\lambda)C_2=0\\
&\{1-\chi_b(J+\lambda)\}C_2-\chi_I(J_s+\lambda)C_1-\chi_I(J+\lambda)C_3=0\\
&\dots\dots\dots\\
&\{1-\chi_b(J+\lambda)\}C_p-\chi_I(J+\lambda)(C_{p-1}+C_{p+1})\quad \text{for } p\geq 3\,.
\end{aligned}\tag{3.83}$$

The bulk-like equations for $p \geq 3$ can be solved by writing

$$C_p = AK^{(p-2)} \quad \text{for } p \geq 2\,. \tag{3.84}$$

Then all equations (3.83) with $p \geq 3$ become identical and give

$$K+\frac{1}{K}=\frac{1}{\chi_I(J+\lambda)}-\frac{\chi_b}{\chi_I}\,. \tag{3.85}$$

This is a quadratic equation for K which leads to pure imaginary solutions with modulus unity for values of λ such that

$$-2\leq\frac{1}{\chi_I(J+\lambda)}-\frac{\chi_b}{\chi_I}\leq +2\,. \tag{3.86}$$

In this interval $K+1/K$ can be rewritten as $2\cos\Phi$ and (3.85) gives

$$\frac{1}{J+\lambda}-\chi_b=2\chi_I\cos\Phi$$

or

$$\lambda=\frac{1}{\chi_b+2\chi_I\cos\Phi}-J\,. \tag{3.87}$$

These are the eigenvalues corresponding to the bulk system. They form a continuum and the condition that the bulk is non-magnetic (i.e. $\lambda > 0$ for these solutions) corresponds to

$$1-(\chi_b-2\chi_I)J\geq 0\,, \tag{3.88}$$

i.e., one recovers Stoner's criterion corresponding to antiferromagnetism between planes. If this is realized then possible magnetic instabilities of the semi-infinite chain described by equation (3.83) must correspond to values of λ falling outside this continuum. The eigenstate must still obey (3.83–85) for which two solutions exist with $K > 1$ or $K < 1$. The only physically acceptable one corresponds to $K < 1$ (i.e. is localized near the surface) for which one gets

$$K(\lambda)=\frac{1}{2\chi_I}\left(\frac{1}{J+\lambda}-\chi_b\right)-\sqrt{\frac{1}{4\chi_I^2}\left(\frac{1}{J+\lambda}-\chi_b\right)^2-1}\,. \tag{3.89}$$

Substituting the form (3.84) of C_p into the second of equations (3.83) and using the relation (3.85) for K we obtain

$$\chi_I(J+\lambda)\frac{1}{K} = \chi_I(J_s+\lambda)C_1 \ . \tag{3.90}$$

If we then eliminate C_1 between this equation and the first equation (3.83) we finally obtain

$$\lambda = \frac{1}{\chi_s + K(\lambda)\chi_I} - J_s \ . \tag{3.91}$$

The onset of magnetism will correspond to λ becoming negative in equation (3.91). The critical value of J_s must be such that $\lambda = 0$ and we find the condition

$$J_s(\chi_s + K(0)\chi_I) \geq 1 \ , \tag{3.92}$$

which is then the equivalent of Stoner's condition but for the existence of a magnetic layer near the metal surface. The situation we have just described corresponds to a ferromagnetic surface situation. Antiferromagnetism could be treated in the same way, with an alternating spin polarization, leading to the corresponding Stoner criterion.

The results obtained here by considering the energetic aspects of the magnetic instabilities are fully equivalent to those of [3.31]. For bulk magnetic materials similar equations may be obtained which describe the change in magnetic moment near the surface. Typical results are shown in Fig. 3.11 where the attenuation factor $K(0)$ of the surface magnetic solution is plotted versus the filling of the

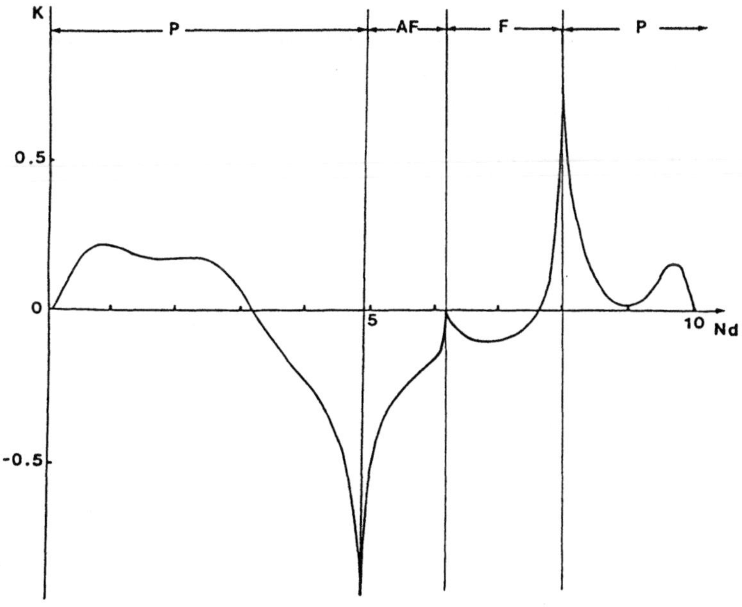

Fig. 3.11. Values of $K(0)$ versus the number of d electrons (calculated for $U = 0.6\,\text{eV}$). P indicates the regions of paramagnetism, AF those of antiferromagnetism and F those of ferromagnetism

d-band. It shows that, except in some narrow regions, K is small so that the change in magnetic moment is localized near the surface.

Exercises

3.1 Repeat the arguments of Sect. 3.1 with a Gaussian shape for the density of states. Show that the qualitative discussion remains unchanged.

3.2 *Zero charge condition*: Show that in the model described by (3.33–35) one can write

$$\delta N_i = \delta N_{i0} - \chi U_i$$

where χ is a local susceptibility. Determine χ and δN_{i0}. Assume that the excess population δN_i creates a local potential

$$U_i = U \delta N_i \ .$$

Determine the final value of U_i and discuss the limits of validity of the zero charge condition.

3.3 Give a detailed argument relating (3.41) to (3.43) by comparing the electrostatic shift in core level state and showing that it is almost the same as for the valence electrons on the same atom.

3.4 *Stoner condition*: Consider the case of ferromagnetism. Starting from the non-magnetic situation, assume that there is a uniform spin polarization ΔN which induces a perturbative potential $U = (U_\uparrow - U_\downarrow)/2$ on each site. Show that one has two relations between δN and U, one of which saturates at large U. Deduce from this the Stoner condition.

4. Electronic States at Covalent Semiconductor Surfaces

The (111) and (100) surfaces of covalent semiconductors are among the most extensively studied systems, particularly in the case of silicon. We do not intend to give here a complete review of the numerous experimental as well as theoretical studies of these surfaces. Instead we try to select what we believe to be the most important results. We discuss them on the grounds of chemical-like models which allow us to obtain very simple pictures of the few essentially different situations.

For this we first apply simple tight-binding arguments to show the basic difference between σ and π bonds and illustrate this with specific examples. We also analyze the unrelaxed, unreconstructed (111) and (100) surfaces showing that the experimental results can be understood in terms of a simple effective Hamiltonian between dangling bonds. This effective Hamiltonian is then applied to discuss possible instabilities, such as spin polarization or charge plus atomic displacement instabilities. We then proceed to a review of the present understanding of relaxation and reconstruction at these surfaces, with specific applications to the Pandey π-bonding chain model of the (111) 2×1 surface, the (111) 7×7 DAS model and the dimer model for the (100) surface. Finally we compare the theoretical results to the available experimental information and try to draw conclusions concerning the possible future evolution of the field.

4.1 Basic Difference Between σ- and π-Bonding

Recent developments in the theoretical understanding of cleaved silicon surfaces require a detailed analysis of the nature of σ- and π-bonding. For this the easiest theoretical treatment ist the tight-binding approximation. As was introduced in Sects. 1.1 and 2.1.2 this method is based on an expansion of the wave function in terms of a restricted "minimal set" of atomic orbitals which are assumed to form an orthonormal basis. The problem then reduces to a diagonalization of the Hamiltonian matrix. In covalent materials the energy bands have a strong molecular character such that this diagonalization can be performed approximately in a simple analytical manner, allowing a very appealing physical understanding in terms of "σ" or "π" bonds. We describe such typical situations in the following.

4.1.1 Covalent σ-Bonds in Group IV Elemental Crystals

We begin, as in Chap. 1, by describing the prototype of the σ-bond which corresponds to perfect tetrahedrally bonded systems. We could write the tight-binding Hamiltonian matrix for such materials in terms of the natural atomic basis set, i.e. one s- and three p-orbitals per atom. However much more physical insight can be obtained from a very simple basis change which allows us to classify the matrix elements by decreasing order of magnitude. For perfect tetrahedral coordination we build sp^3-hybrid orbitals φ_{ij} defined as:

$$\varphi_{ij} = \tfrac{1}{2}(s_i + \sqrt{3}p_{ij}) , \tag{4.1}$$

where s_i is the s-orbital of atom i while p_{ij} is one p-orbital centered on atom i but whose positive lobe points towards its nearest neighbor j. Figure 4.1 shows that the sp^3-orbital φ_{ij} mainly consists of a positive lobe which points from atom i towards atom j.

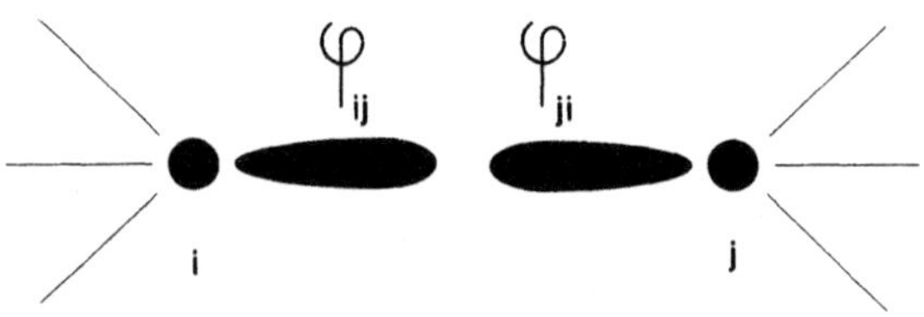

Fig. 4.1. Pair of sp^3 orbitals involved in the same bond of a tetrahedral zinc-blende semiconductor

In this hybrid orbital basis some of the Hamiltonian matrix elements turn out to be much more important than others. These are the matrix elements which connect two sp^3-hybrids φ_{ij} and φ_{ji} involved in the same bond. The corresponding element is:

$$\langle\varphi_{ij}|H|\varphi_{ji}\rangle = \beta . \tag{4.2}$$

Note that, as detailed in Sect. 4.1.3, β can be expressed in terms of the H_{ss}, $H_{s\sigma}$, $H_{\sigma\sigma}$ defined in (2.9) of Sect. 2.1.2. The use of empirical values for these parameters [4.1] always leads to $\beta < 0$. In a first-order approximation we restrict ourselves to the molecular model. Then only β is taken as non-zero, all other interactions being neglected. The system is equivalent to a collection of diatomic molecules with a bonding and an anti-bonding level of energies

$$E = \pm\beta , \tag{4.3}$$

and wave functions:

$$\psi = \frac{1}{\sqrt{2}}(\varphi_{ij} \pm \varphi_{ji}) . \tag{4.4}$$

The degeneracy of the bonding and anti-bonding levels is equal to the number of bonds N in the system. There are four electrons and two bonds per atom so that each bonding state accommodates two electrons with opposite spins. These

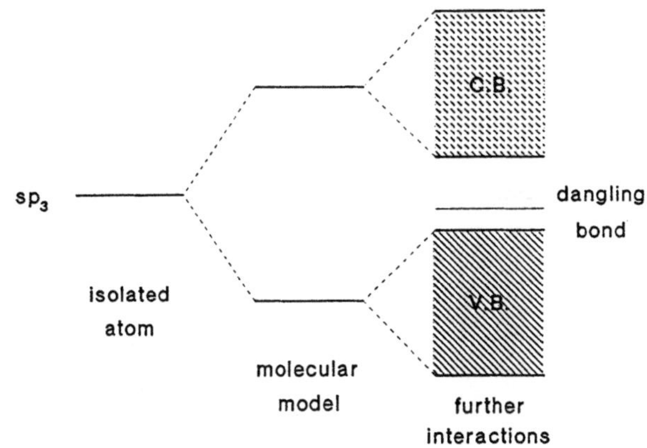

Fig. 4.2. Formation of bands from the molecular model showing the isolated atom sp^3 energy, its splitting into bonding and anti-bonding states and finally the broadening of these states into conduction and valence bands

highly degenerate filled bonding states and empty anti-bonding states represent the simplest crude description of the valence and conduction band of these materials. Incorporation of further interactions will result in a broadening of these bonding–anti-bonding states giving rise to the actual valence and conduction bands as pictured in Fig. 4.2.

These sp^3-bonds which we have just described are the clearest example of σ bonds which involve pairs of orbitals directed along the bond. We now consider other cases involving sp^n and p σ-bonds.

4.1.2 Other Cases of σ-Bonds

We first generalize the notion of sp directional bonding to cases where each atom has equivalent bonds with $n+1$ nearest neighbors ($n = 3$ in three dimensions, 2 in two dimensions, 1 in linear chains, 0 in molecules). For this we follow the approach of *Coulson* [4.2] and on each atom i we build $n+1$ distinct sp^n-hybrid orbitals φ_{ij} of the form:

$$\varphi_{ij} = (1+\lambda^2)^{-1/2}(s_i + \lambda p_{ij}) \ . \tag{4.5}$$

The value of λ is determined in such a way that two φ_{ij} on one atom are orthogonal, i.e.:

$$1 + \lambda^2 \langle p_{ij} | p_{ik} \rangle = 0 \ . \tag{4.6}$$

$|p_{ij}\rangle_{j=1,n+1}$ are $n+1$ unit vectors all equivalent by rotation in an n-dimensional space. The scalar product of two of them is then $-1/n$, leading to:

$$\varphi_{ij} = (1+n)^{-1/2}(s_i + \sqrt{n}p_{ij}) \ . \tag{4.7}$$

Again one can build molecular states from pairs φ_{ij}, φ_{ji}. Usually the number

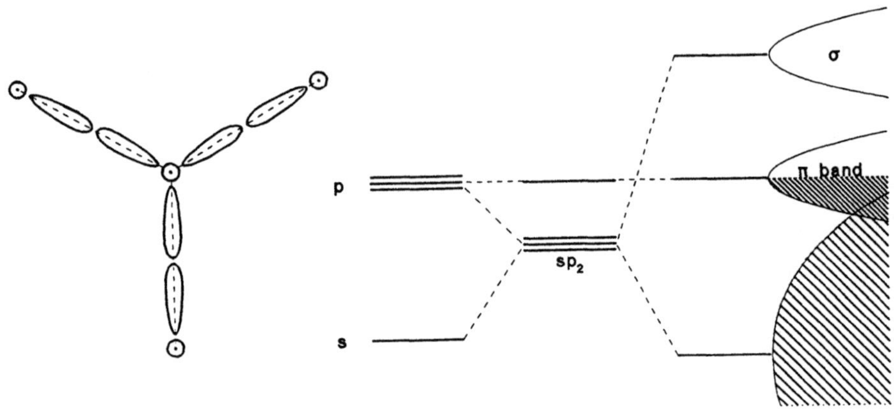

Fig. 4.3. sp^2 bonds – application to graphite. In this particular case a π band is formed between the bonding and anti-bonding states

of electrons is such that there are two electrons per bonding state and none per anti-bonding state. Again such states are broadened into filled valence and empty conduction bands. The best example of this situation is graphite which has a planar hexagonal structure with three bonds per atom, the bonding between planes being much weaker. This corresponds to the formation of sp^2 planar σ-bonds as pictured in Fig. 4.3.

At this stage it is important to come back to the meaning of sp^n-hybrids. Their usefulness is related only to the fact that they yield a quasi-block structure of the Hamiltonian, leading to a simple understanding of the formation of bands. However there is no a priori general prescription for building such hybrids. Only in simple cases is the form of the hybrid completely determined by symmetry (for instance sp^3 in diamond, sp^2 in graphite). In other cases hybrids have to be chosen in such a way that the molecular descriptions are as close as possible to the true band structure.

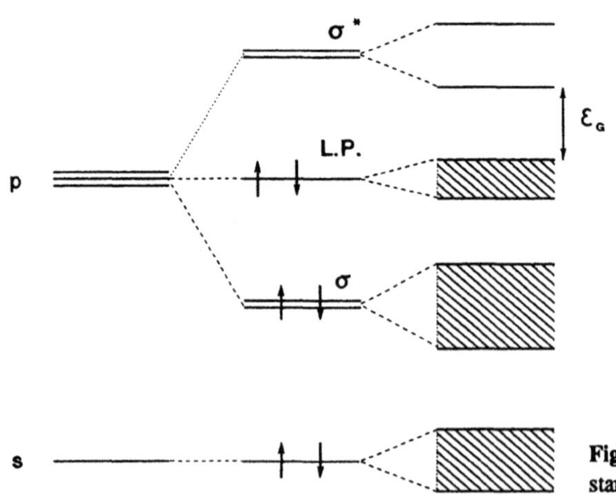

Fig. 4.4. Formation of the Se bands. LP stands for lone pair

We now consider other situations where σ bonds are formed from hybrids which are different from sp^n. This is the case for selenium and tellurium for instance which have a tendency to form chains favoring almost pure "p" σ-bonding. In such systems the difference between the atomic s and p energies is large enough that it is not energetically favorable to build sp hybrids. In fact a more correct picture is to start from the free atom limit shown in Fig. 4.4 and treat, to first order, the s- and p-states independently. We shall see an example of this situation in the case of the O_2 molecule when we discuss π-bonding in Sect. 4.1.3. This is also the case in amorphous As where the three bonds are perpendicular involving pure p atomic orbitals, the s-state being non-bonding.

4.1.3 π-Bonding and Its Strength Relative to σ-Bonding

To discuss the formation of π-bonds we consider two well-known cases: i) the O_2 molecule, ii) graphite. In both situations there is simultaneous formation of σ- and π-bonds and we shall see that the strength of π-bonding is quite generally weaker than that of σ-bonding.

Let us first consider the O_2 molecule. In a tight-binding scheme we have one s- and three p-orbitals per atom, i.e. on the whole a basis set consisting of eight orbitals. There are six independent matrix elements of the Hamiltonian which are: E_s and E_p, the diagonal s and p energies on each atom, and the interatomic matrix elements H_{ss}, $H_{s\sigma}$, $H_{\sigma\sigma}$ and $H_{\pi\pi}$ ($H_{s\pi}$ and H_{sp} being equal to zero in a two-center approximation) in the notation of Fig. 4.5. We use the basis states defined in this figure, leading to the following 8×8 Hamiltonian matrix:

$$
\begin{array}{c|cccccccc}
 & s_A & s_B & \sigma_A & \sigma_B & \pi_A & \pi_B & \pi'_A & \pi'_B \\
\hline
s_A & E_s & H_{ss} & 0 & H_{s\sigma} & & & & \\
s_B & H_{ss} & E_s & H_{\sigma s} & 0 & & 0 & & 0 \\
\sigma_A & 0 & H_{\sigma s} & E_p & H_{\sigma\sigma} & & & & \\
\sigma_B & H_{s\sigma} & 0 & H_{\sigma\sigma} & E_p & & & & \\
\pi_A & & 0 & & & E_p & E_{\pi\pi} & & 0 \\
\pi_B & & & & & H_{\pi\pi} & E_p & & \\
\pi'_A & & 0 & & & & 0 & E_p & H_{\pi\pi} \\
\pi'_B & & & & & & & H_{\pi\pi} & E_p
\end{array}
\qquad (4.8)
$$

The matrix has block-diagonal form. The 4×4 block corresponds to sp mixing. If $E_p - E_s$ were small, it would be interesting to build simple sp-hybrids which would give the dominant matrix elements. In fact this is not at all the case for O_2 since $E_p - E_s$ is fairly large (of order 15 eV). In this limit it is a much better approximation to consider the 2×2 s and p blocks separately and include the effect of $H_{s\sigma}$ by second-order perturbation theory. To zeroth order one gets pure s- and pure p-bonding and -anti-bonding states of energies $E_s \pm H_{ss}$ and $E_p \pm H_{\sigma\sigma}$ respectively.

As regards the 2×2 blocks one gets doubly degenerate bonding and anti-bonding states of energies $E_p \pm H_{\pi\pi}$ corresponding to wave functions

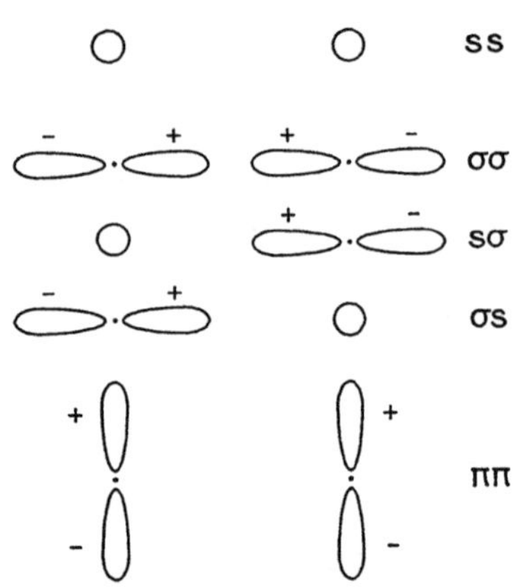

Fig. 4.5. Definition of the two-center interactions with Slater's notation

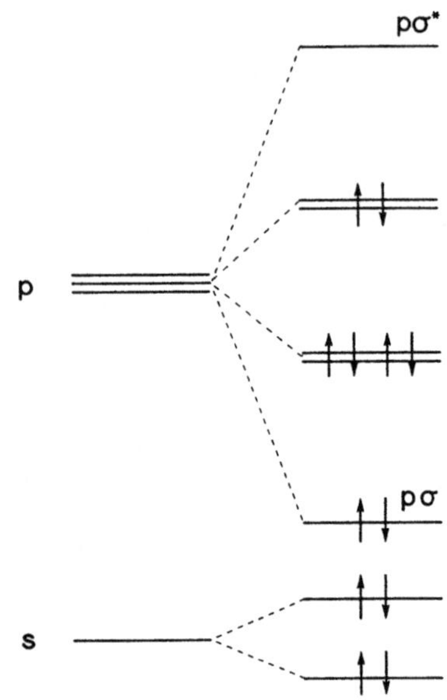

Fig. 4.6. Energy levels of the O_2 molecule

$(\pi_A + \pi_B)/\sqrt{2}$ or $(\pi'_A \pm \pi'_B)/\sqrt{2}$. These form the π-bonds whose essential characteristic is that they are built from pure p atomic orbitals whose axis is perpendicular to the bond. The π-states thus only involve the $H_{\pi\pi}$ interaction. The one-electron energy levels of the O_2 molecule are schematically pictured in Fig. 4.6, corresponding clearly to the situation where $|H_{\pi\pi}| \ll |H_{ss}|$ or $|H_{\sigma\sigma}|$. Comparison with the results of more quantitative one-electron calculations [4.3] demonstrates the validity of these results and shows that the empirical parameters of [4.1] work semi-quantitatively for molecules.

We now discuss another very interesting situation of mixed σ- and π-bonding, namely the case of graphite. The interplanar coupling is very weak so that it is a good starting point to treat each plane separately. As shown in Fig. 4.3 each atom forms three equivalent bonds with its nearest neighbors. As discussed above a convenient basis set is to build three sp^2 directed orbitals pointing towards these neighbors. However this leaves one p_z per atom, where $\mathbf{O}z$ is the axis perpendicular to the plane. These p_z orbitals are only coupled together through nearest neighbor interactions equal to $H_{\pi\pi}$. They do not couple significantly from plane to plane. One thus gets a p_z band centered on E_p and of width equal to $9|H_{\pi\pi}|$. The detailed dispersion relation will be derived later with the similar problem of the (111) surface (Sect. 4.2.1). Figure 4.3 shows the band formation from the σ-bonds as well as the π-band. With four electrons per atom the Fermi level corresponds to a half-filled π-band exactly as given by more sophisticated calculations [4.4].

The respective strength of σ- and π-bonding can be fairly well analyzed by using empirical formulas derived by *Harrison* [4.1] for the nearest-neighbor tight-binding matrix elements. From the fact that actual band-broadening follows closely a free-electron picture, a good approximation is to assume that the matrix elements scale as the inverse squared interatomic distance R. Numerically one gets:

$$\begin{aligned} H_{ss} &= -\frac{10}{R^2}\,\mathrm{eV} \qquad H_{s\sigma} = -\frac{10.8}{R^2}\,\mathrm{eV} \\ H_{\sigma\sigma} &= -\frac{16.9}{R^2}\,\mathrm{eV} \quad H_{\pi\pi} = -\frac{4.8}{R^2}\,\mathrm{eV} \end{aligned} \tag{4.9}$$

with R in Å.

From these values, the bonding–anti-bonding splitting δE_σ, given by 2β in (4.3), is equal to

$$\delta E_\sigma = \frac{|H_{ss}| + 2\sqrt{3}|H_{s\sigma}| + 3|H_{\sigma\sigma}|}{2} = \frac{49}{R^2}\,\mathrm{eV}$$

for sp^3 bonds, and

$$\delta E_\sigma = \frac{|H_{ss}| + 2\sqrt{2}|H_{s\sigma}| + 2|H_{\sigma\sigma}|}{2} = \frac{42.9}{R^2}\,\mathrm{eV} \tag{4.10}$$

for sp^2-bonds. This turns out to be much larger than a typical value for π-bonding splitting which is of the order

$$\delta E_\pi \sim 2|H_{\pi\pi}| = \frac{9.6}{R^2}\,\mathrm{eV}\,. \tag{4.11}$$

This conclusion, which is valid for splitting between one-electron states, is also valid when looking at contributions to the cohesive energy. As discussed in Sect. 2.1.2c the empirical tight-binding total energy is the sum of the one-electron energies of occupied states plus repulsive pair potentials between nearest neighbors. With 2 electrons per π- or σ-bond, we find the corresponding contribution to the cohesive energy to be equal to δE_σ or δE_π. Thus the contribution of π-bonding is about 1/4 of that due to σ-bonding. This will prove to be an important point in the following when analyzing possible reconstructions at the (111) surface.

4.2 The Unrelaxed, Unreconstructed (111) and (100) Surfaces

The first step in calculating the electronic structure of surfaces is to assume that there is no relaxation and no reconstruction. This ideal situation serves as a reference for comparison with experimental data and for understanding the physical origin of eventual modifications in surface electronic or atomic structure. We follow this line of approach and first show how to describe the origin of the most important surface states in terms of dangling bonds. We compare the behavior of the calculated dangling bond surface states to what is obtained for other dangling bond states such as the isolated dangling bond in silicon (amorphous, cristalline or at the Si-SiO_2 interface) or coupled dangling bonds such as the vacancy in silicon. We show that it is possible to describe these dangling bond states by an effective Hamiltonian for which we can get a unique set of parameters. We then apply this idea to show that the (111) surface should be subject to spin polarization as we proposed some time ago and as confirmed by more recent calculations. Such conclusions however can be affected by relaxation and reconstruction as will be shown in the next section.

4.2.1 Description in Terms of Dangling Bonds

The ideal (111) surface of group IV semiconductors with the diamond structure is such that each surface atom loses one of its four nearest neighbors. The corresponding electronic states can again be best understood by using sp^3-hybrid orbitals. These are pictured in Fig. 4.7. In the molecular tight-binding picture (Sect. 4.1.1) each pair of such orbitals involved in the same bond still forms bonding and anti-bonding combinations at energies $\pm\beta$ as in the bulk material. On the other hand each surface atom has a non-bonding sp^3-hybrid which is called a "dangling bond". In the molecular limit its energy is the atomic sp^3-energy, $(E_s + 3E_p)/4$, which we take as the origin of energies. The resulting level structure is given in Fig. 4.2 showing three types of levels: bonding states at $E = -\beta$, dangling bond states at $E = 0$, and anti-bonding states at $E = +\beta$. As shown in Fig. 4.2 the inclusion of further interactions broadens the bonding states into the valence band and the anti-bonding states into the conduction band. Their effect on dangling bond states is the following:

i) to couple each individual dangling bond to the back bonds resulting in a shift of the dangling bond level and a delocalization of the dangling bond wave function;
ii) to couple the dangling bonds together giving rise to a dangling bond surface band.

Both effects will be discussed more quantitatively in Sect. 4.2.2.

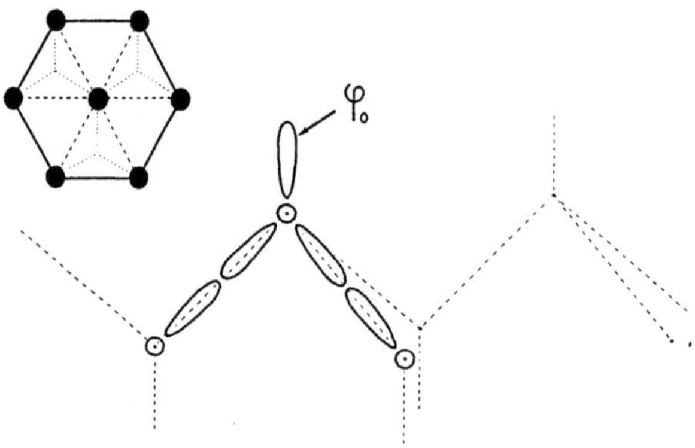

Fig. 4.7. Si(111) surface: Atomic environment and schematic representation of the orbitals

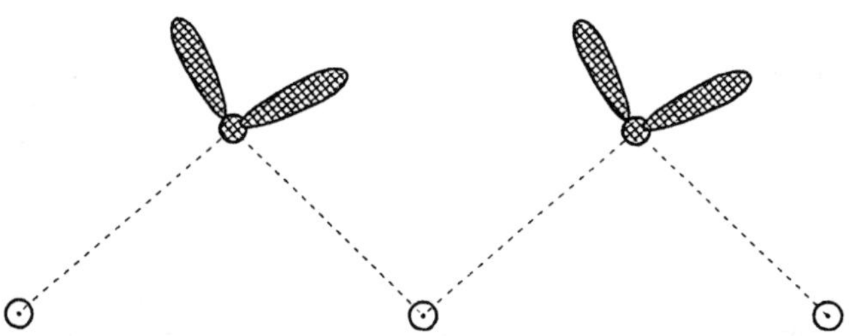

Fig. 4.8. Si(100) surface with its two dangling bonds per atom

A similar reasoning applies to the ideal (100) surface. However each surface atom now has only two covalent bonds with atoms of the first inner plane. In the molecular model there are thus two sp^3 dangling bonds per surface atom (see Fig. 4.8). The level structure in this scheme is the same as for the (111) face except that there are two dangling bond states per atom instead of one. Again further interactions lead to broadening into valence and conduction bands. However their effect on dangling bond states is fundamentally different. It can now be summarized in three steps instead of two:

i) shift of the dangling bond level and delocalization of the dangling bond states
ii) interaction between two dangling bond states on the same atom
iii) interaction between dangling bond states belonging to different atoms.

4.2.2 Effective Hamiltonian for Surface States

Let us now consider the unrelaxed (111) silicon surface. It is characterized in the molecular model discussed above by one sp^3 dangling bond φ_n per surface atom n. All other states are coupled by pairs to form bonding and anti-bonding states at energies $\pm\beta$. When we go beyond the molecular model we have to include the other matrix elements of the Hamiltonian which can be classified into three categories:

i) those connecting the bonding and the anti-bonding states (the back-bond states). We call the eigenstates of the resulting matrix φ_α and their energies ε_α;
ii) the direct interactions between dangling bonds which we write $\langle\varphi_n|H|\varphi_{n'}\rangle = H_{nn'}$;
iii) the coupling of the system of sp^3 dangling bonds with the back-bond states of i) which we write $\langle\varphi_n|H|\varphi_\alpha\rangle = H_{n\alpha}$.

We now expand the total wave function ψ as

$$\psi = \sum_n a_n\varphi_n + \sum_\alpha a_\alpha\varphi_\alpha \,. \tag{4.12}$$

The equations relating the unknown coefficients a_n and a_α are the following, where we take the sp^3 level as the origin of energies:

$$E a_n = \sum_{n'} H_{nn'} a_{n'} + \sum_{\alpha} H_{n\alpha} a_\alpha \ ,$$
$$(E - \varepsilon_\alpha) a_\alpha = \sum_{n'} H_{\alpha n'} a_{n'} \ . \tag{4.13}$$

The second set of equations does not contain terms in $H_{\alpha\alpha'}$ since the φ_α are taken as the true back-bond eigenstates. We now eliminate a_α from equations (4.13) to get:

$$\left(E - \sum_\alpha \frac{H_{n\alpha}^2}{E - \varepsilon_\alpha}\right) a_n = \sum_{n'} \left(H_{nn'} - \sum_\alpha \frac{H_{n\alpha} H_{\alpha n'}}{E - \varepsilon_\alpha}\right) a_{n'} \ . \tag{4.14}$$

This reduces the whole problem to an effective Hamiltonian within the dangling bond subspace. This effective Hamiltonian is built up from the direct interactions between the sp^3 hybrids (the $H_{nn'}$) to which one adds indirect energy-dependent interactions via the back-bonds. This formulation is strictly equivalent to the complete one and, as such, does not bring about any mathematical improvement. However, as in many other situations, the effective Hamiltonian formulation can provide a rather transparent physical picture. For instance if one looks for the surface states in a narrow energy range (in the gap region for example) then one can safely replace $E - \varepsilon_\alpha$ by $\bar{E} - \varepsilon_\alpha$ where $\bar{E}$ is the center of the energy range that is investigated. With this replacement the set of equations becomes an ordinary set of tight-binding equations. If, furthermore, as is quite reasonable, only the nearest neighbor contributions are of importance then one obtains an extremely simple description of the surface band.

For the (111) Si surface, the set of sp^3 orbitals perpendicular to the surface plane behaves exactly as a set of s-like orbitals on the hexagonal lattice of the surface (Fig. 4.7). When truncated to nearest neighbors the corresponding effective tight-binding Hamiltonian contains only two characteristic parameters:

i) the diagonal term or effective dangling bond energy $\varepsilon_d[= \sum_\alpha (H_{n\alpha}^2 / E - \varepsilon_\alpha)$ in (4.14)];
ii) the effective nearest-neighbor interaction β^* $[= \sum_{n'}(H_{nn'} - \sum_\alpha (H_{n\alpha} H_{\alpha n'} / E - \varepsilon_\alpha))$ in (4.14)].

Figure 4.9 gives a striking example of the power and accuracy of this approach. It compares the shapes of calculated dangling bond dispersion curves for Si(111) [4.5] and C(111) [4.6] surfaces with the nearest-neighbor effective Hamiltonian discussed above. This will be seen to apply to other cases as well and we shall find it useful for discussion of a number of cases.

A non-trivial point in the previous formulation concerns the normalization of the states. Let us then write the exact eigenstate ψ of (4.12) as:

$$\psi = c_d \psi_d + c_R \psi_R \ , \tag{4.15}$$

where

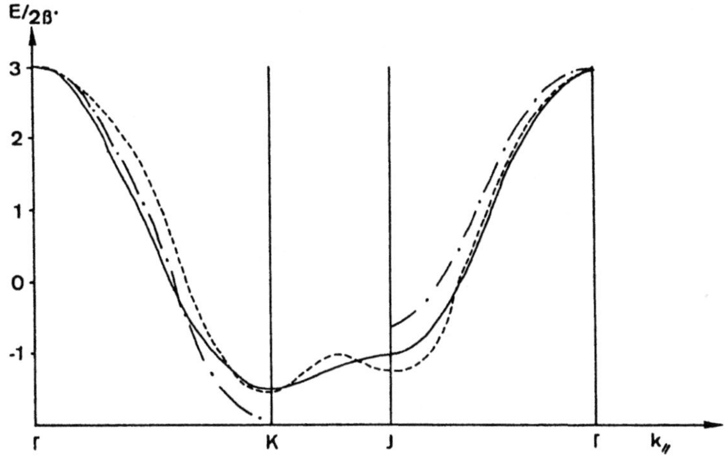

Fig. 4.9. Dispersion of the Si(111) surface band (corresponding to the dangling orbitals: (—) with an effective nearest neighbors interaction; (· – · – ·) *Hirabayashi*'s results [4.6] for C; (- - - -) *Pandey* and *Philips*' results [4.5] for Si

$$\psi_d = \frac{\sum_n a_n \varphi_n}{\sqrt{\sum_n |a_n|^2}}$$

$$\psi_R = \frac{\sum_\alpha a_\alpha \varphi_\alpha}{\sqrt{\sum_\alpha |a_\alpha|^2}} \tag{4.16}$$

are the normalized dangling bond and back-bond parts. With such an expression we have:

$$(E - H_{dd})c_d = H_{dR}c_R \ ,$$
$$(E - H_{RR})c_R = H_{Rd}c_d \ , \tag{4.17}$$

with obvious notation. If one then expresses C_R in terms of C_d in the normalization condition $|c_d|^2 + |c_R|^2 = 1$, one easily obtains

$$|c_d|^2 = \frac{1}{1 + \dfrac{|H_{dR}|^2}{|E - H_{RR}|^2}} \ , \tag{4.18}$$

where E is the exact eigenvalue. This means that, even in the effective Hamiltonian formulation, the dangling bond states are not pure sp^3 states but have delocalized wave functions perpendicular to the surface.

Another way of viewing the same problem is to consider the units of Fig. 4.7 built from one sp^3 hybrid φ_n and its three adjacent back-bonds (i.e. 3 bonding states $\varphi_{Bn\alpha}$ and 3 anti-bonding states $\varphi_{An\alpha}$, with $\alpha = 1$ to 3). Coupling between these leads to a dangling bond state that can be written

$$\psi_n = a_0\varphi_n + a_B \sum_{\alpha=1}^{3} \varphi_{Bn\alpha} + a_A \sum_{\alpha=1}^{3} \varphi_{An\alpha} \tag{4.19}$$

in view of the axial symmetry of the problem. This results in a shift of the dangling bond level as has been calculated in [4.7]. Then one allows for coupling between nearest neighbors ψ_n (on the (111) surface lattice) which will be a sum of a direct coupling between the corresponding φ_n(< 0 with reduced weight $|a_0|^2$) plus extra contributions involving the $\varphi_{Bn\alpha}$ and $\varphi_{An\alpha}$ (giving a positive contribution as shown in [4.7]). On the whole this gives a positive β^*, contrary to the direct negative interaction between parallel sp^3 hybrids. Finally coupling of these states with further back-bonds will not modify the picture very much.

The advantage of this last point of view is that it allows one to define approximate isolated dangling bond states ψ_n which are the building units of other interesting physical situations, such as vacancies [4.8] or trivalent silicon atoms [4.9]. Thus any system of equivalent dangling bonds (with one bond per atom) can be understood as a tight-binding system whose level scheme can be obtained to a good accuracy from the knowledge of two parameters: ε_{db}^* the diagonal term or single dangling bond energy and β_{dd}^* the coupling between dangling bonds which evidently depends upon the relative position of the two dangling bonds.

4.2.3 The Possibility of Surface Electronic Instabilities

For about a decade there was a consensus that the (2×1) structure of the cleaved Si(111) surface was completely explained by the buckling model where one atom of a pair is raised, the other one being lowered. At the same time no really convincing mechanism was proposed to explain the origin of such a reconstruction. However the dangling bond surface state band of the unreconstructed (111) face is narrow since its total width is about 0.9 eV [4.5], corresponding to $\beta^* \sim 0.1$ eV. As this is a half-filled band, the situation is comparable to the case of transition metals near the center of the series (see Chap. 3) and we expect at least some magnetic (or even charge) instabilities in such systems. Such instabilities might be involved in the reconstruction process. Although, at least for Si(111), the situation is now rather well explained in terms of the π-bonding chain model [4.10] (Sect. 4.3.2), *a priori* studies of such instabilities are still relevant even in this particular case. In this section we therefore give an account of the general method of analysis and describe its application to the unreconstructed (111) surface as was done in 1977 [4.11], followed by more recent investigations [4.12–14].

The basis of the treatment is the tight-binding Hamiltonian of the previous section describing the unreconstructed surface state band structure. Apart from the dangling bond level which we take as the origin of energies, the only parameter is β^*, the interaction between nearest neighbor parallel dangling bonds. We explore at the same time the possibility of having either a spin polarization or a charge instability and write the Hamiltonian for electrons with spin σ in the following tight-binding form:

$$H_\sigma = \beta^* \sum_{ij\sigma} |i\sigma\rangle\langle j\sigma| + \sum_i U_{i\sigma}|i\sigma\rangle\langle i\sigma| \,, \tag{4.20}$$

where i, j denote the dangling bond sites and σ the spin. The diagonal terms, in a Hartree–Fock treatment, are given by:

$$U_{i\sigma} = U\Delta N_{i,-\sigma} + \sum_{\substack{i \neq j \\ \sigma'}} K_{ij\sigma'}\Delta N_{j\sigma'} = \sum_{j\sigma'} \gamma_{i\sigma,j\sigma'}\Delta N_{j\sigma'} \,. \tag{4.21}$$

Here $\Delta N_{i\sigma}$ is the change in electron population of spin σ on atom i, U is the intra-atomic Coulomb term, and the $K_{ij\sigma}$ are the interatomic Coulomb terms.

To calculate the instabilities we extend the method used in Sect. 3.4 and express the total energy as:

$$E_{\text{tot}} = \tfrac{1}{2}\sum_{i,\sigma} U_{i\sigma}\Delta N_{i\sigma} + \sum_{\substack{i,j \\ k\text{occ},\sigma}} C^*_{ik\sigma} H^0_{ij} C_{jk\sigma} \,. \tag{4.22}$$

For the same reasons as in Sect. 3.4 we minimize

$$E_{\text{tot}} + \sum_{k,i\sigma} \lambda_{i\sigma}|C_{ik\sigma}|^2 - \sum_{k,\sigma} \varepsilon_{k\sigma} \sum_{i} |C_{ik\sigma}|^2 \tag{4.23}$$

with respect to $C^*_{ik\sigma}$ and get

$$\sum_{j} H^0_{ij} C_{jk\sigma} + (U_{i\sigma} + \lambda_{i\sigma}) C_{ik\sigma} = \varepsilon_{k\sigma} C_{ik\sigma} \,, \tag{4.24}$$

corresponding to the application of the external potential needed to induce the fluctuations $\Delta N_{i\sigma}$. Inserting the last relation into the total energy and following the arguments of Sect. 3.4 we end up with the generalization of (3.71)

$$\Delta E_{\text{tot}} = -\tfrac{1}{2}\sum_{i\sigma} \lambda_{i\sigma}\Delta N_{i\sigma} \,. \tag{4.25}$$

Using the fact that, to first order,

$$\Delta N_{i\sigma} = -\sum_{j} \chi_{ij}(U_{j\sigma} + \lambda_{j\sigma}) \,, \tag{4.26}$$

where χ_{ij} is the tight-binding susceptibility matrix defined by (3.69), one finally gets, from the definition (4.21) of $U_{i\sigma}$:

$$\Delta E_{\text{tot}} = \tfrac{1}{2}\sum_{\substack{i\sigma \\ j\sigma'}} (\chi^{-1}_{ij}\delta_{\sigma\sigma'} + \gamma_{i\sigma,j\sigma'})\Delta N_{i\sigma}\Delta N_{j\sigma'} \,. \tag{4.27}$$

Due to the translational symmetry of its coefficients, this quadratic form can be diagonalized by a two-dimensional Fourier transformation:

$$\Delta N_\sigma(\boldsymbol{q}) = \frac{1}{\sqrt{N}} \sum_{i} e^{i\boldsymbol{q}\cdot\boldsymbol{R}_i} \Delta N_{i\sigma} \,. \tag{4.28}$$

Defining:

$$\chi(\boldsymbol{q}) = \sum_{i} \chi_{0i} e^{i\boldsymbol{q}\cdot\boldsymbol{R}_i} \tag{4.29}$$

and

$$K(\boldsymbol{q}) = \sum_i K_{0i} e^{i\boldsymbol{q}\cdot\boldsymbol{R}_i} \tag{4.30}$$

one can separate in (4.27) the contributions of charge ($\Delta N_i : \Delta N_{i\uparrow} + \Delta N_{i\downarrow}$) and magnetic $\delta N_i = (\Delta N_{i\uparrow} - \Delta N_{i\downarrow})/2$ fluctuations to get

$$\Delta E_{\text{tot}} = \sum_{\boldsymbol{q}} \left\{ \left(\frac{1}{\chi(\boldsymbol{q})} - U \right) |\delta N(\boldsymbol{q})|^2 + \left[\frac{1}{\chi(\boldsymbol{q})} + U + 2K(\boldsymbol{q}) \right] \left| \frac{\Delta N(\boldsymbol{q})}{2} \right|^2 \right\} . \tag{4.31}$$

This gives us directly the criteria for instabilities. For the magnetic type we recover Stoner's condition:

$$U > \frac{1}{\chi(\boldsymbol{q})} . \tag{4.32}$$

We have calculated $\chi(\boldsymbol{q})$ numerically [4.11] and found the result of Fig. 4.10. From this figure the condition for the occurrence of a magnetic instability is given by

$$U > 4\beta^* = 0.4\,\text{eV} \tag{4.33}$$

since $\beta^* \sim 0.1\,\text{eV}$. This result was confirmed in [4.12], the problem being to find a correct estimate of U, the intra-atomic Coulomb term for dangling bond states. This was estimated from studies of the vacancy in Si and a final estimate is $U \sim 1\,\text{eV}$. This estimate is well outside Stoner's condition so that there should be magnetic instabilities. In [4.14] we have computed the total energies due to different magnetic superstructures and have found that the 2×1 antiferromagnetic arrangement is the most stable. There has been some dispute concerning the possibility of the existence of such a ground state for this 2D system [4.15] but we do not reproduce the arguments here. The main point is that spin polarization splits the surface bands for electrons with spin up and down and, if these are separated, one gets an insulator at low temperatures as is observed. Then this could account by itself for some of the experimental results. At high temperature long range ordering disappears (with a Néel temperature $T_N \sim 80\,\text{K}$ [4.14]) but disordered magnetic moments would still subsist leading again to split polarized bands.

This qualitative picture of spin polarization has been confirmed by local density calculations [4.16] whose results correspond, in our model, to $U \sim 0.5\,\text{eV}$. However these authors have also shown that the π-bonding chain model is energetically more favorable than antiferromagnetic polarization [4.16] for the 2×1 Si(111) surface. Spin polarization might even occur in the π-bonding chain model (see Sect. 4.3.2). However the full consequence of magnetic instabilities should apply to the 1×1 surface.

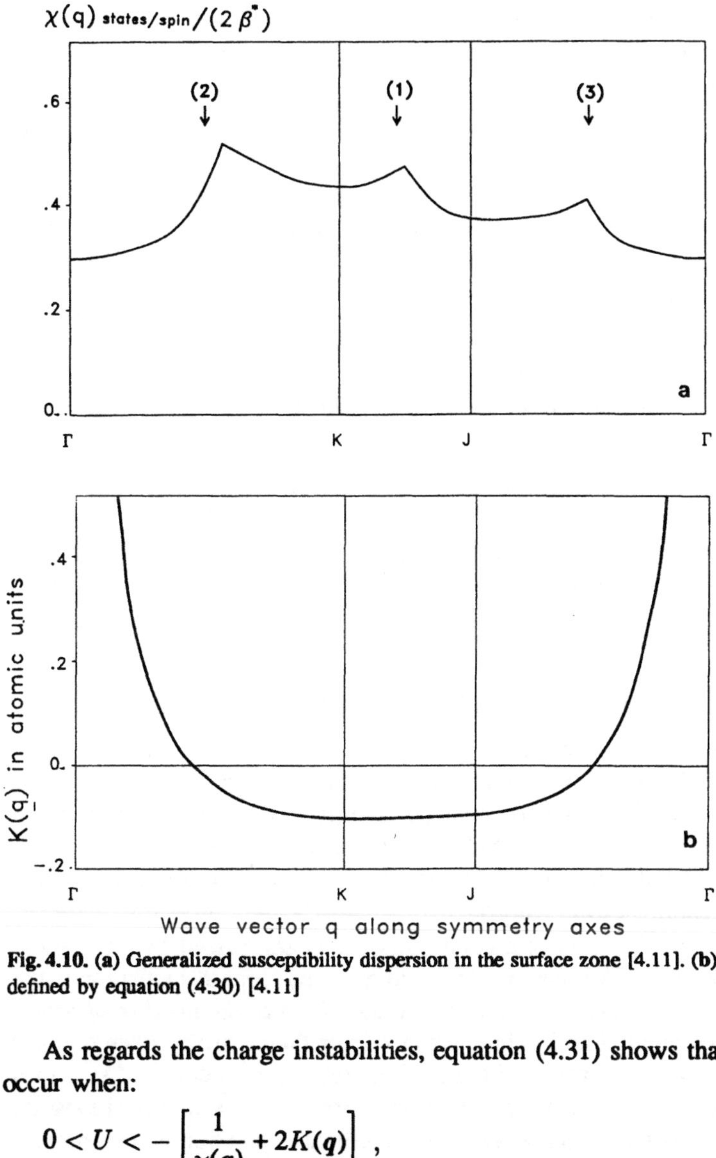

Fig. 4.10. (a) Generalized susceptibility dispersion in the surface zone [4.11]. (b) Lattice sums $K(q)$ defined by equation (4.30) [4.11]

As regards the charge instabilities, equation (4.31) shows that they can only occur when:

$$0 < U < -\left[\frac{1}{\chi(q)} + 2K(q)\right] , \tag{4.34}$$

i.e. only if the lattice sum $K(q)$, defined by (4.30), becomes negative. In [4.11] we have calculated $K(q)$ by approximating the interatomic Coulomb terms K_{ij} by the inverse interatomic distance. The corresponding lattice sum is plotted in Fig. 4.10b and is negative over a large region of q space where its value is about −0.1 a.u. However such electrostatic terms should be screened by a surface dielectric constant for which a rough estimate is $(\varepsilon + 1)/2 \sim 6$ for silicon. The effective value of $2K(q)$ for equation (4.34) is thus of order −0.9 eV. In the same region of q space the maximum value of $2\beta^* \chi(q)$ is ~ 0.5 from Fig. 4.10a. As $\beta^* \sim 0.1$ eV this gives a minimum value for $1/\chi(q)$ of about 0.4 eV. The condition (4.34) then becomes

$$0 < U < 0.5\,\text{eV}\,. \tag{4.35}$$

We have seen previously that U for a surface dangling bond is of the order of 1 eV so that the condition (4.34) for charge instabilities is not likely to be fulfilled. As discussed in Sect. 4.3.2, this is one of the reasons why the buckling model for this surface is not stable.

4.3 Present Understanding of Relaxation and Reconstruction

We first discuss some general principles that govern these phenomena in covalent and partly ionic semiconductors. We then consider the specific cases of Si(111) 2×1 and 7×7, Si(100) 2×1 and finally Si(110). The case of partly ionic semiconductors (III-V and II-VI) will be specifically dealt with in the following chapter.

4.3.1 General Principles Governing Relaxation and Reconstruction

The approach we will follow is to extract the essential energies that govern reconstruction and relaxation phenomena for covalent and partly ionic semiconductors. We will show that some simple energy evaluations account for the observed experimental structures. Although the reasoning we will follow might seem oversimplified, it is justified by the fact that it gives a simple understanding of the result of complex calculations.

Covalent and partly ionic semiconductors are characterized by very strong nearest-neighbor bonds. An obvious principle resulting from this situation is that any relaxation or reconstruction process should *minimize the number of broken bonds* per surface atom. Another interesting rule of thumb arises when one performs a second-order expansion of the total energy δE of a covalent system in terms of the atomic displacements with respect to a set of stable positions. An approximate form is the so-called valence force field expression whose simplest version is

$$\delta E = \frac{k_r}{2} \sum_{\substack{\text{pairs} \\ ij}} (dr_{ij})^2 + \frac{k_\theta r^2}{2} \sum_{\substack{\text{angles} \\ ijk}} (d\theta_{ijk})^2\,, \tag{4.36}$$

where dr_{ij} are the changes in bond length between nearest neighbors i and j, $d\theta_{ijk}$ the changes in bond angles between pairs of nearest neighbors ij and jk, k_r and k_θ are the radial and angular force constants and finally r is the equilibrium nearest neighbor distance. Other simple forms of δE exist such as the *Keating* model [4.17] which uses the crystal symmetry explicitly. However, their use would not change the argument. The strength of the covalent bonds is such that $k_r \gg k_\theta$ as can be seen in Table 4.1. Except for diamond, this holds quite generally and should remain true near surfaces and interfaces. In view of this a very interesting limit to consider is $k_\theta/k_r \to 0$. In this case, the stable geometry of the system, minimizing δE would correspond (when possible) to all $dr_{ij} = 0$. In other words the minimum energy situation *preserves the bond lengths as much as possible*. A good starting point is thus to find a geometrical arrangement where these bond lengths are exactly preserved. One could later allow for a little radial relaxation.

Table 4.1. Different sets of values of k_r and k_θ for Si used in the literature (from [4.20]). These are given in units of 28.935 eV/Å

k_r	k_θ
0.332	0.0066
0.351	0.007
0.261	0.0143
0.261	0.0071

Thirdly we need to understand the behavior of an isolated dangling bond when its electron population changes. This was the basis of *Harrison*'s early discussion of relaxation and reconstruction [4.18], and will be of much help in the following. The isolated dangling bond can in principle have three occupation numbers $n = 0, 1$ or 2. In a symmetric environment the main relaxation corresponds to an axial displacement of the trivalent atom. The corresponding total energy can thus be written

$$E(n,u) = n\varepsilon_0 + \tfrac{1}{2}Un(n-1) - F(n) + \tfrac{1}{2}k\,u^2\,, \tag{4.37}$$

where u is the outwards axial displacement of the trivalent atom from its tetrahedral position (Fig. 4.11), U is the electron–electron interaction, $F(n)$ is the occupation-dependent force and k is the elastic force constant. The simplest

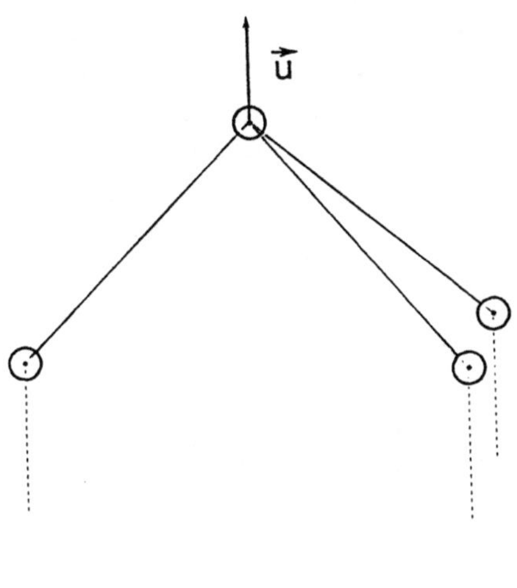

Fig. 4.11. Outwards displacement of the three-fold coordinated Si atom

meaningful approximation is to neglect the slight dependence of k on n and to linearize $F(n)$ in the form

$$F(n) = F_0 + (n-1)F_1 \; . \qquad (4.38)$$

One can minimize $E(n,u)$ with respect to u to get the minimum electronic energy $E_\mathrm{m}(n)$ as a function of the occupation number:

$$E_\mathrm{m}(n) = n\varepsilon_0 + \frac{1}{2}Un(n-1) - \frac{[F_0 + (n-1)F_1]^2}{2k} \; . \qquad (4.39)$$

Let us now suppose that we have a large number of equivalent dangling bonds with an average occupation number equal to 1. In such a case it is interesting to consider the reaction

$$D(0) + D(2) \rightleftharpoons 2D(1) \; , \qquad (4.40)$$

which conserves the number of electrons [$D(n)$ denotes the dangling bond state]. The actual direction in which this reaction occurs depends upon the energy difference

$$U^* = E_\mathrm{m}(0) + E_\mathrm{m}(2) - 2E_\mathrm{m}(1) \; , \qquad (4.41)$$

which, using (4.39), is given by:

$$U^* = U - \frac{F_1^2}{k} \; . \qquad (4.42)$$

This term U^* corresponds to an effective electron–electron interaction, reduced by the effect of relaxation. The sign of U^* will govern the occupation of dangling bonds. In usual situations U^* is positive. Then the stable configuration has all dangling bonds in the state $D(1)$. In negative U situations ($U^* < 0$) one finds the opposite result that dangling bonds can only be in the empty $D(0)$ or completely filled $D(2)$ states.

Let us evaluate U^* in the bulk. The purely electronic contribution U (excluding relaxation) has been calculated by several authors for Si and is of order $U \sim 0.5\,\mathrm{eV}$ [4.9, 19]. From [4.18, 19] one can conclude that $F_1 \sim 1.6\,\mathrm{eV/Å}$. Then the sign of U^* depends sensitively on the value of the force constant k. For a dangling bond in the bulk most of the relaxation can be achieved through angular variations. This leads to a low value of $k \sim 4\,\mathrm{eV/Å^2}$, (see [4.20] for a discussion of this effective force constant). With these values one finds $U^* \sim -0.14\,\mathrm{eV}$ which was the main conclusion of [4.19]. These values should be slightly modified at the surface and we shall discuss this for each specific case. *However the tendency of dangling bonds to be fully occupied or empty must be kept in mind in any analysis of reconstruction processes*. To determine which is the case, one only needs to know the sign of U^*. Its exact magnitude is of less relevance.

The quantity F_1 can be estimated from a simple tight-binding argument due to *Harrison* [4.18]. It requires a knowledge of the variation of the dangling bond level with the axial relaxation u. This will be done for a (111) surface. For a given axial displacement u, we make use of a molecular model based on three equivalent sp^k-hybrids for the three back-bonds of the trivalent atoms, the fourth state of this atom being the dangling bond of energy E_{db}. The values of k are equal to or greater than 2, depending on the axial relaxation amplitude. Each sp^k-hybrid will be engaged in a covalent bond with its nearest neighbors and will thus be occupied by one electron (we neglect the slight ionicity on each back-bond). The contribution of the trivalent atom to the sum of one electron energies will thus be:

$$\Delta E = n\,E_{db} + 3E(sp^k) , \tag{4.43}$$

where n is the number of electrons in the dangling bond state. From trace invariance the quantity $E_{db} + 3E(sp^k)$ is equal to $E_s + 3E_p$ and ΔE reduces to

$$\Delta E = (n - 1)E_{db} + E_s + 3E_p . \tag{4.44}$$

In this expression only the first term depends on u and this molecular model leads to

$$F_0 \sim 0 \quad F_1 = \frac{dE_{db}}{du} . \tag{4.45}$$

One can evaluate the variation of the dangling bond level with the axial relaxation amplitude u from the use of a tight-binding model in two extreme situations (Fig. 4.12):

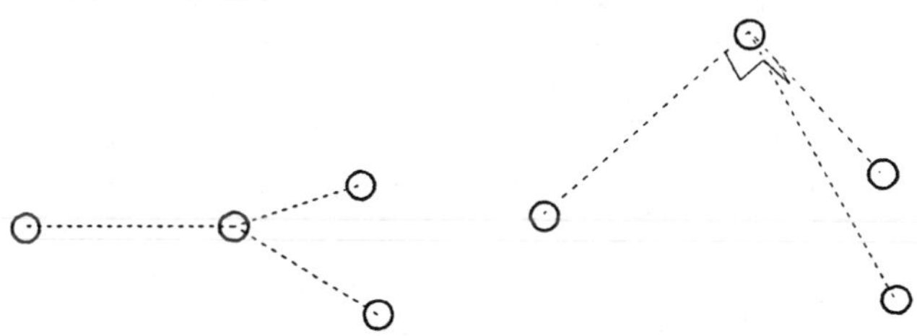

Fig. 4.12. The two extreme situations corresponding to Fig. 4.11. The central atom is the trivalent one

- When the trivalent atom is in the plane of its nearest neighbors (i.e. $u = -(R/3)$ and $k = 2$) the situation is analogous to the case of graphite discussed in Sect. 4.1.3. The dangling bond becomes a pure p-state of energy E_p (atom lowered).
- When the three back-bonds of the trivalent atom become perpendicular (i.e. $u = 2R/3$ and k infinite) then the situation resembles that of amorphous As discussed in Sect. 4.1.2. The dangling bond becomes close to a pure s-state of energy E_s (atom raised).

The general conclusion is that *the dangling bond level is lowered when the atom is raised*. If this simple tight-binding view is applied to a total energy calculation then the occupation dependent force F_1 is given directly by (4.45) and thus reduces to

$$F_1 = \frac{(E_p - E_s)}{R} . \tag{4.46}$$

As $E_p - E_s \sim 4\,\mathrm{eV}$ for Si [4.21,22] one gets $F_1 \sim 1.7\,\mathrm{eV/Å}$ which is remarkably close to the value obtained from the first principles calculations quoted above.

4.3.2 Si(111) 2 × 1: The π-Bonding Chain Model

The metastable Si(111) 2×1 surface is obtained by cleavage. It is thus relatively easy to obtain in ultra high vacuum environments. It has therefore been the subject of extensive experimental and theoretical studies. For many years the most popular model to explain the 2×1 reconstruction has been the buckling model introduced by *Haneman* [4.23]. In this model alternating rows of surface atoms are raised and lowered as shown in Fig. 4.13. At first, this seems quite a natural assumption since the dangling bonds of the (1×1) Si(111) surface give rise to a half-filled metallic band as discussed in Sect. 4.2.3. A 2×1 reconstruction will reduce the size of the Brillouin zone in reciprocal space by a factor of 2; band folding will then occur and will lead to the opening of a gap in the reduced zone. This will result in the lowering of the sum of the one electron energies of occupied states. The opening of a gap is also related to the fact that the dangling bond energy is raised when the atom is lowered while it is lowered when the atom is raised (Sect. 4.3.1). However such a view is oversimplified since it completely neglects electrostatic effects due to the charge redistribution at the surface. In fact we have seen before (Sect. 4.2.3) that the coupling between the parallel dangling bonds at the (111) surface is small ($\sim 0.1\,\mathrm{eV}$) so that a useful starting point is to neglect this coupling and treat the dangling bonds independently. Our general analysis of Sect. 4.3.1 can then be applied. We first analyze the quantity $U^* = U - (F_1^2/k)$. The buckling reconstruction can be stable only if U^* is found negative. To test the validity of the simple arguments that we intend to use, we first evaluate U and k for a bulk dangling bond. Then we turn to the surface U^* and discuss the 2×1 reconstruction of the (111) surface in this perspective.

The bare intra-atomic U is of the order of 10 eV but, for a dangling bond, delocalization effects roughly reduce this by a factor of 2 [4.8]. Furthermore dielectric screening will again reduce it by roughly the dielectric constant $\varepsilon \sim 10$ in the bulk. This leads to a final value for the bulk dangling bond U of 0.5 eV in agreement with numerical calculations [4.19].

At the (111) surface however the situation is slightly different since for the 2×1 buckled situation there are chains which have an occupation number 0 and

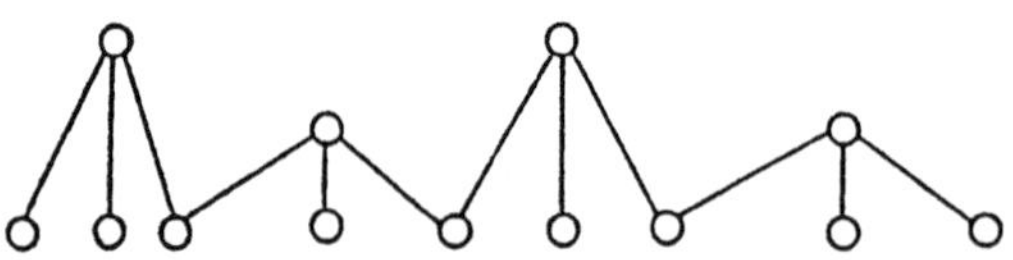

Fig. 4.13. Schematic representation of the 2×1 reconstruction in the buckling model

2 respectively. Thus the bare U is the atomic value (reduced by delocalization i.e. 5 eV) minus a lattice sum $\sum_n[(\pm)e^2/R_n]$ (due to charges on other atoms as in Sect. 4.2.3) whose magnitude has been evaluated in [4.11] to be -2.7 eV for Si. The resulting bare U is thus 2.3 eV. This has to be reduced by the dielectric screening which, however is less efficient at the surface than in the bulk. A reasonable estimate is obtained by replacing ε by $(\varepsilon+1)/2$, i.e. an extra reduction factor of about 6, leading to a final value for U of about 0.4 eV.

A similar situation holds for the effective force constant which cannot be reduced as much at the surface as in the bulk since there is much less freedom for interbond angles to relax. Following the arguments developed in Chap. 8 a good value of k is obtained by only allowing atoms of the surface plane to move. The corresponding calculation leads to a value of k of order 6 eV/Å^2. Taking $U \sim 0.4$ eV, $F_1 \sim 1.6$ eV/Å and $k \sim 6$ eV/Å^2 one gets a positive U^* of order 0.13 eV. This shows, as confirmed by LDA calculations [4.16], that the buckled model should not be stable. Furthermore the order of magnitude of U^* is an indication that the neglect of π type overlap between dangling bonds involving energies of order 0.1 eV is not a good approximation.

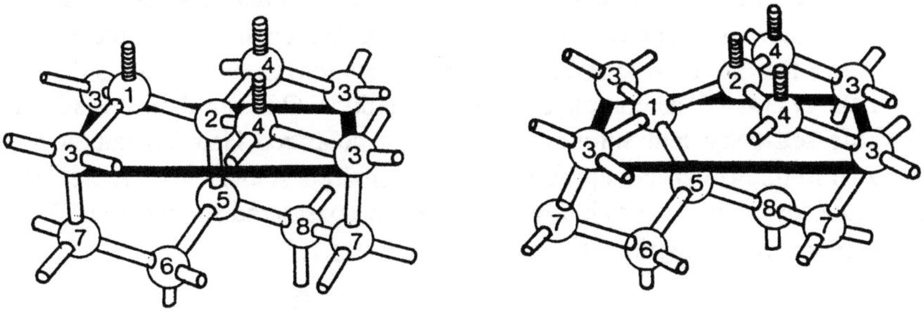

Fig. 4.14. Atomic structure of the Si(111) surface [4.10]. Unreconstructed surface (*left*); Pandey's model (*right*). The solid line represents the 2*d* unit cell

However this conclusion only became clearer with the work of *Pandey* who proposed a new surface topology, the π-bonding chain model [4.10]. The corresponding lattice rearrangement is shown in Fig. 4.14 where one sees that the atomic structure of the first two layers has been completely modified. The basic principles of this reconstruction are the following:

1) keep the number of dangling bonds per surface atom constant
2) keep bond lengths practically equal to the bulk value
3) group the dangling bond atoms along chains of nearest neighbors.

It is clear that condition 1) and 2) will not strongly modify the surface energy per atom except for some small elastic energy associated with bond-length or bond-angle distortions. The main change arises from 3) which allows for π-interactions between the parallel dangling bonds of nearest neighbors. This coupling leads to a relatively broad band which is split into two subbands by the inequivalence of the two dangling bond atoms in the unit cell. With one electron per dangling bond only the lowest subband, whose states have bonding character, is filled. This leads to some π-bonding gain in energy that tends to stabilize this structure.

The energy gain in the π-bonding configuration has been found to be rather large [4.10] $\sim 0.3\,\text{eV}$ for an optimied chain configuration. Another interesting point is that the change in topology can be obtained easily from the ideal 1×1 structure by simultaneously depressing a surface atom (No. 1 in Fig. 4.14) into the second layer and by raising a second layer atom (No. 2) into the surface layer. This breaks a subsurface bond (2-5) and establishes a new bond (1-5). The barrier for this process has been calculated and is smaller than 0.03 eV [4.16]. This is much smaller than the normal Si-Si bond strength ($\sim 2.3\,\text{eV}$). This explains why this structure could be created easily during the cleavage process and also why it transforms easily to a 1×1 during chemisorption.

One might wonder whether the π-bonded chain model is stable with regard to spin polarization. This was done in [4.24] by using a nearest-neighbor tight-binding effective Hamiltonian for the dangling bonds as defined by expression (4.20). However this Hamiltonian now applies to the chains which furthermore have two inequivalent atoms per unit cell. The spin unpolarized Hamiltonian is then characterized by two parameters: β^* the nearest neighbor interaction along the chain and $\pm U_0$ the diagonal terms on the two inequivalent atoms. Figure 4.15 clearly shows that the corresponding dispersion relation accounts quite satisfactorily for the results of the complete LDA calculations. One can then investigate the possible self-consistent solutions of the spin polarized Hamiltonian. It has been shown in [4.24] that the antiferromagnetic chain is stable for $U \gtrsim 0.85\,\text{eV}$, which is just lower than the value which we have considered for $U (\sim 1\,\text{eV})$.

Comparison with experiment is given in Sect. 1.3.1.

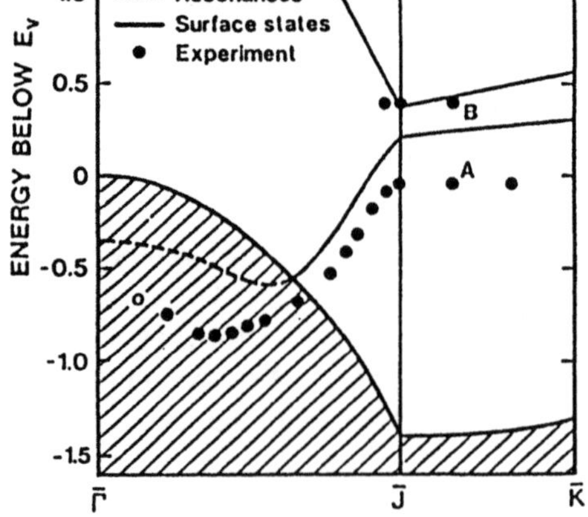

Fig. 4.15. Experimental and theoretical dispersion curve (Pandey's model) [4.16]

4.3.3 Si(111) 7 × 7: The Dimer–Adatom–Stacking-Fault Model

Considering the complexity of the 7×7 Si(111) unit cell (Sect. 1.3.1) which contains 103 atoms, it is probably not possible to draw conclusions with arguments as simple as those used in the preceding section. We may however notice that the proposed dimer–adatom–stacking-fault (DAS) model does greatly reduce the number of dangling bonds.

The non-reconstructed (111) cell corresponding to the 7×7 unit cell contains 49 atoms which are all in the surface plane. This can be seen in Fig. 4.16. In the DAS model the dimerization leads to only 42 atoms in the first surface plane. Of these atoms, 36 have their dangling bond saturated by the bond with the adatom. There are thus 6 atoms left in the first surface plane with a dangling bond. Each of the 12 adatoms also has one dangling bond and each corner atom (whether it is a vacancy or not) contributes to one dangling bond per unit cell. The number of dangling bonds has thus been reduced from 49 to 19.

Furthermore, one notices that the reconstruction is accompanied by very large angular distortions. The dangling bonds saturated by the adatoms, and the dimer bonds are such examples. These distortions extend in depth to the third layer below the surface. It is not certain in such cases whether or not bond length variations should contribute to a decrease in the large angular distortion energy.

The unit cell is too large to be approached by ab initio techniques. An empirical tight-binding energy minimization scheme has been used to gain some insight

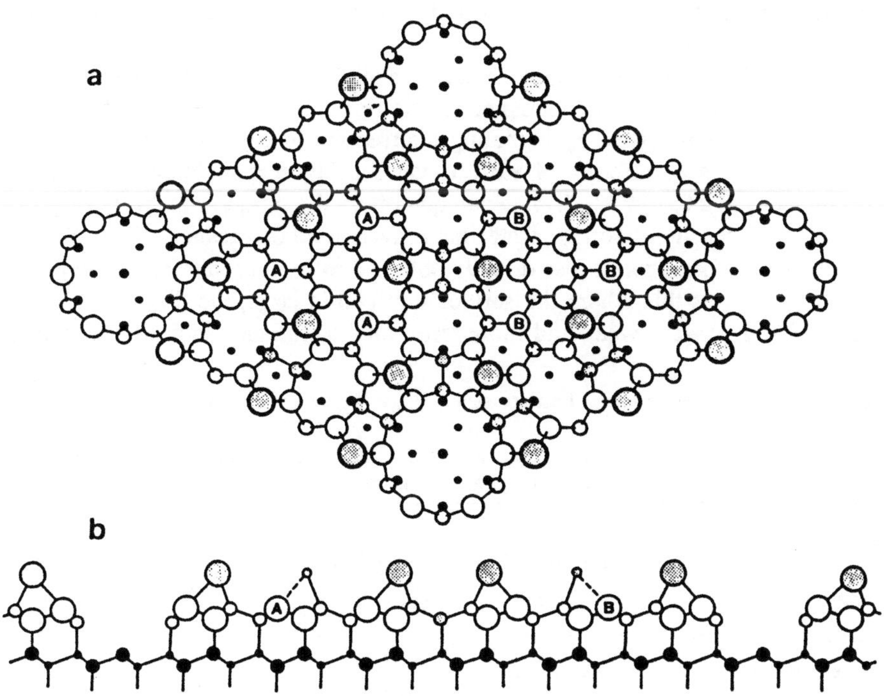

Fig. 4.16. The dimer–adatom–stacking-fault reconstruction model; see Chap. 1

into the stability of the DAS model [4.25]. A simple four-orbital Hamiltonian is used to calculate the band structure. The change in total energy with atomic displacement and the number of bonds can be evaluated from the calculated one-electron band structure using a semi-empirical form (see [4.25] for more details). This form is generalized to give a better account of the large angular distortion energy due to the adatom geometry.

Two geometries are considered for the calculation of the surface energy: the ideal surface and the DAS model. The atomic positions of the adatoms and the atoms of the three last layers in the DAS model were allowed to converge towards equilibrium. It must be stressed that, to keep the size of the system manageable for the diagonalization of the Hamiltonian, the study was restricted to a slab of four layers of silicon. Hydrogen saturates the dangling bonds of the back face. Thus the ideal case contains 196 silicon atoms and 49 hydrogen atoms per unit cell. The DAS model contains 200 silicon atoms and 49 hydrogen atoms per unit cell. The requirement that the back face of the slab be sufficiently far from the surface is thus barely satisfied.

The first important finding is that some bonds are indeed stretched, probably to reduce some of the angular distortion energy. The adatom–first-layer bond length is 5 to 5.7% longer than the bulk Si-Si bond length. The dimer-dimer bond length is also stretched compared to the bulk one by about 3%. Furthermore, the positions of the atoms in the third layer are significantly affected. The slab should thus be thicker in order to accurately describe the bonding to bulk Si. In the final geometry, the reconstructed surface is more stable than the ideal one by 0.435 eV per surface unit cell.

The new surface spectroscopic techniques made possible by the tunneling microscope are particularly well adapted to this very open surface. Also the picture of atomic positions given by tunneling microscopy [4.26], can now be supplemented by the spectroscopy of surface states using current imaging tunneling spectroscopy (CITS) [4.27]. This technique offers the unique capability of probing the local density of surface states on a given atom or vacancy of the surface. This type of information is also directly obtained from tight-binding calculations. In a case such as this where the surface unit cell is large, the dispersion of the surface bands is very weak and more traditional techniques such as photoemis-

Table 4.2. Measured and calculated surface state eneriges for the DAS model of the Si(111) 7×7 surface

	Experimental data	Theoretical calculation
Adatoms	$E_F - 0.2\,eV$ $E_F + 0.5\,eV$	E_F $E_F + 0.8\,eV$
Surface free atoms	$E_F - 0.8\,eV$	$E_F - 0.9\,eV$
Adatom back bonding	$E_F - 1.7\,eV$	$E_F - 1.5\,eV$
Corner atom with dangling bond	$E_F - 1.0\,eV$ to $E_F - 0.6\,eV$	$E_F - 0.7\,eV$

sion are ill-adapted. Table 4.2 gives a comparison between the measured [4.27] and the calculated [4.25] surface states. The agreement is remarkable considering both the experimental uncertainties and the theoretical limitations due to the size of the slab and the simplicity of the Hamiltonian.

4.3.4 Si(100) 2 × 1: The Dimer Model

The geometrical situation is pictured in Fig. 4.17. For each surface atom (1) and (2) two covalent bonds have been broken, a fact which tends to make the ideal geometry highly unstable. This can be understood very easily from the general arguments we have developed in Sect. 4.3.1. The surface should rearrange in such a way that the number of dangling bonds is minimized and also the length of the covalent bonds is close to their bulk value. In the limit where the elastic energy associated with distortions in bond angles is negligible, it is easy to move two second nearest neighbor surface atoms 1 and 2 (Fig. 4.17) in such a way that all distances d_{1A}, d_{1D}, d_{2B}, d_{2C} are kept equal to the bulk bond length and also that $d_{12} = R$. This means that there is now a covalent bond between atoms 1 and 2. This greatly stabilizes the surface since it reduces the number of dangling bonds to one per surface atom instead of two in the unreconstructed situation.

Such a reconstruction where pairs of surface atoms become covalently bonded corresponds to the dimer model of the (100) surface. The local situation is depicted in Fig. 4.18 which corresponds to the more general asymmetric dimer model. However, to understand the origin of this asymmetric reconstruction it is best to consider first the symmetric case where $\delta x_1 = -\delta x_2$ and $\delta z_1 = \delta z_2$ (Fig. 4.18). In this case the condition that all distances be equal to R leads to

$$2\delta x_1 = d_{AB} - R = R\left(\sqrt{\tfrac{8}{3}} - 1\right) . \tag{4.47}$$

which gives $\delta x_1 = 0.75$ Å for silicon. This is just the value obtained for $(\delta x_1 - \delta x_2)/2$ in all asymmetric or symmetric dimer models quoted in [4.28] which

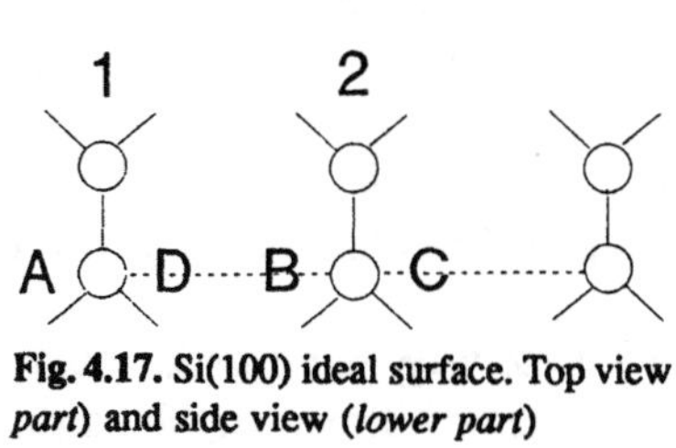

Fig. 4.17. Si(100) ideal surface. Top view (*upper part*) and side view (*lower part*)

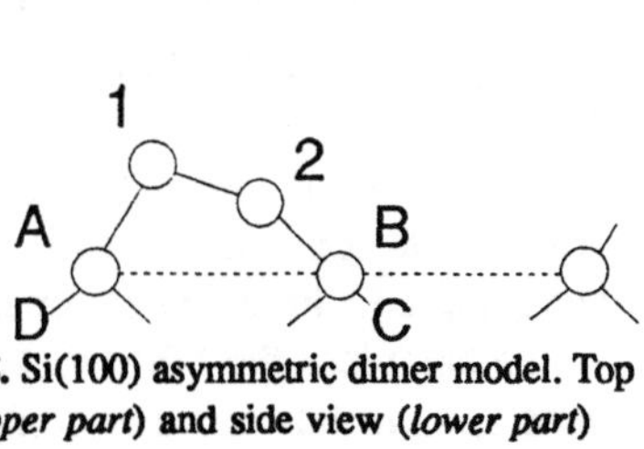

Fig. 4.18. Si(100) asymmetric dimer model. Top view (*upper part*) and side view (*lower part*)

range from 0.7 to 0.8 Å. This means that the symmetric situation just described and preserving all distances is a correct starting point for understanding the formation of the asymmetric dimer.

Let us begin with the symmetric case and discuss the possible lowering of symmetry. There again, we have two competing phenomena:

1) the interaction V between the two dangling bonds of atoms 1 and 2 which gives rise to a bonding and an anti-bonding state; this would lead to a gain in energy equal to $2|V|$;
2) a possible upwards motion of atom 1 and a corresponding downwards motion of atom 2; this again involves an effective Coulomb energy U^* as discussed in Sect. 4.3.1.

To lowest order one can treat the two situations separately and the condition for the existence of an asymmetric dimer is

$$U^* < -2|V| \,. \tag{4.48}$$

Here U^* is still different from the values discussed before for the isolated dangling bond and for the Si(111) 2×1 case. The bare value of U now corresponds to electron transfer from atom 2 to atom 1 and involves the intra-atomic value ($\sim 6\,\mathrm{eV}$) minus the interatomic term, roughly equal to e^2/R ($\sim 6\,\mathrm{eV}$). This has still to be reduced by an effective dielectric constant $\varepsilon^* \sim 6$ and the screened U is therefore negligible. This means that one has now $U^* \sim -F_1^2/k \sim 0.5\,\mathrm{eV}$ which would represent the gain due to the lowering of symmetry. Various calculations as well as experimental evidence reported in [4.28] lead to the conclusion that V is small enough for the inequality (4.48) to be satisfied.

Exercises

4.1 Calculate the energy levels of the O_2 molecule as given by the eigenstates of (4.8), by making use of the symmetry properties, i.e. by building symmetric and antisymmetric states. Determine their numerical value from (4.9). Compare with more sophisticated values as given in [Ref. 2.16, p. 28].

4.2 *Electronic structure of graphite*: Graphite is a lamellar crystal and one can, as a good first approximation, treat each plane separately. The atoms in each plane form a hexagonal planar lattice.

Find the unit cell of this lattice. Determine the electronic structure in a nearest neighbor tight-binding approximation with one "s" and three "p" orbitals per atom.

Show that, if the z-axis is perpendicular to the plane, the p_z orbitals of each atom are decoupled from the other s, p_x, p_y orbitals. Calculate the dispersion curve $E_z(\boldsymbol{k})$ (the π band) for this system.

To calculate the other dispersion curves in a simple way, build three equivalent sp^2 orbitals on each atom, pointing towards its three nearest neighbors (in the same way as for the sp^3 orbitals of Sect. 4.1.1). Calculate $E(\mathbf{k})$ in a description which includes a coupling β between two hybrids involved in the same bond and a coupling Δ between two hybrids on the same atom. Discuss the formation of the bands.

4.3 *Effective coupling for dangling bonds*: Use equation (4.14), starting from the molecular model where the α are the bonding and anti-bonding states. Determine: (i) the self-energy of each individual dangling bond and (ii) the effective coupling between dangling bonds on the Si(111) surface. Apply this numerically with Harrison's parameters.

4.4 *Si(111) surface dispersion relation*: Calculate the (111) surface dangling bond dispersion relation in a model containing the following effective tight-binding interactions: (i) the dangling bond self-energy Σ and (ii) the nearest neighbor effective interaction β^* between dangling bonds. Show that one recovers Fig. 4.9.

4.5 *Surface dielectric constant*: Using classical image-charge theory calculate the dielectric constant at the interface between two dielectrics with dielectric constants ε_1 and ε_2. For a charge e in medium 1 calculate its image charge e'. Determine the potential at a distance r from charge e and r' from its image charge. From its value at the interface (where $r = r'$) determine the dielectric constant at the interface.

4.6 *Energetic considerations for dangling bonds*: One can obtain surprisingly good values of total energies from the molecular model of exercise 2.1 applied, for instance, to silicon. Calculate the neutral vacancy formation energy within this model and compare it to the local density value ($\sim 4\,\text{eV}$). Deduce the energy involved in the creation of a neutral isolated dangling bond.

4.7 *Dangling bond U*: We want to estimate the electron–electron interaction U at a bulk dangling bond. Start with the free atom sp^3 value $U \sim 10\,\text{eV}$ and calculate the reduction due to the dielectric constant $\varepsilon = 10$ and the fact that the dangling bond is only 70% localized on the original sp^3 orbital.

4.8 *Improved calculation of F_1*: Describe how one can improve the calculation of F_1 in Sect. 4.3.1 by making use of a cluster tight-binding total energy calculation. In this the perfect crystal is simulated by a central atom connected to the sp^3 hybrids of its neighbors while one of these is removed to simulate the trivalent atom. Write down the various steps of the calculation allowing one to derive F_0 and F_1.

5. Surfaces of Compound Semiconductors

5.1 General Conditions for Self-Consistency: The Local Neutrality Condition

Compound semiconductors are characterized by the fact that there is electron transfer between the electropositive and the electronegative atom. This means that self-consistency effects are very important, even for bulk properties such as the splitting between LO and TO phonon modes. Self-consistency will become even more important in the treatment of surfaces and interfaces. For this reason the first section is devoted to an analysis of such effects. This begins with general arguments separating long-range from short-range effects and showing the physical basis of the so-called "local neutrality condition". Then a tight-binding formulation of self-consistency effects will be presented for further application to the prediction of core-level shifts and band offsets. Finally we give the values of the cation and anion dangling bond levels determined from the zero-charge condition which will help us in the understanding of surface and interface properties.

5.1.1 General Arguments

We begin by discussing bulk screening. Let us consider a general defect characterized by a bare (i.e. unscreened) perturbation potential $V_b(\boldsymbol{r})$. This perturbation induces a modification in the electron density that tends to reduce the strength of the perturbation and leads to a screened potential $V(\boldsymbol{r})$ which, in a first order treatment, can be expressed as:

$$V(\boldsymbol{r}) = \int \varepsilon^{-1}(\boldsymbol{r}, \boldsymbol{r}') V_b(\boldsymbol{r}') d\boldsymbol{r}' , \qquad (5.1)$$

where $\varepsilon^{-1}(\boldsymbol{r}, \boldsymbol{r}')$ is the inverse dielectric function. One often considers the Fourier transform of (5.1) which can similarly be written:

$$V(\boldsymbol{q}) = \sum_{\boldsymbol{q}'} \varepsilon^{-1}(\boldsymbol{q}, \boldsymbol{q}') V_b(\boldsymbol{q}') . \qquad (5.2)$$

Due to the translational periodicity of the perfect bulk crystal it can be shown [5.1, 2] that the dielectric matrix (and its inverse) have non-vanishing elements only when

$$q' = q + G\,, \tag{5.3}$$

where G is a reciprocal lattice vector. The most commonly discussed situation corresponds to a macroscopic bare potential, i.e. when $q' \simeq 0$ (potential with an infinitely slow variation on the microscopic scale). In such a case we obtain from (5.2)

$$V(q + G) = \varepsilon^{-1}(G, 0)V_b(q)\,, \tag{5.4}$$

where q is small enough compared to the dimensions of the first Brillouin zone to have

$$\varepsilon^{-1}(q + G, q) = \varepsilon^{-1}(q + G, 0) \sim \varepsilon^{-1}(G, 0)\,. \tag{5.5}$$

From (5.4) one obtains the macroscopic component of the screened potential:

$$V(q) = \varepsilon^{-1}(0, 0)V_b(q)\,. \tag{5.6}$$

This is simply the definition of the static macroscopic dielectric constant ε which is thus obtained as

$$\varepsilon = \left[\varepsilon^{-1}(0, 0)\right]^{-1}\,. \tag{5.7}$$

However, (5.4) shows quite generally that a bare macroscopic perturbation not only generates a macroscopic ($q \sim 0$) screened potential, but also microscopic components $V(q + G)$ each of them varying as $V(q + G)\,\mathrm{e}^{\mathrm{i}(q+G)\cdot r}$ in real space. Such microscopic components correspond to local-field effects. It has been argued in [5.3] that the microscopic components are an order of magnitude smaller than $V(q)$ itself. Furthermore, if one averages the screened potential over a unit cell of the crystal in real space then the average $\langle \mathrm{e}^{\mathrm{i}(q+G)\cdot r}\rangle$ vanishes in the limit of small q. Then when $V_b(r)$ varies slowly in space one can write

$$\langle V(r)\rangle_R = \frac{1}{\varepsilon}\langle V_b(r)\rangle_R \tag{5.8}$$

where R is the center of the unit cell. This is just the local relation between the screened and bare macroscopic potentials in a dielectric medium.

To illustrate this let us consider a point defect characterized by a perturbative potential consisting of a short-range part (roughly localized in an atomic volume around the defect) and a long-range Coulomb part behaving as $-Qe^2/r$ (with Q positive or negative). At large distances from the defect the Coulomb potential varies slowly in space and can be considered as a macroscopic potential. Thus (5.8) becomes valid and the potential is screened by ε giving $-Qe^2/\varepsilon r$. The important problem is to establish at what distance from the defect this becomes valid. This question can be answered in the context of linear screening theory. However the full theory described above in terms of $\varepsilon^{-1}(q, q')$ is complex and can only be worked out numerically. One can answer this question analytically in a much simpler way by using an approximate dielectric response. One first neglects the non diagonal parts of the dielectric matrix so that one has

$$V(\boldsymbol{q}) = \frac{V_b(\boldsymbol{q})}{\varepsilon(\boldsymbol{q})} , \tag{5.9}$$

where $1/\varepsilon(\boldsymbol{q})$ is the diagonal part of $\varepsilon^{-1}(\boldsymbol{q}, \boldsymbol{q}')$. This approximation neglects local field effects which are not very important in these materials. A good estimate of $\varepsilon(\boldsymbol{q})$ can be obtained from the following very simple considerations. Considering the sreening of a bare charge $+Qe$ one knows that the long range screened potential is equal to $-Qe^2/\varepsilon r$, which means that an amount $Q(1 - 1/\varepsilon)$ of the bare charge is completely screened as in a metal while the remaining part Q/ε is left unscreened. The way in which the part $Q(1 - 1/\varepsilon)$ is screened can thus be analyzed along the lines of Thomas–Fermi theory, which is valid here since the electron density in systems like Si, GaAs, etc. is relatively high. From these considerations the screened potential takes the form of (5.9) with

$$\frac{1}{\varepsilon(\boldsymbol{q})} = \frac{1}{\varepsilon} + \left(1 - \frac{1}{\varepsilon}\right) \frac{1}{1 + \lambda^2/q^2} , \tag{5.10}$$

where q is the modulus of $\boldsymbol{q}$ (the model is isotropic) and λ is the inverse Thomas–Fermi screening length (equal to $4k_F/\pi$, where k_F is the Fermi wave number). Equation (5.10) gives a realistic expression for $\varepsilon(\boldsymbol{q})$ as can be judged from Fig. 5.1 where it is compared to the result of [5.4]. In real space the screened potential is analytical and is given by

$$V(\boldsymbol{r}) = -\left[\frac{Qe^2}{\varepsilon r} + \left(1 - \frac{1}{\varepsilon}\right) \frac{Qe^2}{r} \mathrm{e}^{-\lambda r}\right] . \tag{5.11}$$

Typical values are $\varepsilon \sim 10$ and $\lambda = 1$ a.u. This means that the second correction term can be neglected when $r \gtrsim 2.5$ Å, i.e. at a nearest neighbor distance from the defect site. In other words the bare charge Qe is reduced to Qe/ε, i.e. becomes an order of magnitude smaller within the "atomic sphere" surrounding the defect.

This finding is very important. It means that for a neutral defect ($Q = 0$) the self-consistent solution is that the total charge within the atomic sphere is zero. For charged defects it is Qe/ε which can again be approximated by zero if ε is

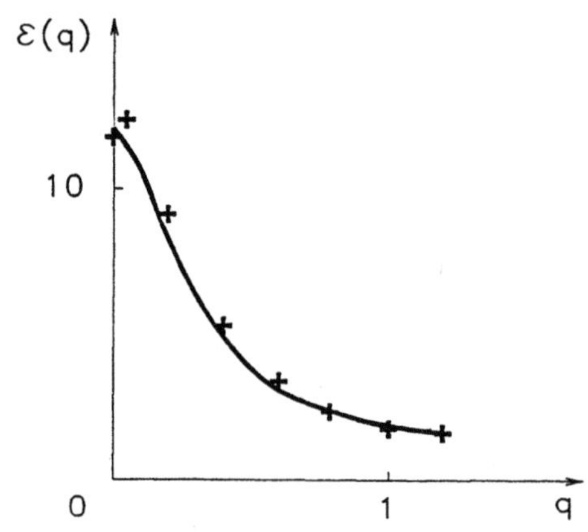

Fig. 5.1. Wavevector-dependent dielectric function for Si. Crosses are taken from [5.4]; full line corresponds to (5.10)

large enough. This leads to the local charge neutrality approximation which has produced a lot of interesting simple results. We shall see later numerous cases where the charge neutrality criterion can be applied with fair success.

5.1.2 Tight-Binding Formulation

As already discussed briefly in Sect. 3.2.2 the tight binding method can be made self-consistent in a relatively simple manner. For this one assumes that the interatomic matrix elements, usually limited to nearest neighbors [5.5] or next-nearest neighbors [5.6] remain unaltered by the charge redistribution. This is not strictly correct but can be justified at least approximately [5.7]. Thus the only consequence of charge transfer is to shift the diagonal terms, i.e. the atomic energies. We work here with *sp*-systems for which the shift of the s and p-levels E_s and E_p is taken to be identical and is given for atom i by

$$\Delta E_i = \sum_j \gamma_{ij} \delta N_j \ , \tag{5.12}$$

where δN_j is the excess number of electrons on atom j and the γ_{ij} are the corresponding Coulomb terms. These fall into two categories: the intra-atomic term $\gamma_{ii} = U_i$ and the interatomic terms γ_{ij} which, in the tight-binding limit, are taken to be proportional to the inverse interatomic distances $1/R_{ij}$. As explained in Sect. 3.2.2, the δN_j are functions of these shifts ΔE_i which can be determined by integrating the partial densities of states defined in Sect. 2.3.3. Usually this is done numerically by Green's function techniques but this can also be achieved by means of simpler molecular models as will be discussed in Chaps. 6 and 7. In several cases one can linearize this relation and write

$$\delta N_i = \delta N_i^0 + \sum_j \chi_{ij} \Delta E_j \ , \tag{5.13}$$

where the χ_{ij} are the linear susceptibilities. Equations (5.12) and (5.13) can be written in matrix form

$$\begin{aligned} \Delta E &= \gamma \delta N \\ \delta N &= \delta N^0 + \chi \Delta E \end{aligned} \tag{5.14}$$

and can be solved for the ΔE to give

$$\Delta E = (I - \gamma\chi)^{-1} \gamma \delta N^0 \ . \tag{5.15}$$

The matrix $(I - \gamma\chi)^{-1}$ is the tight-binding inverse dielectric matrix. The formulation leading to (5.15) allows us to understand easily the zero charge approximation. As the susceptibility matrix is essentially negative and γ is positive (electron repulsion), the eigenvalues of the dielectric matrix in normal situations are larger than unity. In cases of strong screening the eigenvalues of $\gamma\chi$ are much larger than unity and (5.15) simplifies to

$$\Delta E \sim -\chi^{-1} \delta N^0 \ . \tag{5.16}$$

This approximate expression is just what is obtained by imposing $\delta N = 0$ in (5.14). This justifies the use of the local neutrality condition for determining the ΔE_i. Such a condition will be used with much success in the following sections. It has also been the basis of midgap level theories of Schottky barriers and heterojunction band offsets as will be discussed in Chap. 7.

In practical tight-binding self-consistent calculations the parameters are taken from fits to bulk bandstructure [5.5, 6]. Such fits give the interatomic parameters and also the diagonal energies except for an additive constant which shifts the overall band structure. This constant is determined by imposing the condition that the average atomic level of the bulk system be equal to the corresponding free atom value. For instance, in an AB compound, this level E is given by:

$$\bar{E} = \frac{1}{2}\left(\frac{E_{SA} + 3E_{PA}}{4} + \frac{E_{SB} + 3E_{PB}}{4}\right) . \tag{5.17}$$

Such values can be obtained either from experiment [5.8] or from Hartree–Fock calculations [5.9]. Another point is that the fitted values of the bulk E_{si} and E_{pi} must result from a bulk self-consistent treatment and are thus different from the non-self-consistent values E^0_{si} and E^0_{pi}. These last values must be deduced from the expression (5.14) by calculating the δN_i with the fitted prameters and by writing:

$$\begin{aligned} E^0_{si} &= E_{si} - \Delta E_i \\ E^0_{pi} &= E_{pi} - \Delta E_i \end{aligned} \tag{5.18}$$

A final point concerns the values of the intra-atomic Coulomb terms U_i. These can be obtained either empirically from atomic spectra [5.10] or from other more or less empirical procedures [5.11]. They can also be calculated directly, for instance by using a simple method detailed in the appendix to Chap. 6. In self-consistent tight-binding calculations (such as those concerning band offsets in Sect. 7.2) it is better to choose the intra-atomic terms U_A and U_B to obtain the best description of bulk dielectric screening. This is achieved by fitting the inverse dielectric matrix in reciprocal space to give the correct known value at $\boldsymbol{q} = \boldsymbol{0}$ [i.e. $1/\varepsilon(0) \sim 1/10$] and also at the Brillouin zone boundary [where, in general, $1/\varepsilon(G_0) \sim 1/2$].

The electrostatic contributions of (5.12) for surfaces or interfaces involve the calculation of lattice sums over planes, having the following general form:

$$I(\boldsymbol{q}, \boldsymbol{r}) = \sum_{\boldsymbol{r}_1} \frac{\exp(\mathrm{i}\boldsymbol{q} \cdot \boldsymbol{r}_1)}{|\boldsymbol{r} - \boldsymbol{r}_1|} \tag{5.19}$$

where the sum is taken over identical atoms belonging to a given plane, and the site $\boldsymbol{r}$ may or may not lie in the plane. General methods for evaluating such sums are described in [5.12].

5.1.3 Zero-Charge Dangling Bond Levels and Work Function

The creation of a surface in covalent, III-V or II-VI materials with the zinc-blende structure involves the breaking of a number of covalent bonds and thus the creation of dangling bonds. These dangling bonds are quite reactive and they are directly involved in the rebonding at surfaces which occurs in chemisorption processes and also at the formation of interfaces. It is thus of interest to determine the properties of intrinsic idealized dangling bonds that would occur at unrelaxed and unreconstructed surfaces, in order to serve as a reference for systematic analyses of rebonding properties.

The method which is followed here is to work in the context of tight binding where such entities are easier to define. As described in [5.13] the easiest way to study dangling bond properties is to start from the bulk crystal and remove one atom to create the corresponding vacancy. Such a vacancy is a system of four interacting dangling bonds resulting in two characteristic energy levels: a non degenerate one of A_1 symmetry and a threefold degenerate one of T_2 symmetry [5.14]. These states are built from equivalent combinations of the four dangling bond states. The center of gravity of these levels $(E_{A_1} + 3E_{T_2})/4$ is a measure of the individual dangling bond energy E_{db} and their splitting $(E_{T_2} - E_{A_1})$ is a measure of their interaction in the vacancy. This is illustrated in Fig. 5.2.

The first important task is to determine the position of the dangling bond energy E_{db} as a function of charge. Let us call Z the number of electrons in the

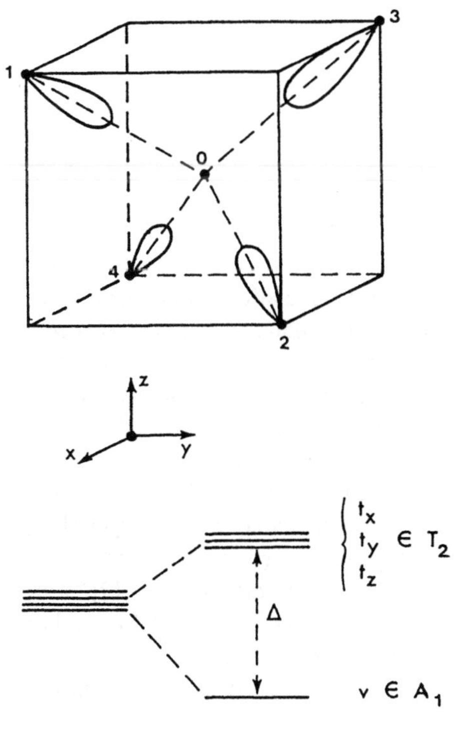

Fig. 5.2. Top: vacancy molecule in the zinc-blende structure with its four sp^3 dangling bonds. Bottom: A_1 and T_2 vacancy levels

dangling bond levels A_1 and T_2 of the vacancy. This corresponds to $Z/4$ electrons per dangling bond and we want to calculate the quantity $E_{db}(Z/4)$ characteristic of the individual dangling bond. For neutral dangling bonds we have: $Z_0 = 4$ in covalent systems (2 electrons in A_1, 2 in T_2), $Z_0 = 3$ for anion vacancies and $Z_0 = 5$ for cation vacancies in 3-5 compounds. The quantities $E_{db}(/Z_0/4)$ have been calculated in [5.15] by putting a perturbative potential on the neighbors of the vacancy, adjusted to get local neutrality on these atoms. The corresponding results are given in Table 5.1. It is also interesting to consider other charge states by changing the value of Z by an amount ΔZ. In such cases the potential on the neighbors must be such that the number of excess electrons with respect to the neutral atom is reduced to $\Delta Z/\varepsilon$ according to our previous analysis. As ε is large one makes little error by again using a local charge neutrality condition. This allows one to obtain $E_{db}(Z/4)$ as a function of Z. From this one obtains the effective Coulomb repulsion parameter U, defined as the shift in E_{db} per excess electron, i.e. $[dE_{db}/d(Z/4)]_{Z_0}$. These are also given in Table 5.1

Table 5.1. Dangling bond energies and effective Coulomb repulsion parameters of anions and cations for IV-VI and III-V compounds

	E_{db} anion	E_{db} cation	U_d anion	U_d cation
AlAs	0.47	1.37	0.96	0.96
			0.4	
GaSb	0.08	0.59	0.52	
GaAs	0.10	0.83	0.52	0.64
GaP	0.28	1.34	0.86	1.12
InP	0.48	1.08	0.8	0.68
InAs	0.27	0.41	0.48	
InSb	0	0.53	0.32	
Si	0.11	0.11		
Ge	−0.10	−0.10		
C	1.48	1.48		

Another interesting by-product of tight-binding calculations concerns the estimation of the ionization potential I_p of the semiconductor, i.e. the absolute position of the top of the valence band. Experimental values of this quantity have been collected in [5.16] together with the absolute position of the average dangling bond level $I_p - \varepsilon_B$ (where ε_B is Tersoff's neutrality level [5.17] which, as will be shown in Chap. 7 must be identified with the average dangling bond energy, i.e. half the sum of the anion and cation dangling bond energies of Table 5.1). These are given in Table 5.2 where they are compared with tight-binding predictions. For this we have positioned the tight-binding band structures with respect to the vacuum by determining the average atomic level $\bar{E}$ of (5.17) from the free atom value. We take $\bar{E}$ from the Hartree–Fock values of [5.11] but shifted by an amount $\bar{U}/2$ where $\bar{U}$ is the average intra-atomic Coulomb term also discussed in Sect. 4.2.3. The physical reason for this shift of $\bar{E}$ between the

Table 5.2. Comparison of the measured ionization potential (I_p) with the energy difference ($-E_v$) between the top of the valence band and the vacuum level. Also included is the average dangling bond level ($I_p - \varepsilon_B$) and its tight-binding estimation ($-E_v - \bar{E}_{db}$)

	I_p [eV]	$-E_v$ [eV]	$I_p - \varepsilon_B$ [eV]	$-E_v - \bar{E}_{db}$ [eV]
AlAs		5.79		4.87
GaSb	4.91	5.06	4.84	4.72
GaAs	5.56	5.55	5.06	5.09
GaP	6.01	5.92	4.84	5.11
InP	5.85	6.07	5.11	5.29
InAs	5.42	5.47	4.98	5.02
InSb	4.90	4.95	4.89	4.70
Si	5.33	5.60	4.97	5.50
Ge	4.80	5.23	4.67	5.34
C		8.33		6.85

free atom and the solid is explained in the appendix to Chap. 6. Once this is done the top of the valence band E_v is automatically positioned with respect to the vacuum and $-E_v$ gives the ionization potential I_p directly. Comparison with the experimental values shows very good agreement, within a few tenths of an eV for all IV and III-V systems. This is within the uncertainties of experiment and theory (experimental values include dipole layers). Finally we give the quantity $-E_v - \overline{E_{db}}$ where $\overline{E_{db}}$ is the average dangling bond energy with respect to the top of the valence band. This again agrees well with $I_p - \varepsilon_B$.

5.1.4 Separation of Long-Range and Short-Range Effects

A fairly general problem that occurs at the surfaces of semiconductors is that surface (or interface) states, when they exist, can accommodate an extra charge. From the macroscopic point of view this leads to the well known situation of Fig. 5.3 where a space charge region develops within the semiconductor while there is a compensating two-dimensional superficial charge density σ. Of course

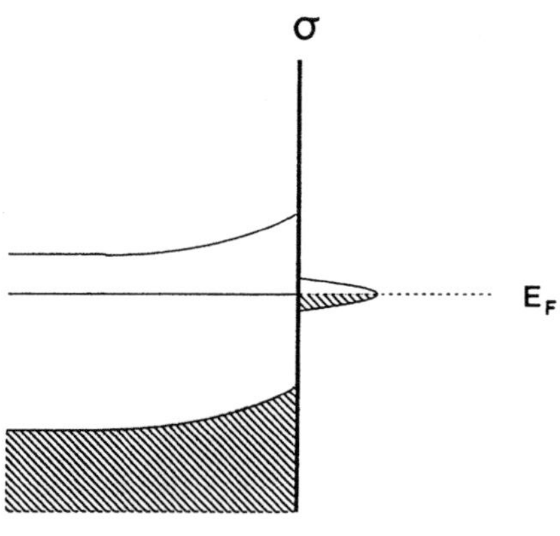

Fig. 5.3. Surface states and space charge region. The bulk semiconductor is on the left and σ is the surface charge density

there is macroscopic band bending. One usually calculates the position of the surface states for the flat band situation, thus assuming that band bending does not modify appreciably their shape and position with respect to the bulk bands in the vicinity of the surface, apart from a rigid shift of the bands. Our aim here is to investigate to what extent this is valid.

To discuss this point let us consider a surface made up of identical atoms, the superficial charge density σ corresponding to an excess δN of electrons on each of these atoms. Macroscopic screening will, according to the discussion of Sect. 5.1.1, reduce this to $\delta N/\varepsilon$ practically localized on these surface atoms. The macroscopic situation corresponds to a uniform distribution of this excess charge, such that one has:

$$\frac{\sigma}{\varepsilon} = -\frac{e\delta N}{\varepsilon s} \tag{5.20}$$

where s is the area per surface atom. In practice this is not true since the excess electron density has the periodicity of the two-dimensional surface lattice. This will introduce some corrections to the surface potential which might in principle affect the position of the surface states. The extreme case is the tight-binding limit where each atomic excess charge $-e\delta N$ is considered as a point charge. Let us then calculate the potential in the vicinity of one surface atomic site (taken as the origin) due to this excess charge on every other site. As discussed in Sect. 5.1.2 this is given by a lattice sum of the general form (5.12) which is written

$$\gamma_0(z) = \frac{e^2\delta N}{\varepsilon} \sum_j \frac{1}{\sqrt{R_j^2 + z^2}} \tag{5.21}$$

where $\boldsymbol{R}_j$ are the lattice vectors along the surface and z is the coordinate perpendicular to it (Fig. 5.4). The special case of the lattice site located at the origin is handled by considering that e^2/R_0 is the intra-atomic Coulomb term, U, as it should be when $z \to 0$. Expression (5.21) can be estimated to a fairly good approximation by isolating the term $j = 0$ and replacing the remaining sum by an integral. This leads to:

$$V_0(z) = \frac{e^2\delta N}{\varepsilon s} \left(\frac{1}{\sqrt{R_0^2 + z^2}} + \frac{2\pi}{s} \int_{\sqrt{\frac{s}{\pi}}}^{R_M} \frac{R\,dR}{\sqrt{R^2 + z^2}} \right) \tag{5.22}$$

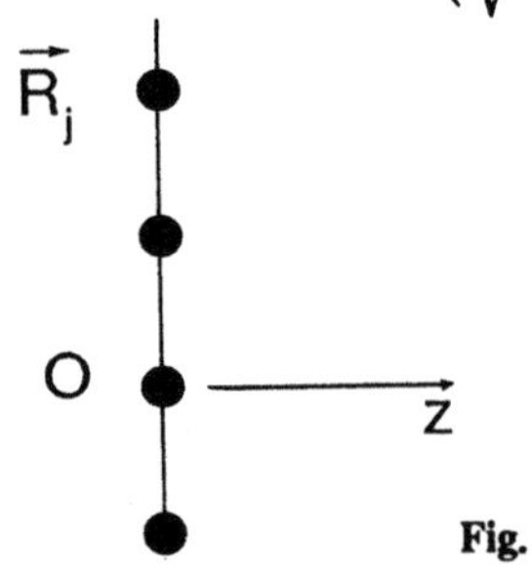

Fig. 5.4. Local configuration used in (5.21)

where we have considered a circular surface of radius R_M (ultimately tending to infinity) and where the lower bound in the integral corresponds to the excluded surface of the central site (circle of radius $\sqrt{s/\pi}$). Evaluation of (5.22) gives

$$V_0(z) = \frac{e^2 \delta N}{\varepsilon} \left(\frac{2\pi}{s} R_M + \frac{1}{\sqrt{R_0^2 + z^2}} - \frac{2\pi}{s} \sqrt{\frac{s}{\pi} + z^2} \right) . \tag{5.23}$$

The first term corresponds to the total surface charge which is exactly compensated by the bulk space charge in the macroscopic treatment. The correction to the macroscopic potential at $z = 0$ is thus

$$\Delta V_0(z = 0) = \frac{\delta N}{\varepsilon} \left(U - 2e^2 \sqrt{\frac{\pi}{s}} \right) . \tag{5.24}$$

It is easy to show that (5.23) leads to the correct macroscopic potential when z becomes larger than the atomic size, i.e. larger than R_0 and $\sqrt{s/\pi}$. In this limit one gets

$$\Delta V_0(z) \sim -\frac{2\pi \delta N e^2}{\varepsilon s} |z| \tag{5.25}$$

corresponding to a macroscopic field $(1/e)(d/dz)\Delta V_0(z)$ equal to $\pm(\sigma/(2\varepsilon_0 \varepsilon)$ as it should be (since $\varepsilon_0 = 1/4\pi$).

We can now estimate the shift in position of the surface states due to this surface charge density σ. This shift ΔE must be given by

$$\Delta E = \theta \Delta V_0(0) \tag{5.26}$$

where θ is the weight of the surface atomic orbitals in the surface state wave function since this wave function is somewhat delocalized. Thus $\Delta V_0(0)$ represents an upper bound for ΔE. An order of magnitude of U is about 7 to 8 eV (Sect. 4.3.2). On the other hand $\sqrt{s}$ should be of order 3 Å. This means that $U - 2e^2\sqrt{\pi/s}$ is of order -7 eV. However this is reduced by $\varepsilon \sim 10$ to give

$$\Delta V_0(z = 0) \sim -0.7 \delta N \text{ in eV} . \tag{5.27}$$

We shall see in Sect. 6.1.2 that the value of δN corresponding to a bulk concentration of ionized impurities of $10^{17}\,\text{cm}^{-3}$ is about 10^{-3} electrons per atom. This means that ΔV_0 and thus ΔE are smaller than 10^{-3} eV which is completely negligible.

5.2 The (110) Surface

We shall consider here the example of the GaAs(110) face which is the most studied case. In the past there has been uncertainty concerning the existence of surface states in the gap region. Some of the early experiments [5.18, 19] did conclude that there were indeed such states in the gap region. However, it is now clear that intrinsic surface states do not exist within the band-gap [5.20, 21]. In fact it has been proved that such states were due to defects pinning the Fermi level near midgap [5.20–24]. The situation was also somewhat ambiguous on the theory side due to the uncertainty in the predictions of the energy position of the localized states. In fact the most important contribution was due to *Chadi* [5.25] who showed that the (110) surface is subject to substantial relaxation, the effect of which is to remove the intrinsic surface states from the gap for reasons which we discuss below.

We first consider the ideal unrelaxed surface. Then we analyze Chadi's results for the relaxation and their effect on the dangling bond surface states. Finally we proceed to a comparison between experiment and theory for the core level shifts.

5.2.1 The Ideal Unrelaxed Surface

The atomic structure of this surface is shown schematically in Fig. 5.5 which displays the corresponding unit cell together with the bonds with atoms belonging to the first inner plane. There have been a large number of calculations on this surface, either using empirical tight-binding theory or the pseudopotential method (for a summary of these before 1979, see [5.26]).

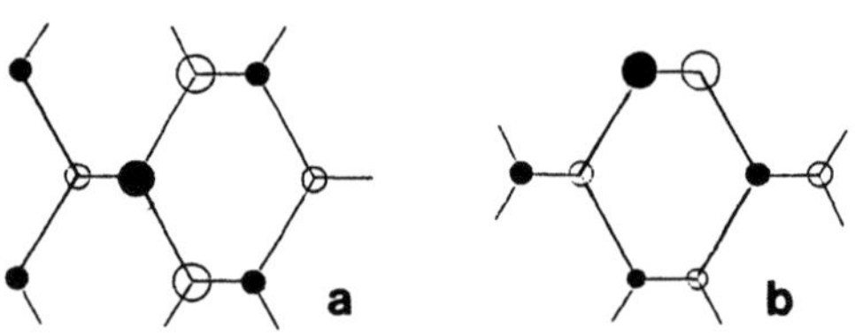

Fig. 5.5a,b. GaAs(110) ideal surface atomic configuration: (a) top view, (b) side view

Most tight-binding calculations have been non-self-consistent and they predict two tight-binding surface bands: the lower one corresponds to the anions and the upper one to the cations. This general view is confirmed by the pseudopotential calculations of [5.27] which concludes that the As dangling bond bond is centered at $-0.25\,\mathrm{eV}$ from the top of the valence band, while the Ga derived one is at $+1\,\mathrm{eV}$, i.e. well within the band-gap, in contradiction to all recent experimental data. We shall see that this contradiction will be removed by including the effect of relaxation.

In fact a good idea of the behavior of intrinsic surface states can be obtained from the values of the anion and cation dangling bond energies. Non-self-consistent values were determined in [5.28] for a number of III-V compounds and, for GaAs, they are in fairly good agreement with the pseudopotential results. This is also the case with the dangling bond levels which we have determined from the zero charge condition in Table 5.1. However the average splitting between the cation and anion dangling bond bands should be $\sim 0.73\,\mathrm{eV}$ from Table

5.1 instead of the 1.25 eV from the pseudopotential calculation. This is due to the fact that the dangling bonds of the (110) face belong to nearest neighbors and cannot really be considered as isolated. Their interaction opens up the gap between the two subbands. There is also an effect due to electron transfer since the anion band will now be populated by 2 electrons whereas the cation one will be empty. This formally corresponds to a transfer of 3/4 electron between the two dangling bonds. This point will be analyzed further in Sect. 5.2.3 when considering core level shifts.

5.2.2 The Relaxed Surface

We have seen in Sect. 4.3.1 on covalent systems that the equilibrium position of a threefold coordinated atom depends on the electron population of its dangling bond. Roughly speaking, when it possesses two electrons the atom is raised and the dangling bond energy is lowered; when it is empty the situation tends to be planar, the dangling bond state being more like a p-state. Thus one would expect the surface anion to be raised and its energy lowered, the reverse being true for the cation.

In [5.25] *Chadi* has performed a tight-binding total energy calculation of the type described in Sect. 2.1.2 by adding short range repulsive terms to the sum of one-electron energies. Using a non self-consistent version of the method he was able to show that the anions were indeed raised and the cations lowered. In a more recent treatment [5.29] a charge dependent version has been applied to several III-V compounds and systematically compared with experiment. The stable situation is pictured in Fig. 5.6. We reproduce here the trends of two basic parameters $\Delta_{1\perp}$ and Δ_{1Y} defined in this figure. Their theoretical values are plotted in Figs. 5.7 and 5.8 versus the experimentally determined values. The agreement is extremely good and confirms the overall picture quite convincingly. Again let us stress that this is in full agreement with the simplified view in which the cations tend to form a planar system with their three nearest neighbor anions.

These atomic displacements tend to lower the anion energies and to raise the cation energies. Detailed calculations [5.26] show that the corresponding dangling

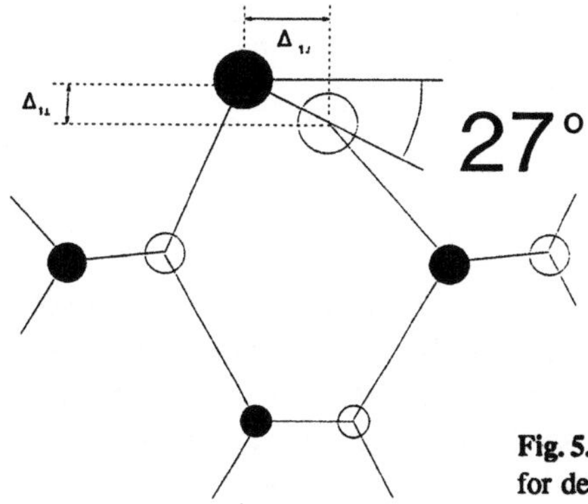

Fig. 5.6. Relaxed GaAs(110) surface, side view. See text for details

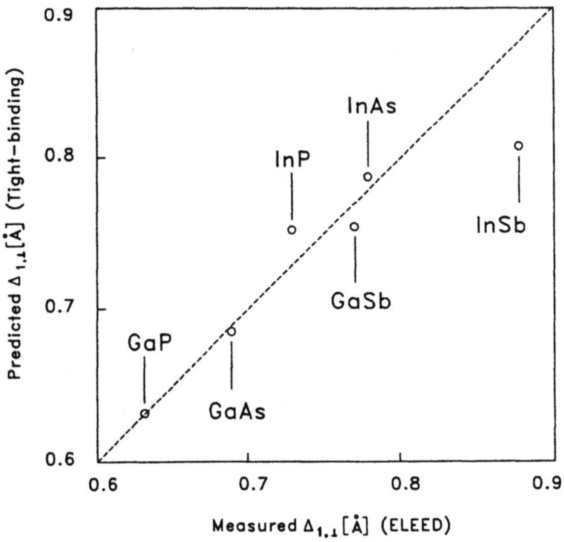

Fig. 5.7. Comparison between predicted and measured $\Delta_{1\perp}$ displacement parameters of the (110) surface of III-V compounds [5.29]

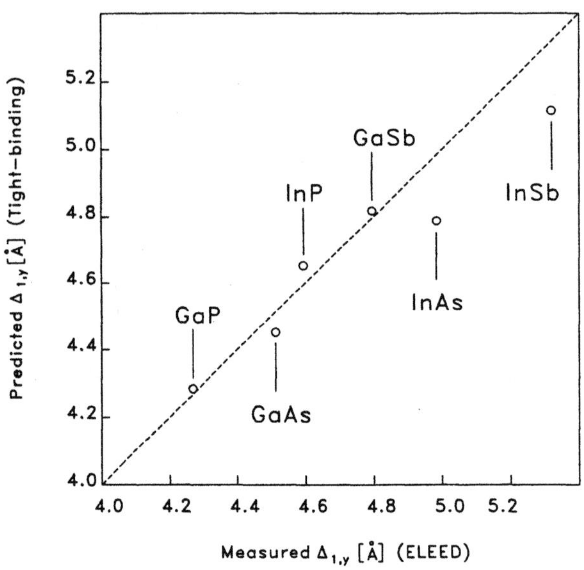

Fig. 5.8. As for Fig. 5.7 but for the Δ_{1y} parameter

bond bands are pushed out of the band-gap in agreement with experimental information. Finally, recent LDA calculations [5.30] fully confirm the picture and have yielded a value of 1.22 eV per unit cell for the surface energy compared with 1.21 eV obtained experimentally, as well as a photoelectric threshold of 4.94 eV compared with 5.15 to 5.75 eV obtained experimentally.

5.2.3 Core Level Shifts

Some years ago photoemission experiments made it possible to determine the core level binding energies of IV [5.31, 32] and III-V [5.33–35] semiconductor surfaces. The observed shifts for covalent systems (Si) are very sensitive to surface orientation and reconstruction. For the (110) surfaces of III-V compounds experiments yielded shifts of 0.2 to 0.4 eV towards higher (lower) binding energies for surface cations (anions). The purpose of this section is to present a calculation of these core level shifts. We consider eight different III-V compounds and study the influence of surface reconstruction. The theoretical approach is based on the tight binding approximation. We first discuss a simple molecular model prior to giving a more complete picture based on a self-consistent tight-binding calculation [5.36].

First let us describe the physical origin of the sign and order of magnitude of the core level shift. For this we use the molecular or bond orbital picture of these materials [5.37, 38]. The model consists of pairs of sp^3-hybrids of energies E_C (cation) and E_A (anion) belonging to the same bond and coupled by a hopping integral V. In this way one gets the usual bonding and anti-bonding states. The contribution of each bond to the electron population of the atoms is $1+f$ on the anion and $1-f$ on the cation where f characterizes the ionicity and is given by:

$$f = \frac{(E_C - E_A)/2}{\sqrt{[(E_C - E_A)/2]^2 + V^2}} \tag{5.28}$$

(we note that f is always positive and less than unity).

In the same picture the creation of the nonpolar (110) surface corresponds to breaking such bonds, leaving an array of alternate cation and anion dangling bonds. As $E_C > E_A$ these two electrons will populate the anion dangling bonds so that the excess population δn_A on the anions relative to the bulk will be equal to $2-(1+f)$, i.e.:

$$\delta n_A = 1 - f\ , \tag{5.29}$$

with the opposite $\delta n_C = -\delta n_A$ on the cations.

The value of δn_A given by (5.29) is positive and will induce a positive shift $U_A \delta n_A$ of the anion valence intra-atomic levels, where $U_A > 0$ is the corresponding Coulomb potential. Similarly the cation shift $U_C \delta n_C$ will be negative. From electrostatic arguments one can show that the core levels will experience a shift of the same order of magnitude, i.e. $U_A \delta n_A > 0$ for the cations [5.39]. Comparison with experiment shows that these predicted shifts have the correct sign. The order of magnitude is fixed by the value of U_A and U_C. As in Sect. 4.2.3, these effective Coulomb terms are deduced from their free atom values, which are on the order of 8 eV, by taking the effect of screening into account. With $\varepsilon \sim 10$, the values obtained are $U_A \sim U_C \sim 0.8$ eV. As $\delta n_A \sim 0.6$, one finds a positive shift of about 0.5 eV for the anions and a negative one ~ -0.5 eV for the cations.

This simple treatment thus provides the basic understanding of the chemical shift for the (110) faces. It shows clearly that these shifts are not directly related to the net charge on the anion and cation in the bulk material. In the simple molecular model these are $1-4f$ and $-(1-4f)$ respectively, with a sign change for $f = 0.25$, while equation (5.29) always keeps a constant sign in agreement with experiment. This conclusion contrasts with that of ***Mönch*** [5.40] who related chemical shifts to bulk interatomic charge transfers. Our conclusion is confirmed by the more complete calculations to be described below where we have checked that two different tight-binding models, leading to bulk net charges of opposite sign do in fact predict practically identical chemical shifts at the (110) surface.

We now describe the complete tight-binding treatment performed for eight semiconductor compounds. We use the first nearest neighbor tight-binding parametrization of *Vogl* et al. [5.5]. To study the effect of surface reconstruction we use a d^{-2} scaling law (where d is the interatomic distance) for the interatomic parameters, as proposed by *Harrison* [5.41]. We choose the atomic geometries of zinc blende semiconductor surfaces reviewed in [5.42] and, for simplicity, only consider displacements of atoms belonging to the surface plane. We also assume that the Coulomb potential due to charge redistribution has only one diagonal element on each atom. Finally we calculate the atomic charges near the surface by using a decimation technique (Sect. 2.3) which is the one best suited for the required accuracy (better than 0.1 e).

To determine the surface core level shifts we start from a non self-consistent calculation in which unmodified bulk parameters are used. The charges calculated on each type of atom of the surface and the first two inner planes are plotted in Fig. 5.9. One finds that the charge disturbance is mainly localized in the surface plane and that it is reduced by the reconstruction. Furthermore the net excess charge relative to the bulk for surface anions and cations keeps a constant sign independent of the sign and magnitude of the bulk net atomic charges, confirming the simple picture discussed above.

In order to calculate the core level shift we use the "local charge neutrality condition" introduced in Sect. 5.1 [5.43]. In principle, this approximation works best when the screening is strong, i.e. essentially for long-wavelength perturbations. This does not seem to be the case here since the charge perturbation is localized on the surface plane and, furthermore, is changing sign from anions to cations. However, it is periodic along the surface and can be expanded in a Fourier series involving the reciprocal lattice vectors K of the surface plane. In view of the rapid decrease of the atomic orbital form factors the dominant terms in such an expansion correspond to the smallest reciprocal vectors, i.e. $\pm(2\pi/a)$ (100). This screening effect will mainly correspond to a reduction by a q-dependent dielectric function $\varepsilon(q)$ with q close to $2\pi/a$ (100). This effective dielectric constant turns out to be of order 3 to 4, much smaller than the long-wavelength value. To estimate the accuracy of a local-charge-neutrality criterion under these conditions one can use the arguments given in [5.43] to show that the true shift is equal to the zero-charge value minus a corrective term $\Delta_b/\varepsilon(\varepsilon-1)$, where Δ_b is the "bare" or non self-consistent shift and ε the effec-

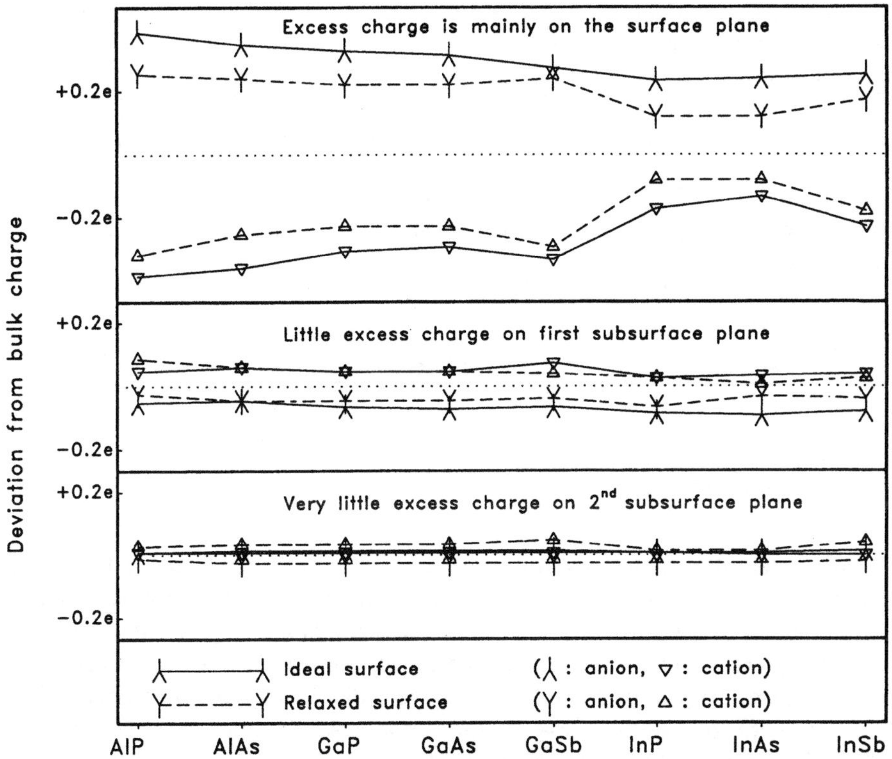

Fig. 5.9. Atomic charge of the first three planes of the (110) surface of III-V compounds. This shows that the charge transfer between anion and cation is mainly confined to the first plane

tive dielectric constant discussed above. For the anions Δ_b is given by $U_A \delta n_A$. With $\delta n_A \sim 0.2$ (Fig. 5.9) and the values of U_A and ε quoted above we get a correction term smaller than 0.1 eV. Thus we can conclude on general grounds that the zero-charge approximation will overestimate the core shifts by an amount smaller than 0.1 eV.

The predicted core level shifts are reported in Table 5.3 for the unreconstructed and reconstructed surface and compared to available experimental data. Except for InP and AlP the theoretical values are roughly equal and opposite as suggested by the simple model. For reconstructed surfaces the calculated cation shift is almost independent of the material (~ -0.3 eV for Al and In, -0.4 eV for Ga). This is not so for the anion for which the shift decreases when the cation is varied in the sequence Al, Ga, In. The overall agreement with experiment is quite correct not only for the sign but also for the order of magnitude. The general conclusion is that the core shifts do not depend strongly on the material, their absolute values always lying in the range 0.2 to 0.4 eV (except for one theoretical prediction of 0.69 eV for AlP).

Table 5.3. Calculated surface core level shifts for anions ΔE_A and cations ΔE_C. Results of photoemission experiments are also reported

Compound		AlP	AlAs	GaP	GaAs	GaSb	InP	InAs	InSb
Theory ideal surface	ΔE_A	0.68	0.53	0.54	0.43	0.28	0.11	0.06	0.19
	ΔE_C	−0.63	−0.47	−0.60	−0.58	−0.40	−0.56	−0.58	−0.46
Theory reconstructed surface	ΔE_A	0.69	0.40	0.43	0.40	0.41	0.00	0.28	0.22
	ΔE_C	−0.40	−0.33	−0.41	−0.38	−0.35	−0.34	−0.31	−0.28
Experiment	ΔE_A	–	–	–	0.37	0.30	–	–	0.29
	ΔE_C	–	–	−0.28	−0.28	−0.36	−0.3	−0.26	−0.22

5.3 Polar Surfaces

It was recognized almost twenty years ago by *Nosker* et al. [5.44] that polar faces of ionic compounds can lead to a divergence of the macroscopic potential as a function of layer thickness. They concluded that a charge accumulation (or depletion) was needed in the immediate vicinity of each surface to prevent such a divergence. They proposed that this charge redistribution could be obtained by reconstruction or even facetting of the surface. A few years later *Masri* et al. [5.45] investigated the same problem for the (111) faces of the zinc-blende compounds and proposed that the required surface charge could be obtained simply in some cases by suitably populating the corresponding surface dangling bond states. Finally, *Harrison* et al. have discussed this problem again in the context of interfaces [5.46] and have proposed modifications in the surface composition analogous to the earlier proposal of *Nosker* et al. [5.44].

In this section we first give a general discussion of the electrostatic problems arising at these surfaces. We then discuss the (100) and (111) faces showing that there are now well established consequences of this general phenomenon.

5.3.1 General Electrostatic Stability Condition

Let us first consider the general situation of Fig. 5.10 where the planes parallel to the surface contain only one type of atom which alternatively bears a charge equal to $-q$ or $+q$. We suppose that the interplanar distances are alternatively R and R'.

An idea of the behavior of the macroscopic potential in such a system can be obtained easily by smearing out the point charges within each plane parallel to the surface. This leads to uniformly charged planes with a superficial charge density $\pm\sigma$, where σ is related to q by

$$\sigma = \frac{q}{s}, \tag{5.30}$$

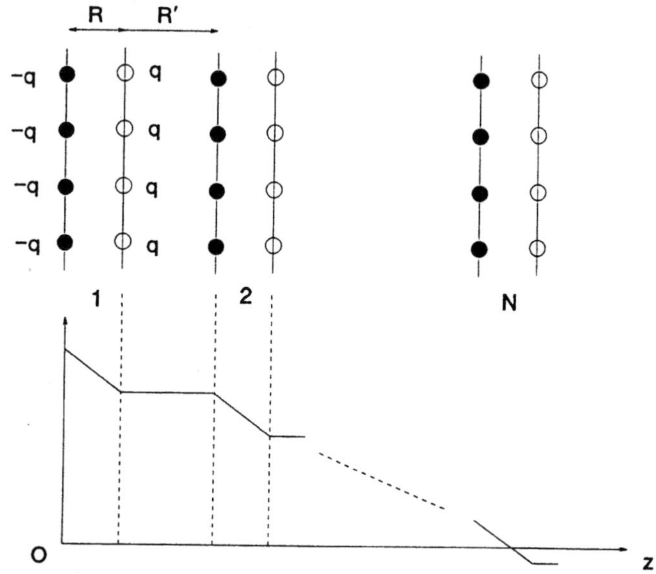

Fig. 5.10. Average over individual planes of the electrostatic potential for alternating layers with opposite charges

where s is the atomic surface. Under such conditions the electron potential V will vary with distance z as shown in Fig. 5.10, since one has a succession of double layers. The potential drop δV in each double layer is given by

$$\delta V = 4\pi e\sigma R\,. \tag{5.31}$$

As the potential is constant between the double layers the total potential drop ΔV is

$$\Delta V = 4\pi e\sigma RN\,, \tag{5.32}$$

where N is the number of double layers. The total length is $(R+R')N$, corresponding to a non-vanishing macroscopic field E, equal to $\Delta V/[(R+R')Ne]$ i.e.

$$E = 4\pi\sigma\frac{R}{R+R'}\,. \tag{5.33}$$

Such a situation is not stable when the thickness of the slab increases to normal values. As emphasized in [5.46] and calculated in Chap. 6 δV is of order of magnitude 30 eV. This value has to be multiplied by N to get the total drop ΔV and for usual situations this would lead to huge values, i.e. to an unacceptable situation. The only way to recover a normal situation corresponding to a zero bulk electric field is to impose on each surface an excess superficial charge density $-\sigma'$ on the left surface and $+\sigma'$ on the right. This will induce an extra potential drop of

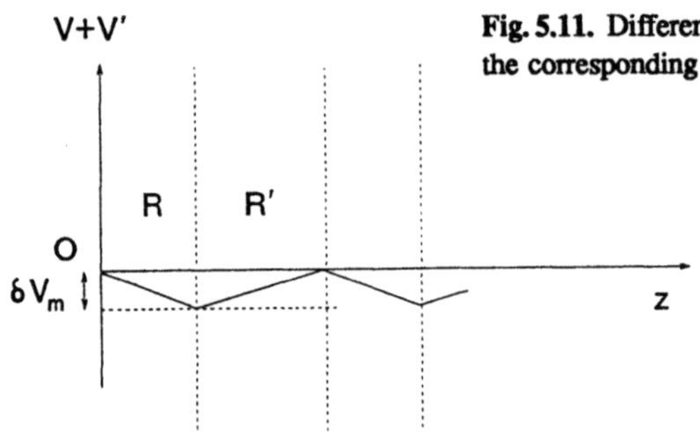

Fig. 5.11. Difference between the potential of Fig. 5.10 and the corresponding macroscopic potential

$$\Delta V' = +4\pi e\sigma'(R+R')N \ . \tag{5.34}$$

The stability condition requires that $\Delta V + \Delta V'$ be zero. This determines σ' which turns out to be

$$\sigma' = -\frac{R}{R+R'}\sigma \ . \tag{5.35}$$

The surface electronic or atomic structure must then be such that this condition be realized. For (111) faces where $R = d/3$ and $R' = d$ (d being the interatomic distance) one has $\sigma' = -\sigma/4$. For (100) faces on the other hand R and R' are equal so that $\sigma' = -\sigma/2$.

Once the potential of Fig. 5.10 is compensated by V' due to charges $\pm\sigma'$ located exactly in the surface planes it behaves as shown in Fig. 5.11 along the slab. This corresponds to double layers, $\mp(\sigma+\sigma')$ separated by R and $\pm\sigma'$ separated by R', as pictured in the figure. The peak δV_m in the periodical potential drop is

$$\delta V_m = 4\pi e\sigma \frac{RR'}{R+R'} \ . \tag{5.36}$$

It is clear that the macroscopic field has now vanished but, from Fig. 5.11, there is still a dipole layer of magnitude $\delta V_m/2$. The simplest way of suppressing this dipole layer while still allowing it to be localized in the vicinity of the surface is to put an extra σ_d on the surface layer and $-\sigma_d$ on the first inner plane. The corresponding potential drop is then $-4\pi e\sigma_d R$ which, to compensate $\delta V_m/2$, requires that

$$\sigma_d = \frac{R'}{2(R+R')}\sigma \ . \tag{5.37}$$

Once this is done the potential behaves as in Fig. 5.12, i.e. it oscillates in the unit cell around a zero average value, as it should in the bulk system.

Now that we have derived the general arguments let us consider the zinc-blende compounds along the lines given by *Harrison* [5.46]. We start from a slab of covalent material (say Ge) with (111) or (100) faces and see what happens when it is transformed into GaAs. We can do this by first freezing the electron

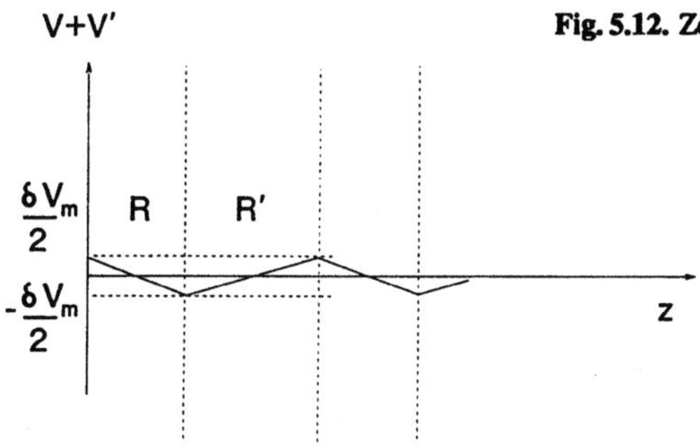

Fig. 5.12. Zero average potential obtained from Fig. 5.10

density and changing the nuclear charges by $+e$ or $-e$. This situation would thus correspond to the case of Fig. 5.10 with $q = +e$ with all the consequences described above. If one allows for electronic redistribution in the framework of linear screening theory then two effects arise independently:

1) screening of the macroscopic component of the field which simply reduces it by the macroscopic dielectric constant,
2) charge redistribution within the unit cell which only alters the oscillating part.

Thus all the previous discussion on instabilities remains valid since the σ, σ' and σ_d are simply divided by ε.

Thus we conclude that the compensating charges near the As(111) or (100) faces are $-e/4$ and $-e/2$ per atom with the opposite on the Ga faces, and that this is completely general. This result can be obtained in different ways. For the (111)As face we require an extra population of 1/4 electron per atom in a dangling bond surface band that already contains 5/4 electron. Similar conclusions hold for other faces. Another possibility is to modify the atomic structure in the vicinity of the surface. For instance, if on the As(111) face one replaces one As atom in four by a Ge atom (which is possible at the Ge-GaAs interface), or if one creates one As vacancy in four surface atoms then the electrostatic condition is automatically fulfilled. We shall see in the next two sections that this is in complete agreement with a number of experimental results. However we have seen that even if these conditions are fulfilled there is still a dipole layer and this is still important. However, it can be compensated for various modifications in the structure, for instance by creating pairs of vacancies or anti-sites between the first two planes.

A final point that is worth mentioning refers to the fact that, at first sight, tight-binding descriptions of these compounds seem to contradict the above conclusions. For instance, as a result of a bulk tight-binding calculation, As atoms appear to have a total net charge $-q$ and Ga atoms $+q$ with $q > 0$. We could apply the previous reasoning, replacing e/ε by q, and obtain an apparently different condition for preventing the electrostatic instability. This is not true however since the net charge $-q$ on As implies an electron population equal to $5 + (q/e)$, which in turn corresponds to a contribution of $(1/4)[5 + (q/e)]$ per bond. At the As surface the stability condition requires an excess charge of $+q/4$ on the As atoms with respect to their bulk value $-q$. This means that the number of electrons on the surface As atoms must be $5+[(3/4)(q/e)]$. Each of the three back bonds contributes $(1/4)[5+(q/e)]$, i.e. a total number of $(3/4)[5+(q/e)]$. The difference is 5/4 electron that must be placed in dangling bonds as in the previous discussion.

5.3.2 GaAs(111) Surfaces

Polar surfaces cannot be obtained by cleavage. They require the use of highly elaborate growth techniques. Although chemical vapor deposition (CVD) and metal-organic CVD (MOCVD) techniques are often recognized as the most promising for actual technological application, molecular beam epitaxy (MBE) is more often used for the physical study of the surface. This is due to various factors. Parameters such as fluxes or substrate temperatures are thought to be more independent of each other in MBE, and also CVD processes involve high partial pressures of undesired chemical species in the vicinity of the growing surface. Perhaps the most important factor is that conventional characterization methods using electron diffraction (LEED or RHEED) can be used to characterize the MBE surface during growth. As a result, MBE has found a very wide application in the growth of III-V compound surfaces. As we saw in Sect. 5.3.1, along a polar axis such as (111), GaAs can be seen as a succession of double layers. It then terminates as a (111) Ga surface on one side and a $(\bar{1}\bar{1}\bar{1})$ As surface on the other.

In the case of the (111) Ga surface of GaAs, a wide variety of diffraction patterns has been seen: (2×2), $(\sqrt{3}\times\sqrt{3})$-$R30°$, (3×3), $(\sqrt{19}\times\sqrt{19})R23.4°$ [5.47]. The different reconstructions are obtained by changing the surface preparation conditions. Early papers tried to explain the reconstruction using a completely stoichiometric surface (buckling model) [5.48]. This model involves a large intra-atomic Coulomb repulsion on the raised atoms [5.49]. However, there exists experimental evidence that the reconstruction is non-stoichiometric [5.49, 50]. The most studied case is the (2×2) reconstruction. The generally accepted model is that of Ga-vacancy buckling, originally proposed by *Tong* et al. on the basis of LEED intensity analysis [5.49]. The corresponding top view and side view of the three top layers taken from [5.49] are shown in Figs. 5.13 and 5.14. Capital letters denote the surface Ga atoms, small letters the As atoms of the second layer. Arrows indicate the relaxation of the As atoms around the Ga vacancy.

The vacancy buckling model was further shown to be stable compared to the ideal surface by *Chadi* [5.51]. A total energy calculation has also been made using a tight-binding scheme very similar to that used for the Si(111) 7×7 reconstruction (Sect. 4.3.3) and the GaAs(110) relaxation [5.52]. However, no electrostatic terms were included in the calculation and no attempt was made to get a self-consistent solution. The final charge transfers found by *Chadi* are small ($\pm0.14\,\bar{e}$) but even so, they correspond to atomic potentials far from the self-consistent values (Sect. 5.1). However, more extensive calculations using self-

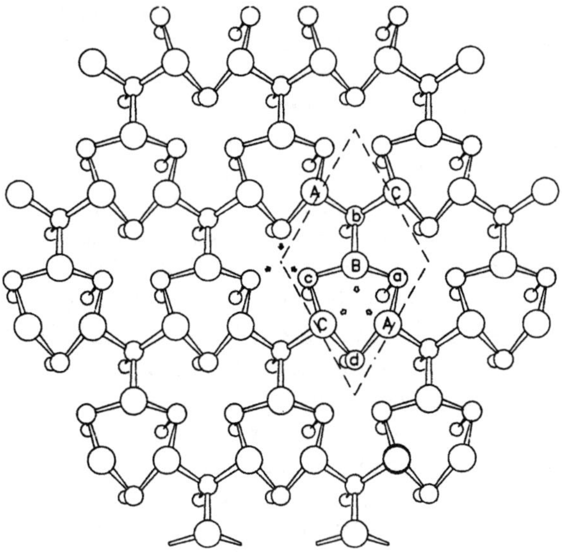

Fig. 5.13. Top view of the (2 × 2) vacancy buckling (VB) model. See text for details

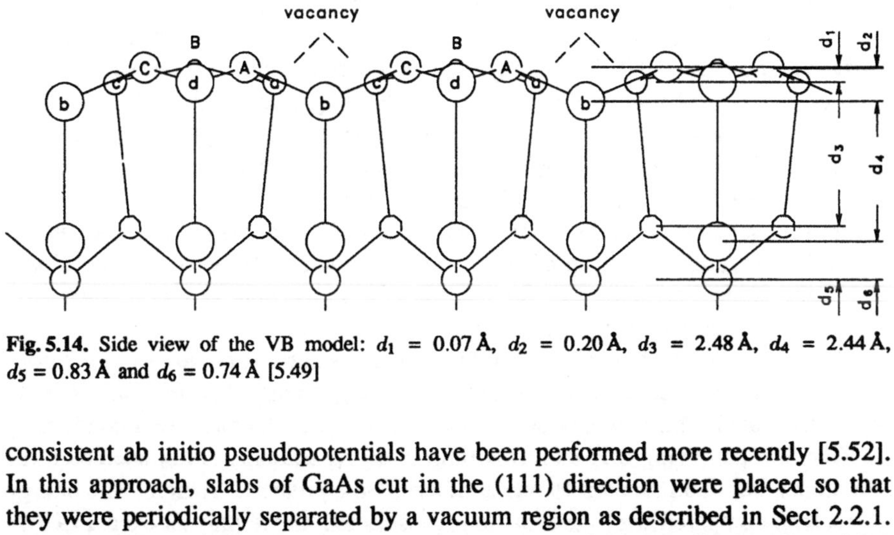

Fig. 5.14. Side view of the VB model: $d_1 = 0.07$ Å, $d_2 = 0.20$ Å, $d_3 = 2.48$ Å, $d_4 = 2.44$ Å, $d_5 = 0.83$ Å and $d_6 = 0.74$ Å [5.49]

consistent ab initio pseudopotentials have been performed more recently [5.52]. In this approach, slabs of GaAs cut in the (111) direction were placed so that they were periodically separated by a vacuum region as described in Sect. 2.2.1. However, if one does this without care the nonzero field described in Sect. 5.3.1 will create a spurious field in the vacuum region, and this vacuum field can have an influence on the reconstruction. To avoid such an effect, *Kaxiras* et al. used the trick of joining slabs of GaAs which were equivalent through an inversion operation. Details of the exact matching condition at the middle plane (between the two symmetric slabs), where the charge state is a little tricky, may be found in [5.53]. Each slab was eight atoms thick and the vacuum region corresponded to four GaAs bond lengths. The total energy was calculated using the LDA approximation for the exchange-correlation term.

A further feature in this ab initio calculation is an approximate calculation of the chemical potential of the Ga and As atoms. Indeed in comparing surfaces reconstructions of different stoichiometry it is necessary to take into account the energy involved in adding or removing one atom from the surface region. Since the experimental setup is usually MBE which uses atomic Ga and most often As_2, both these forms (as well as the stable As_4) were considered as reservoirs of atoms to adjust the stoichiometry. Another possibility is to take GaAs itself as the reservoir. The energy of a single atom can be evaluated in the pseudopotential approximation by integration of the different terms of the single atom Hamiltonian. The binding energy of the atoms in the molecular-like reservoirs (As_2, As_4) is taken from experiment. The binding energy in GaAs is calculated. By changing the reservoir, different growth conditions can be simulated from a stoichiometric situation to one with an excess of one of the species.

Different initial configurations can be chosen before beginning the energy minimization procedure. The choice is guided by the fact that Ga tends to adopt an sp^2 hybridization whereas As tends towards an sp^3-configuration. However, although this happens almost ideally on the (110) surface, it cannot be the case for all the atoms of a (111) or a $(\bar{1}\bar{1}\bar{1})$ surface.

It is remarkable that the results of both models, the simple tight-binding scheme of *Chadi* and the elaborate procedure of *Kaxiras*, are very similar over a wide range of growth conditions. Both show that the Ga-vacancy-buckled surface is the most stable. The energy gain compared to the ideal unreconstructed surface is 2.3 eV/(2 × 2) unit cell in the tight-binding case and 3.3 eV for a GaAs reservoir in the (111) ab initio calculation. The atomic positions in the fully relaxed situation where the energies have been calculated are very close to those determined experimentally from LEED [5.49, 51].

In this geometry, the coverage in Ga is 3/4 of a monolayer and the surface is semiconducting with a gap of approximately 0.44 eV. Thus the surface adopts the charge state that suppresses the electric field in the bulk as was predicted on the basis of a simple electrostatic model (Sect. 5.3.1).

However, in the ab initio calculations, the As-saturated range of growth conditions shows a tendency towards a more stable geometry than the Ga-vacancy-buckling model. This new model has also a coverage of 3/4 and a semiconducting surface, the gap being 0.16 eV. This time three As adatoms are bonded to three Ga atoms and lie in relation to one another in an equilateral triangle. The energy difference compared with the ideal unreconstructed lattice plus an excess of As_2 is 5.0 eV/(2 × 2) unit cell. This surface may only exist in growth conditions where the excess As_2 can maintain this otherwise unstable reconstruction. It is nevertheless interesting to note that in this case also, simple electrostatic arguments give correct predictions for the coverage.

In contrast to the (111) Ga surface, the $(\bar{1}\bar{1}\bar{1})$ As surface does not favor a vacancy formation as is shown by endothermic energies obtained from both tight-binding and ab initio calculations [5.51, 54]. This is mainly related to the ionicity of the GaAs bond. Due to the high energy of a Ga dangling bond and the comparatively low energy of the As dangling bond, it is energically favorable to remove a Ga atom from the surface but not an As atom. It is found in the ab initio calculations that triangle adatom geometries as the one discussed for the As-saturated (111) surface are most stable. The adatom can be Ga or As depending on the growth conditions. Typical energies are $-3.5\,\mathrm{eV}/2 \times 2$ unit cell for Ga triangle compared to the ideal unreconstrcted case and the Ga reservoir case, and $-4.5\,\mathrm{eV}/2 \times 2$ unit cell for As triangles, with an As_2 reservoir. Another surface reconstruction which is found to have a comparable total energy is then 3-As, 9-Ga (puckered) hexagon (see [5.54] for details). All these reconstructions correspond to a coverage of 3/4 in agreement with the electrostatic model. Nevertheless, as argued by *Kaxiras* et al. [5.54], it is doubtful that the As-triangle surface actually occurs experimentally due to the very low sticking coefficient of As on this surface. Thus for As-rich growth conditions it is possible that a metastable reconstruction such as a staggered As-vacancy or a single Ga adatom are kinematically favored.

Given these reconstructions, the Ga triangle and the staggered As-vacancy, the $(\sqrt{19} \times \sqrt{19})$ reconstruction may be easily explained as a case of intermediate stoichiometry, in agreement with experiment.

5.3.3 GaAs(100) Surfaces

The GaAs(100) surface is by far the most studied polar semiconductor surface, because it is the one most used for technological applications. Extensive photoemission experiments [5.55–58], RHEED diffraction patterns [5.59], early Auger measurements [5.60] and recent scanning tunneling microscopy measurements (STM) [5.61] have been used to study it.

Depending on the growth conditions, it is possible to observe different reconstructions. Correlated Auger and diffraction experiments on the surface grown under varying conditions show some indication that the different reconstructions may be related to different coverages [5.60]. The sequence of reconstructions going from a Ga-rich surface to an As-rich surface is: 4×6, 1×6, $C(8 \times 2)$, $C(2 \times 8)$, $C(4 \times 4)$. The $C(2 \times 8)$ and $C(2 \times 4)$ reconstructions are the most extensively studied cases and we will concentrate on them here.

From photoemission experiments *Larsen* et al. were able to identify the main surface related peaks in the valence bands with calculations for a 2×1 reconstruction [5.55]. The dispersion of the bands could be accounted for by letting two dangling bonds of neighboring As atoms interact and form dimers. This is very similar to the 2×1 reconstruction of Si (see Sect. 4.3.4). The formation of dimers was found to be stable compared to the ideal surface using a tight-binding calculation of the total energy [5.62]. Such a dimerization can account for a 2×1 reconstruction, but the four-fold periodicity is more difficult to infer from these photoemission measurements.

The two-fold periodicity is much less prominent in the RHEED patterns [5.56] where no spots, but only streaks are observed [5.63, 64]. This was interpreted as a disorder in the two-fold $[1\bar{1}0]$ direction [5.65]. From a microscopic point of view the four-fold periodicity could be easily obtained by a sequen-

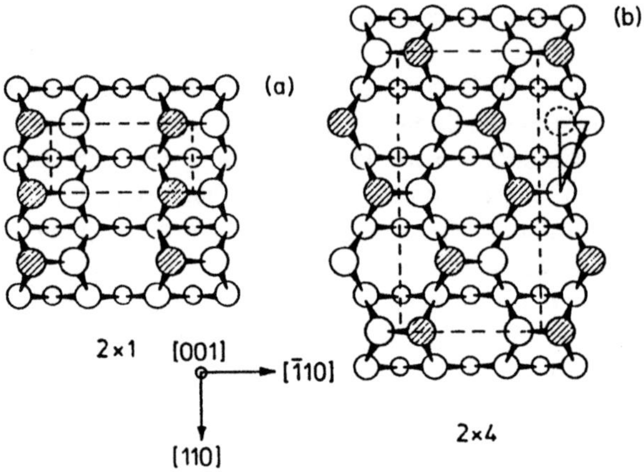

Fig. 5.15. 2 × 1 asymmetric dimer structure (**a**) and the 2 × 4 fully reconstructed asymmetric dimer model (**b**). For details see [5.55]

cing of tilted dimers as in Fig. 5.15. The sequencing of dimers in the two-fold periodicity could then be made random to account for the observed streaks in the diffraction pattern. Thus the 2 × 4 reconstruction could be explained using a full monolayer coverage. This assumption was supported by core level intensities. However other coverages extending down to 0.5 monolayers of As have also been reported [5.66]. The model proposed links the four-fold periodicity observed to the existence of dimers, and thus to the two-fold periodicity. This is in contradiction with the result of exposing a GaAs(100) (2 × 4) reconstructed surface to a H_2 plasma and to excited H_2 [5.67]. Even for very low exposures, hydrogen breaks the two-fold periodicity but remarkably the four-fold periodicity remains present in the RHEED patterns. Thus the two-fold and four-fold reconstructions seem to have different origins. Furthermore, valence band measurements on these surfaces have shown surface peaks that could only be attributed to the bonding of hydrogen both to surface As and to surface Ga atoms [5.68]. It is thus very probable that both species exist on the surface.

Ab initio self-consistent pseudopotential calculations show that the full layer surface coverage is metallic [5.69]. This is contradicted by photoemission measurements that indicate semiconducting behavior. Different reconstructions corresponding to different As coverages between 0.5 and 0.75 have been investigated by *Chadi* [5.56, 70] using the tight-binding approach. Three different models (all leading to a semiconducting surface) have been investigated and are shown in Fig. 5.16. Two involve missing As dimers (a) and (c), and one, missing Ga atoms, so that the As in the third layer can dimerize. They all correspond to a (2 × 4) reconstruction since it was shown that a (2 × 2) reconstruction is energetically less favorable [5.70]. Figure 5.16a corresponds to a coverage of 1/2 whereas Fig. 5.16c corresponds to a coverage of 3/4. Both (b) and (c) are found to have a comparable energy and to be more stable than (a) by 0.2 eV/(1 × 1) unit cell. However kinetic effects may play a significant role on these surfaces which are

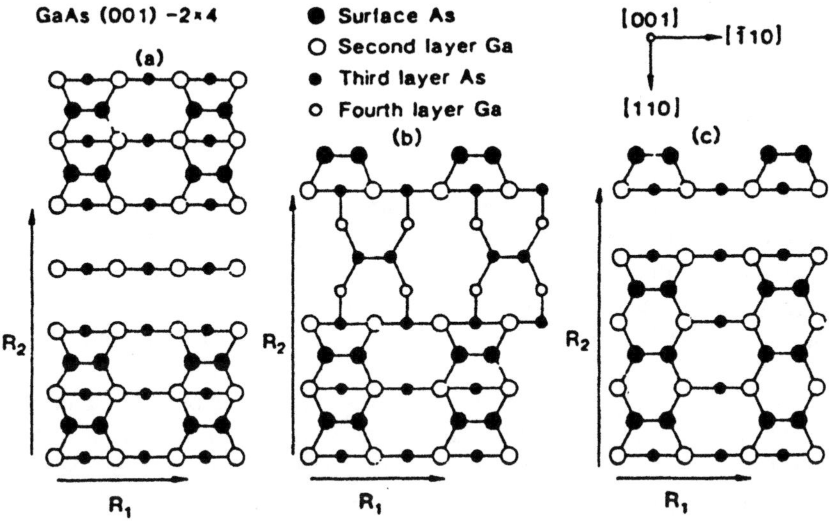

Fig. 5.16. Top view of three models of 2 × 4 structures involving As dimers and missing dimers (a), (c) and also missing Ga atoms (b) [5.56]

grown in conditions far from equilibrium. Thus it is possible that by varying the temperature conditions at the end of the growth (when the Ga flux is turned off, but not the As flux) the stoichiometry may vary. A very nice confirmation that model (c) is a possible configuration is given by the STM pictures, confirming the scheme of Fig. 5.16c. A disorder in the two-fold periodicity is compatible with the streaks seen in RHEED. We should however remark that this picture was obtained on a surface grown by MBE, and covered with excess As as a protective layer to bring the sample into the STM apparatus. The As was then removed thermally. It would be very interesting to obtain the equivalent data in situ in the MBE apparatus for varying growth conditions.

Exercises

5.1 *Thomas–Fermi screening*: In the Thomas–Fermi approximation the electron density $n(\boldsymbol{r})$ at point $\boldsymbol{r}$ is taken to be locally the same as for a free electron gas for which the potential has its local value $V(\boldsymbol{r})$. Express $n(\boldsymbol{r})$ in terms of the Fermi energy E_F and $V(\boldsymbol{r})$. Linearize this expression for weak $V(\boldsymbol{r})$. Use this and Poisson's equation to express the q-dependent dielectric function in the form (5.10) and find the value of λ. Apply this numerically to the case of silicon.

5.2 *Susceptibility matrix in the molecular model*: Generalize the molecular model of Sect. 4.1.1 to atoms with different intra-atomic sp^3 energies. Calculate the electron population on a single atom as a sum of contributions from the four bonds. Use this to calculate the susceptibilities defined by (5.13).

5.3 *Electrostatic lattice sums over planes*: Suppose one has to evaluate an electrostatic lattice sum of the form

$$S = \sum_j \frac{e^2}{\sqrt{R_j^2 + d^2}}$$

where the $\boldsymbol{R}_j$ are atomic positions within a plane and d is the distance to the plane. Show that one can calculate S by including explicitly a few terms of the discrete summation corresponding to the smaller $|\boldsymbol{R}_j|^2$ and replacing all others by an integral from a value $R_{\min}$ to a value $R_{\max}$ assuming circular symmetry. Relate $R_{\min}$ to the number of atoms explicitly included in the discrete sum and $R_{\max}$ to the total area of the plane.

Apply this to the (111) planes of an AB zinc-blende compound. Evaluate the sum S for $d = 0$, at a lattice point, and excluding the term $\boldsymbol{R}_j = 0$ from the summation. Compare the results with no discrete correction to the case where the first shell of neighbors is included in the discrete sum.

5.4 *Core level shifts*: From the results of exercise 1.1 determine the core level shifts at the (110) surfaces of III-V compounds according to the molecular model of Sect. 5.2.3, with a screened Coulomb parameter $U_A = U_B = 0.8\,\mathrm{eV}$.

5.5 *Macroscopic potential for polar surfaces*: Following the approach of Sect. 5.3 determine the numerical value of the macroscopic potential drop that would develop across a (111) oriented slab of GaAs of thickness 1 μm, if there was no surface compensating charge. For this take the GaAs lattice parameter (2.35 Å) and assume that the net charge on the atoms is $\pm 0.2\,e$.

6. Chemisorption on Semiconductor Surfaces

There have been many studies of chemisorption on metals and semiconductors and it is impossible to treat all aspects of this problem here. For this reason we have focussed our interest on two particular cases of chemisorption on semiconductors. The first one concerns the growth of a metallic layer for which there have been recent studies at low temperatures. This is directly connected to the Schottky barrier height which will be discussed in Chap. 7. The second system which is developed in Sect. 6.2 is the As layer on the (111) face of covalent semiconductors which represents an extremely well defined problem from the structural point of view and can be considered as a precise test for the theory.

6.1 Chemisorption of Metal Atoms on Semiconductors: The Formation of the Schottky Barrier

We shall see in Chap. 7 that the Schottky barrier height seems to be determined in many cases by the pinning of the Fermi level at the neutrality level E_B which is also the average dangling bonds energy. To achieve deeper understanding of what happens at the interface several studies of the behavior of the Fermi level position with respect to metal coverage have been performed (for recent reviews of the subject, see [6.1,2]). A very intriguing feature of the results obtained at room temperature (*RT*) is that pinning is achieved for very low coverages, of the order of 0.2 ML (monolayers). Another point is that, for thick layers, it seems that one can observe several pinning positions and this observation has led to the "Unified Defect Model" of *Spicer* et al. [6.3]. The problem is that room temperature (*RT*) is sufficient for the adatoms to rearrange themselves on the surface in which case they tend to form metallic islands, for which the Fermi level pinning is analogous to that of a thick metallic layer. In this regard much progress has been achieved recently through low temperature (*LT*) studies by several authors [6.4–8]. In such cases, at low coverages, the metal adatoms tend to remain distributed on the surface, the barrier height for motion being large enough to prevent adatom diffusion and clustering. One thus gets a uniform distribution of adatoms which represents an ideal situation for a better understanding of the bonding at these interfaces.

A major difference between the *RT* and *LT* results for the Fermi level position on n- and p- type GaAs is schematically pictured in Fig. 6.1. At *RT*

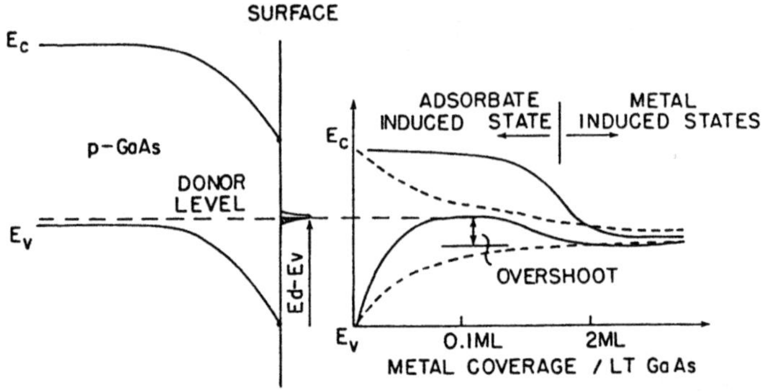

Fig. 6.1. Left: band bending near the surface. Right: schematic representation of the variation of the surface Fermi level for low-temperature (*full line*) and room-temperature (*dotted line*) adsorption of metals on the (110) face of GaAs. Upper curves (starting from E_c) are for n-type, lower curves (starting from E_v) for p-type

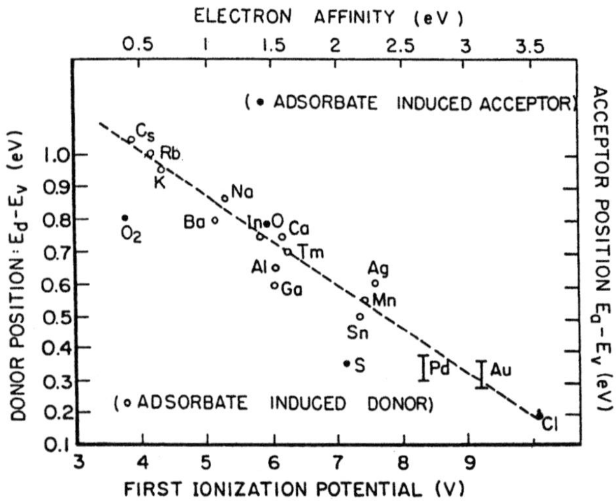

Fig. 6.2. Plot of the adsorbate induced donor (∘) or acceptor (•) levels versus the first adatom ionization potential

the Fermi level E_F moves symmetrically with coverage for n- and p- type GaAs and tends towards the thick layer position. At *LT* the situation is drastically different. E_F remains constant on n- type materials and moves abruptly for a relatively large coverage ($\sim$ 1 ML to 2 ML) which probably corresponds to the onset of metallization. On p- type materials E_F moves rapidly, then saturates and finally shifts smoothly downwards towards its limiting value. This *LT* behavior is characteristic of donor states induced by adatom chemisorption. *Khan* et al. [6.1] as well as *Monch* [6.9] have identified the saturation position with the adatom induced donor energy and have plotted this energy against the first atomic ionization potential finding a linear dependence (Fig. 6.2). Similar results are obtained for other adatoms such as O, S, Cl which induce acceptor states and are also plotted on Fig. 6.2 but with respect to the atomic electron affinity.

In the following sections we demonstrate that these very typical results can be quantitatively described for GaAs (110) by the formation of adatom-Ga and adatom-As single bonds. We first analyze these bonds in terms of a Hartree–Fock model and show that the donor states must be identified with metal-Ga bonding states. We obtain in the same model a linear correlation with the atomic term values that quantitatively agrees with Fig. 6.2. The second section deals with the very low coverage situation, describing how the Fermi level reaches ist saturation value. The third section is concerned with higher coverages (~ 0.1 to 1 ML) where dipole-dipole interactions become important and we demonstrate that their inclusion explains the lowering of E_F, i.e. the fact that there is an overshoot. Finally we discuss other cases and the onset of metallization.

6.1.1 The Isolated Chemisorbed Molecule

We want do deal here with the LT low coverage situation where each adatom-substrate bond can be treated independently of the others. We consider first adatoms with one s electron, i.e. the alkali atoms but also Au and Ag. In a LT situation each adatom, when reaching the surface, tends to achieve the closest possible local bonding configuration which does not necessarily correspond to an absolute minimum in energy. For alkali atoms on GaAs(110), for instance, it is likely that single bonds are formed either with one Ga or one As dangling bond. This will result in filled bonding states and empty anti-bonding states, representing a favorable situation as regards the bonding strength. Statistically one expects an equal number of adatom-Ga and adatom-As bonds even if one of them ist slightly stronger than the other. Experimental evidence for the existence of two adsorption sites at low coverages has been obtained recently for Cs on GaAs and InP(110) from the splitting of core levels [6.10].

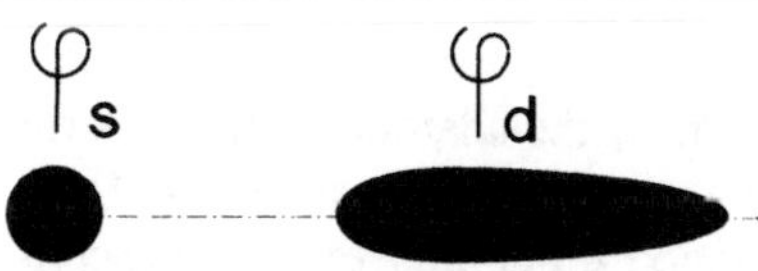

Fig. 6.3. Coupling of the adatom s (φ_s) orbital to an sp^3 (φ_d) dangling bond

Let us then analyze the electronic states obtained in the formation of one such bond treated individually. To obtain a very simple description we shall make use of an approximation equivalent to the defect molecule model used with much success in the theory of point defects [6.11]. Strictly speaking this involves treating the subsystem of Fig. 6.3, which consists of the s-orbital φ_s of the adatom and the sp^3 orbital φ_d corresponding to the Ga or As dangling bond. This trivially leads to a 2×2 matrix of the form:

$$\begin{bmatrix} E_s & V \\ V & E_d \end{bmatrix} \tag{6.1}$$

where E_s is the s adatom energy, E_d the dangling bond energy and V the covalent coupling strength between the two. Diagonalization of this matrix leads to bonding and anti-bonding states whose energies are given by:

$$E_{\substack{A\\B}} = E_d + \delta \pm \sqrt{\delta^2 + V^2}\,, \tag{6.2}$$

with

$$\delta = \frac{E_s - E_d}{2}\,, \tag{6.3}$$

the + or − sign corresponding to anti-bonding and bonding states respectively. However such a simplified model, if applied literally, cannot give good results. As discussed in [6.11] this is due to the fact that when allowing the molecule to interact with the surrounding system the sp^3 dangling bond delocalizes over its back-bonds and even further away. This can be taken into account simply by "renormalizing" the defect molecule parameters, i.e. by replacing E_d by the true dangling bond energy (and not its sp^3 approximation) and by reducing V with respect to ist sp^3 value by the square root of the corresponding delocalization factor. To be more precise the true dangling bond state wave function ψ_d can be written as $\alpha\varphi_d + \sqrt{1-\alpha^2}\varphi'_d$ where φ_d is the pure sp^3 state while φ'_d is a combination of the states of the surrounding crystal. The weight of the sp^3 state is $|\alpha|^2$ and the covalent coupling V between φ_s and ψ_d will be reduced by the factor α.

As there is some amount of charge transfer between the adatom and the semiconductor one needs to perform a self-consistent treatment. For this we make use of a tight binding Hartree–Fock approximation in which the diagonal terms are linear functions of the atomic electron populations. This leads us to write, as in Sect. 5.1.2,

$$E_s = E_{s0} + \frac{U_s}{2} n_s + \sum_j \frac{e^2}{R_j} \delta n_j\,, \tag{6.4}$$

where n_s is the electron population of the adatom, U_s the adatom Coulomb term (one has to take $Us/2$ because of the exchange term which cancels half the Hartree term), δn_j is the excess electron population on atom j of the semiconductor and R_j is the corresponding distance to the adatom.

In principle a similar expression should be written for the shift in dangling bond level E_d. However the corresponding self-consistent calculation would be relatively cumbersome and one can derive a fairly accurate answer from very simple physical arguments that are related to the charge neutrality criterion discussed in Sect. 5.1.2. As we are considering semiconductors with high values of the dielectric constant ($\varepsilon \sim 12$), the true self-consistently screened solution will be close to the mathematical limit $\varepsilon \to \infty$. in other word one can replace the semicondutor by a metal of the same average electron density. In this limit the shift in E_s can be readily estimated since the sum $\sum_j (e^2/R_j)\delta n_j$ is the self-consistent response of a metal to an external point charge equal to $-e\delta n_s$ (where $\delta n_s = (n_s - 1)$ is the net excess of electrons on the adatom). It has been

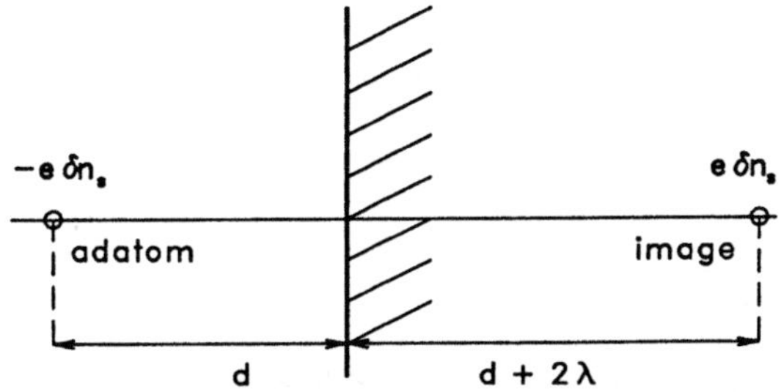

Fig. 6.4. Corrected image of a charged adatom where λ is the Thomas–Fermi screening length (for a full discussion see [6.12])

shown in [6.12] that even for small distances d between the adatom and the surface, $\sum_j (e^2/R_j)\delta n_j$ can be approximated by a modified image-charge theory (Fig. 6.4). This gives for E_s the extremely simple result

$$E_s = E_{s0} + \frac{U_s}{2} + \left(\frac{U_s}{2} - \frac{e^2}{2(d+\lambda)}\right)\delta n_s \, , \tag{6.5}$$

where d is the distance of the adatom to the semiconductor surface and λ is the Thomas–Fermi screening length. From the same type of arguments the shift of the dangling bond energy should be imposed by a local charge neutrality criterion, i.e. should be written as in Sect. 5.1 in the form

$$E_d = E_{d0} + U_d \delta n_d \, , \tag{6.6}$$

where E_{d0} is the position of the neutral dangling bond (corresponding to a filling $Z = 3/4$ or $Z = 5/4$ for the Ga and As dangling bonds) and U_d is the effective Coulomb term, equal to 0.6 eV for both dangling bonds (see Sect. 5.1 for a more detailed discussion).

We can now proceed to the self-consistent calculation. Let us fill the bonding state with n_b electrons. Then we can express δn_s as:

$$\delta n_s = \frac{n_b}{2}\left(1 - \frac{\delta}{\sqrt{\delta^2 + V^2}}\right) - 1 \tag{6.7}$$

and δn_d as:

$$\delta n_d = \frac{n_b}{2}\left(1 + \frac{\delta}{\sqrt{\delta^2 + V^2}}\right) - Z \tag{6.8}$$

with $Z = 3/4$ or $5/4$ for Ga or As dangling bonds. From (6.3, 5, 6) we get

$$\delta = \delta_0 + \frac{1}{4}\left(U_s - \frac{e^2}{d+\lambda}\right)\delta n_s - \frac{U_d}{2}\delta n_d \, , \tag{6.9}$$

with

$$\delta_0 = \frac{\left(E_{s0} + \dfrac{U_s}{2} - E_{d0}\right)}{2} \, . \tag{6.10}$$

This leads to the self-consistency condition for δ:

$$\delta = \frac{\delta_0 - \frac{1}{4}\left(1 - \frac{n_b}{2}\right)\left(U_s - \frac{e^2}{d+\lambda}\right) - \left(\frac{n_b}{2} - Z\right)\frac{U_d}{2}}{1 + \frac{\frac{n_b}{8}\left(U_s - \frac{e^2}{d+\lambda} + 2U_d\right)}{\sqrt{\delta^2 + V^2}}} . \tag{6.11}$$

This equation has to be solved iteratively and gives a different solution for coupling with Ga or As dangling bonds and for varying occupation numbers n_b of the bonding states. We have thus to specify now the physical situation of interest and also how we estimate the parameters, especially the adatom-surface distance d.

Le us consider the physical situation first. We want to get an idea of the ground state configuration of the neutral surface. We have assumed that in the LT deposition there is statistically an equal number of adatom-Ga or adatom-As bonds. As there is one electron per adatom and on the average one electron per dangling bond the neutral surface will correspond to each bonding state being occupied by two electrons. The corresponding level scheme is pictured in Fig. 6.5. The highest occupied level is the adatom-Ga bonding state and will be the first to be ionized. Its ionization level $E(2,1)$ is given by the difference in the total energy $E_{tot}(n_b = 2)$ of the system with two electrons and $E_{tot}(n_b = 1)$ with one electron. We find

$$E(2,1) = E_{tot}(n_b = 2) - E_{tot}(n_b = 1) , \tag{6.12}$$

which can be estimated by using Slater's transition-state argument (see Sect.3.2.3), i.e. by expanding this expression to second order in the difference $[n_b - 3/2]$. We then find

$$E(2,1) = \left(\frac{\partial E_{tot}}{\partial n_b}\right)_{n_b=3/2} . \tag{6.13}$$

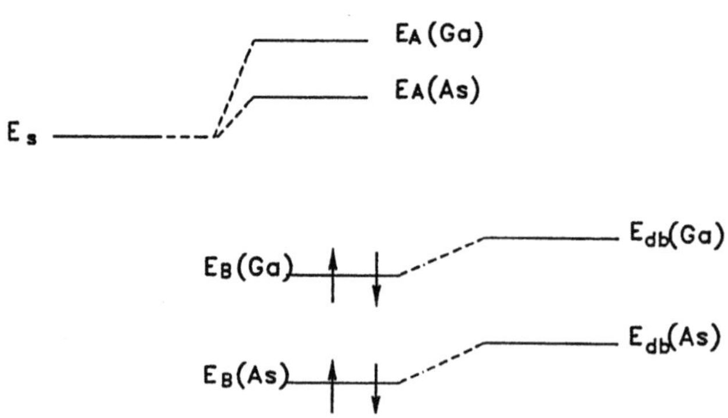

Fig. 6.5. Adatom level scheme showing the formation of bonding and anti-bonding states resulting from the coupling of the s adatom state to the Ga or As dangling bonds

As the derivative of E_{tot} with respect to n_b is the one-electron eigenvalue [6.13] this means that the single donor level $E(2,1)$ is simply given by the bonding adatom-Ga level calculated at the intermediate occupation number $n_b = 3/2$. It thus requires the solution of (6.11) for δ with this value of n_b.

Let us now describe how one obtains numerical values for the parameters. The first of these is the interatomic distance d between the adatom and its neighbor (Ga or As). We empirically fix it as the sum of the atomic radii tabulated by Clementi [6.14], a procedure that has been shown to be quite efficient in a number of cases [6.15]. Once we have d we calculate the interatomic coupling between the adatom state and the sp^3 state from Harrison's rules (see [6.16] and Sect. 4.1.3) and reduce it by the delocalization factor α which we have estimated to be 70%. The energy E_{s0} is taken to be the Hartree-Fock energy of the neutral atom since, in the neutral atom, there is complete cancellation between the Coulomb and the self-exchange term. The absolute position of the top of the GaAs valence bands is obtained in the same way as in section 5.1.3 i.e. by referring it first to the average sp^3 level $\bar{E}$ in this material (using for this the tight binding parametrization of [6.17]. We then determine $\bar{E}$ from a Hartree-Fock free atom calculation, giving $\bar{E}_0$, and adding to this half the Coulomb term $\bar{U}/2$ (see Appendix) to compensate for the change in self-exchange term between the atom and the solid. The values of the Coulomb terms U_s and $\bar{U}$ are taken from Harrison [6.18], and the Hartree-Fock values are given in Table 6.1.

Table 6.1. Parameters for the model of the isolated chemisorbed molecule

	R_s [Å]	E_{s0} [eV]	U [eV] [6.17][a]	V [eV]
Cs	2.98	−3.37	5.05	−0.38
Rb	2.65	−3.75	5.02	−0.45
K	2.43	−4.01	5.56	−0.50
Na	1.90	−4.96	6.17	−0.68
Ag	1.65	−5.99	6.34	−0.79
Au	1.74	−6.01	6.75	−0.75
Ga	1.36		6.61	
As	1.14		8.31	

[a]) With U from [6.17] E_d (Ga) = −4.69 eV, $E_{d\,(As)}$ = −5.42 eV

The results for the bonding of adatom-Ga chemisorbed molecules are plotted in Fig. 6.6 versus the atomic term value E_{s0}, for $U_d = 0.6$ eV as discussed before. One gets for the Ga bonding state a straight line with a slope of about 0.2. This is in remarkable agreement with the experimental plot of Fig. 6.2, which has a comparable slope. Even the absolute position of the donor level in the gap is in very good agreement with the experimental value. This is is not too surprising since it was shown in [6.11] that the molecular model used for transition metal impurities in bulk GaAs induces a systematic error of about 0.2 eV with respect to the full Green's function calculation. Finally, a further confirmation of the

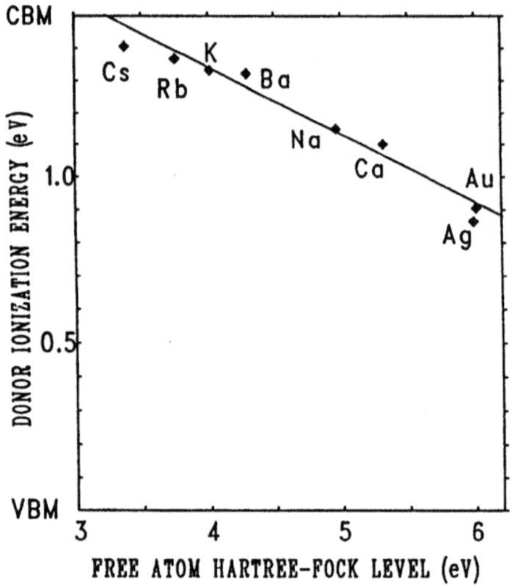

Fig. 6.6. Predicted adatom energy levels versus the free atom values [6.21]

validity of the model comes from the fact that the anti-bonding states fall high in the conduction band so that there are no available acceptor states.

Having found the origin of the donor level we proceed to explain the LT evolution of the Fermi level as a function of metal coverage. We first discuss how the Fermi level reaches its pinning value at very low coverages and, in the following section, we discuss its behavior between 0.1 and 1 ML coverages where electrostatic interactions between chemisorbed molecules begin to play a role.

6.1.2 The Very Low Coverage Limit

In this section we discuss the behavior of the Fermi level at very low coverages, i.e. before pinning can occur [6.9]. We examine the situation described before where there are donor states and consider the cases of n- and p-type materials. The case of n-type materials is very simple. In our previous model both adatom-Ga and adatom-As anti-bonding states lay in the conduction band. This means that there cannot be any accumulation of negative charge (i.e. excess electron density) on the surface states to compensate for a positive space charge due to ionized bulk shallow donors. There will thus be no band bending and the Fermi level will remain constant versus coverage, at least at low coverages.

The situation for p-type materials is drastically different since the negative bulk space charge due to ionized acceptors can only be compensated by ionized surface donors which, as we have shown in the previous sections, correspond to adatom-Ga bonding states. The fraction of ionized surface donors N_s^+ to the total number of donors is given by Fermi-Dirac statistics:

$$N_s^+ = \frac{N_s}{1 + \frac{1}{2}\exp\left(\frac{E_F - E(2,1)}{kT}\right)} , \qquad (6.14)$$

where the factor 1/2 accounts for degeneracy and $E(2,1)$ is the donor level calculated in Sect. 6.1.1.

On the other hand the space charge region can be treated in the conventional depleted zone approximation. For a homogeneous distribution of acceptors in the space-charge region, the total band bending ΔV depicted in Fig. 6.7 is easily evaluated and is given by

$$\Delta V = \frac{e^2 N_A X^2}{2\varepsilon} , \tag{6.15}$$

where N_A is the density of bulk shallow acceptors and X is the width of the space charge region. The latter can be obtained from the global charge neutrality condition

$$N_s^+ = N_A X . \tag{6.16}$$

From the three previous relations one can obtain an equation for the difference $E_F - E(2,1)$

$$E_F - E(2,1) = \varepsilon_A - \varepsilon(2,1) + \frac{e^2 N_A}{2\varepsilon} \left(\frac{N_s}{N_A}\right)^2 \left(1 + \tfrac{1}{2} \exp \frac{E_F - E(2,1)}{kT}\right)^{-2} , \tag{6.17}$$

where all the quantities are defined in Fig. 6.7, ε_A being the bulk acceptor ionization energy and $\varepsilon(2,1)$ the distance in energy between the surface donor level and the top of the valence band at the surface. Equation (6.17) can be solved numerically but it is better to find analytical solutions valid over some range of values of N_s. To be precise we take the usual value of $N_A \sim 10^{17}\,\mathrm{cm}^{-3}$ and express N_s in fractions of a monolayer [one monolayer on GaAs(110) corresponds to $N_s = 0.88\,10^{15}\,\mathrm{cm}^{-2}$]. We also take $\varepsilon_A \sim 0$ and $\varepsilon(2,1) \sim 0.5\,\mathrm{eV}$, a reasonable average value. We can then consider several possibilities:

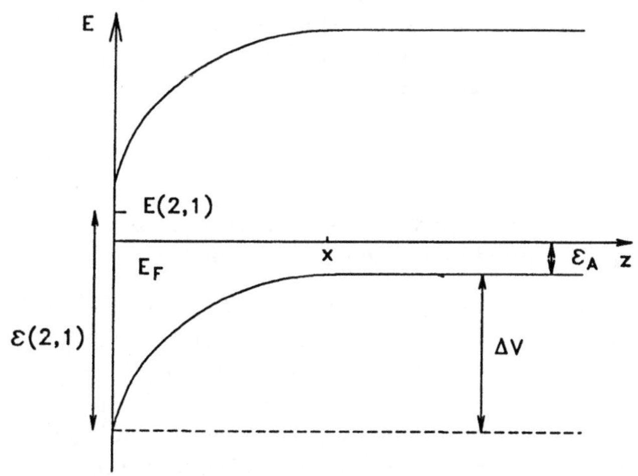

Fig. 6.7. Definition of the important parameters describing the metal–semiconductor interface

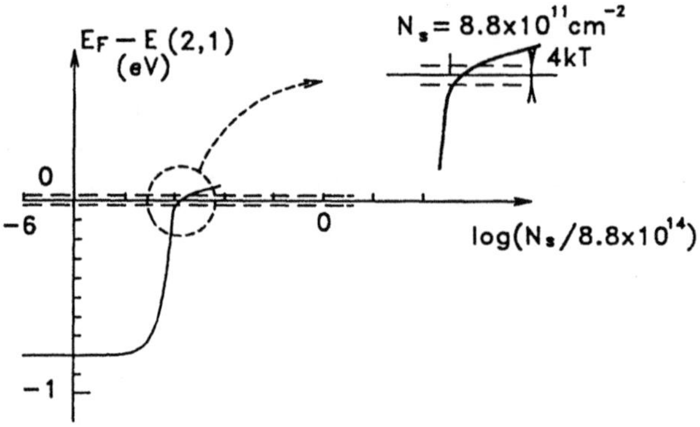

Fig. 6.8. Evolution of the surface Fermi level with the surface density of adatoms (values used for this figure are those given in the text for room temperature)

1) $N_s = N_{s0}$. This is the value at which $E_F - E(2,1)$ strictly vanishes. This critical value is given directly from (6.17) by

$$N_{s0} = \frac{3}{2}\sqrt{\frac{2\varepsilon N_A}{e^2}[\varepsilon(2,1) - \varepsilon_A]}\ , \tag{6.18}$$

and with the numbers given above, $N_{s0} \sim 10^{12}\,\text{cm}^{-2}$, i.e. 10^{-3} ML.

2) $N_s < N_{s0}$. Here $E_F - E(2,1)$ is negative and, as soon as it is lower than $-2kT$, a very good approximate value is

$$E_F - E(2,1) = \varepsilon_A - \varepsilon(2,1) + \frac{e^2}{2\varepsilon}\frac{N_s^2}{N_A}\ , \tag{6.19}$$

which shows that at very low coverage E_F rises parabolically with N_s. Expression (6.19) in principle crosses zero at $N_s = 2/3 N_{s0}$ but is not valid in the range $-2kT < E_F - E(2,1) < 0$ where it is almost flat as shown in Fig. 6.8.

3) $N_s > N_{s0}$. Here $E_F - E(2,1)$ is small compared to $|\varepsilon_A - \varepsilon(2,1)|$ and as soon as it becomes equal to $2kT$ one gets the approximate expression

$$E_F - E(2,1) \sim kT \ln\frac{3N_s}{N_{s0}}\ . \tag{6.20}$$

This variation is very slow at low temperatures, amounting to $2.3\,kT$ (i.e., $\sim$ 20 meV at 80 K) when N_s is increased by one order of magnitude.

The main result of this section is that pinning occurs at $N_s \sim 10^{-3}$ ML and that beyond this value E_F varies very little, which should translate experimentally into some kind of plateau as shown in the curve of Fig. 6.8. Note that this value of N_s corresponds to the adatom-Ga bonding states discussed earlier, corresponding to a total coverage of 2×10^{-3} ML. The conclusion that one needs $\sim 10^{12}$ states $\times\ \text{cm}^{-2}$ for $N_A \sim 10^{17}\,\text{cm}^{-3}$ to pin the Fermi level is identical to that found in [6.20].

6.1.3 Electrostatic Interaction Between Adsorbed Species

When increasing the metal coverage below 1 ML the interactions between adatoms will become important. As long as the chemisorbed bonds are not in nearest-neighbor positions one can neglect the Hamiltonian matrix elements between them. However this is not true of the long range electrostatic interactions between the corresponding dipoles which can modify the self-consistent treatment and eventually shift the donor level position with respect to the low coverage value derived in Sect. 6.1.1.

We thus consider one particular metal-substrate bond for which we generalize equation (6.5) in the form

$$E_s = E_{s0} + \frac{U_s}{2} + \left(\frac{U_s}{2} - \frac{e^2}{2d'}\right)\delta n_s + \sum_i \left(\frac{e^2}{R_i} - \frac{e^2}{\sqrt{R_i^2 + 4d'^2}}\right)\delta n_{si} , \quad (6.21)$$

where the sum over i corresponds to all other adatom-substrate bonds. For each of them we have used the same simple treatment which includes the excess charge $-e\delta n_{si}$ and its image. (In this equation we take $d' = d + \lambda$ as discussed above.)

In the sum over i we have to distinguish three types of bonds: adatom-As with two electrons, adatom-Ga with two electrons and ionized adatom-Ga with one electron. We assume that each bond belonging to one given category has the same δn_s which we label $\delta n_{s,\mathrm{As}}$, $\delta n_{s,\mathrm{Ga}}$ and $\delta n^+_{s,\mathrm{Ga}}$ respectively (this corresponds to an average approximate crystal). From this, expression (6.21) simplifies to

$$E_s = E_{s0} + \frac{U_s}{2} + \left(\frac{U_s}{2} - \frac{e^2}{2d'}\right)\delta n_s + U'(x_{\mathrm{As}})\delta n_{s,\mathrm{As}} + U'(x_{\mathrm{Ga}})\delta n_{s,\mathrm{Ga}} + U'(x_{\mathrm{Ga}^+})\delta n^+_{s,\mathrm{Ga}} , \quad (6.22)$$

where the U' denote the lattice sums and x represents the fraction of monolayer of the corresponding kind of chemisorbed bond.

A straightforward method to evaluate the lattice sums U' is to replace them by an integral, which gives

$$U'(x) = N \int_{1/\sqrt{\pi N}}^{R} \left(\frac{1}{R} - \frac{1}{\sqrt{R^2 + 4d'^2}}\right) 2\pi R dR , \quad (6.23)$$

where N is the surface density of the corresponding adatoms. The lower bound in the integral corresponds to the surface occupied by the central site which has to be excluded from the integral and the upper bound R will be ultimately made equal to infinity. Integrating (6.23) yields

$$U' = \frac{4\pi N d'}{\dfrac{1}{2d'\sqrt{\pi N}} + \sqrt{\dfrac{1}{4\pi N d'^2} + 1}} . \quad (6.24)$$

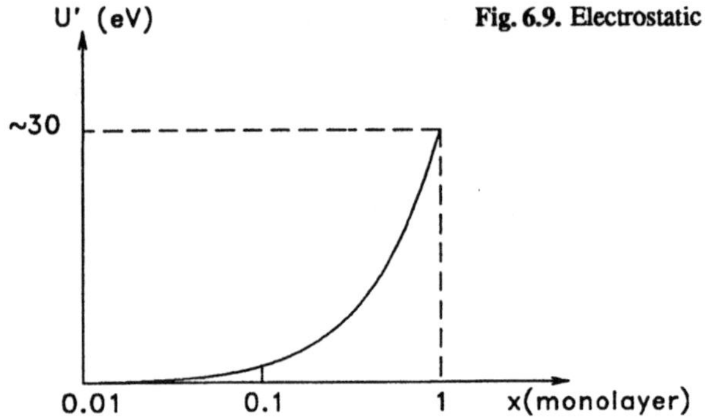

Fig. 6.9. Electrostatic lattice sum as obtained from (6.25)

From Table 6.1 a typical value of d' is 3.7 Å. Taking into account the fact that one ML corresponds to $N \sim 10^{15}\,\mathrm{cm}^{-2}$ one finds numerically

$$U'(x) \simeq \frac{40x}{\dfrac{1}{\sqrt{10x}} + \sqrt{1 + \dfrac{1}{10x}}}\,\mathrm{eV}\,, \tag{6.25}$$

where x is the fraction of monolayer. Of course the replacement of the discrete sum by an integral is an approximation. However one does not know the exact distribution of adatoms and furthermore, comparing the integral approximation to the exact result on a square lattice, there is little difference. Thus expression (6.25) is plotted in Fig. 6.9 showing that it has non negligible values for coverages x lying between 0.01 and 1 ML. In this range there will be a substantial correction to the donor level position calculated in Sect. 6.1.1.

The first conclusion that one can draw stems from the fact that the concentration N_s^+ remains fixed at the value $10^{12}\,\mathrm{cm}^{-2}$, i.e. $x_{\mathrm{Ga}^+} \sim 10^{-3}$ ML. In this case $U'(x_{\mathrm{Ga}^+})$ is much smaller than 0.1 eV, i.e. it remains negligible. Furthermore there is an equal concentration of bonds with As and Ga atoms, i.e. we have $x_{\mathrm{Ga}} = x_{\mathrm{As}} = x/2$. Thus the atomic level E_{s0} of (6.22) can be written

$$E_s = E_{s0} + \frac{U_s}{2} + \left(\frac{U_s}{2} - \frac{e^2}{2d'}\right)\delta n_s + U'\left(\frac{x}{2}\right)(\delta n_{s,\mathrm{Ga}} + \delta n_{s,\mathrm{As}})\,. \tag{6.26}$$

This is in principle valid for any bond. We first apply it to filled ($n_b = 2$) adatom-Ga and adatom-As bonds. This would lead to two coupled equations for the corresponding value of δ. We replace this by an equation for an average value $\bar{\delta}$ which is thus given by

$$\bar{\delta} = \overline{\delta_0} - \frac{1}{2}\left[\frac{U_s}{2} - \frac{e^2}{2d'} + 2U'\left(\frac{x}{2}\right)\right]\frac{\bar{\delta}}{\sqrt{\bar{\delta}^2 + \bar{V}^2}} \tag{6.27}$$

i.e.,

$$\bar{\delta} = \frac{\overline{\delta_0}}{1 + \dfrac{\left[U'\left(\dfrac{x}{2}\right) + \dfrac{U_s}{4} - \dfrac{e^2}{4d'} + \dfrac{U_d}{2}\right]}{\sqrt{\bar{\delta}^2 + \bar{V}^2}}}\,. \tag{6.28}$$

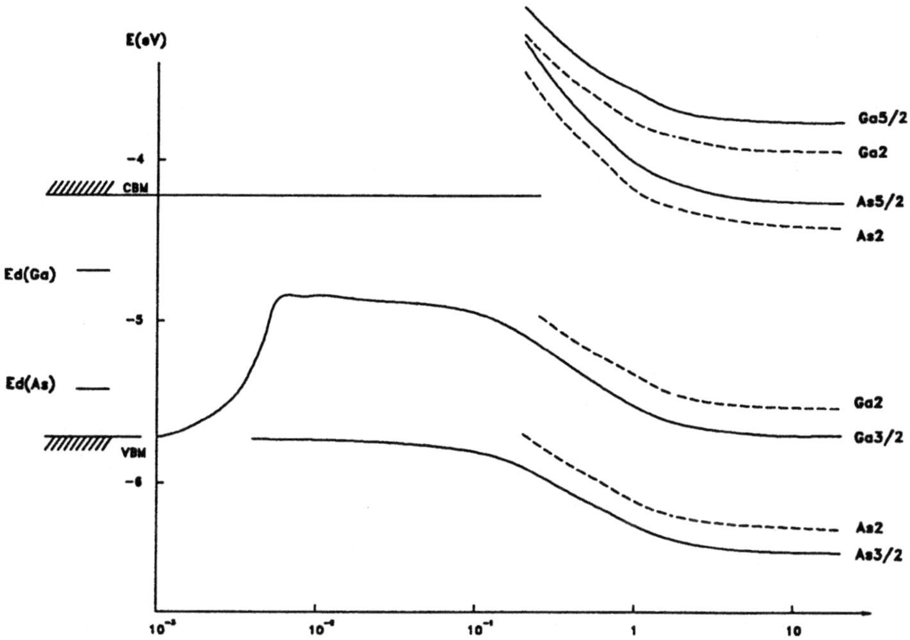

Fig. 6.10. Coverage dependence of the adatom bonding and anti-bonding levels as obtained from the calculation described in the text

At relatively high coverages $x/2 \sim 0.1$ to 0.5 ML $U'(x/2)$ takes values ranging from 1.65 to 13 eV and $\bar{\delta}$ will be substantially smaller than $\bar{\delta_0}$.

We can now calculate the donor level $E(2,1)$ for adatom-Ga bonds exactly in the same way as in Sect. 6.1 but taking into account the last corrective term in (6.26). We thus obtain the new equation for δ:

$$\delta = \frac{\delta_0 - \frac{1}{16}\left(U_s - \frac{e^2}{d'}\right) - U'\left(\frac{x}{2}\right)\frac{\bar{\delta}}{\sqrt{\bar{\delta}^2 + \bar{V}^2}}}{1 + \frac{3}{16}\left(U_s - \frac{e^2}{d'} + 2U_d\right)\frac{1}{\sqrt{\delta^2 + V^2}}} . \tag{6.29}$$

This expression differs from that for the isolated molecule [given by (6.11) for $n_b = 3/2$] by the corrective term $-U'(x/2)(\bar{\delta}/\sqrt{\bar{\delta}^2 + \bar{V}^2})$ which is always negative and thus tends to reduce δ with respect to its low coverage limit. This behavior is clearly shown in Fig. 6.10 where the numerical results obtained for the case of Ag are plotted. We see that all bonding levels calculated for $n_b = 3/2$ experience a substantial drop and finally saturate but at coverages for which one can no longer neglect the overlap between the adatoms.

In the same figure the curves representing the anti-bonding states are also drawn. When these fall into the gap they should behave as acceptor states. From the transition state argument their ionization energy must be calculated with 1/2 electron in the anti-bonding state, i.e., for a total number of electrons in the bond $n_b = 5/2$. Thus the two upper curves of Fig. 6.10 are calculated with that occupancy. It is interesting to note that the As anti-bonding state does not fall within the gap for $x < 1$ which means that E_F should remain constant for n-type materials. These curves are in fairly good correspondence with experiment and they explain fairly well the drop of E_F for p- type materials. However they are likely to become incorrect for coverages of order unity because of two-dimensional broadening effects due to the interaction between adatoms and at still higher coverages, to the onset of metallization. Such effects are discussed in the next section.

6.1.4 Broadening Effects and the Onset of Metallization

We now consider the broadening of the molecular level due to the overlap between the wave functions of the adatoms. In the spirit of the tight-binding method, two adatoms interact when they are nearest neighbors, defined to be such that the corresponding atomic spheres of radius R_{at} are practically in contact. Then their interaction is determined by Harrison's expression [6.16] which gives (Sect. 4.1.3):

$$\beta = \frac{10\,\mathrm{eV}}{(2R_{at}\ \mathrm{in}\ \text{Å})^2} \tag{6.30}$$

with R_{at} taken from [6.14]. For Cs, $R_{at} \sim 3$ Å and, assuming for instance a square lattice, one gets a full monolayer for a maximum coverage θ_M of $\sim 1/3$ on GaAs(110) and $\theta_M \sim 1$ for Ag ($R_{at} = 1.65$ Å). For $0.1\,\theta_M < \theta < \theta_M$ we have a more-or-less random array of adatoms on Ga or As dangling bonds. We treat this in the virtual crystal approximation which consists in replacing the Ga and As dangling bonds by an average dangling bond. To go further we consider that the adatoms are located on the sites of a two-dimensional lattice with a random distribution. We generate this random distribution numerically and calculate the eigenvalues of the corresponding tight-binding matrix (including atom–dangling-bond and atom–adatom interactions). Self-consistency is achieved in the same way as before but, for delocalized states, Slater's transition state now corresponds to 2 electrons instead of 3/2 or 5/2. We have performed calculations for a square lattice and obtain a filled bonding band and an empty anti-bonding band for the coverages of interest. The position of the Fermi level is then taken to be the top of the bonding band for p-type materials and the bottom of the anti-bonding band for n-type materials. The corresponding results are plotted in Fig. 6.11 for Cs and Ag.

Figure 6.11 also illustrates two other regimes: i) at very low coverages where, as discussed in Sect. 6.1.2, the application of Fermi statistics gives a parabolic rise of the Fermi level before there are enough donors to pin it; ii) at $\theta > \theta_M$ where the second adatom layer is found to immediately produce metallic behavior, the Fermi level becoming identical in n- and p-type materials. To calculate this situation we have simply extended the square lattice of the first layer into a cubic lattice with the same interactions between metal atoms as given by (6.30). Of course this represents an idealized model but it contains the

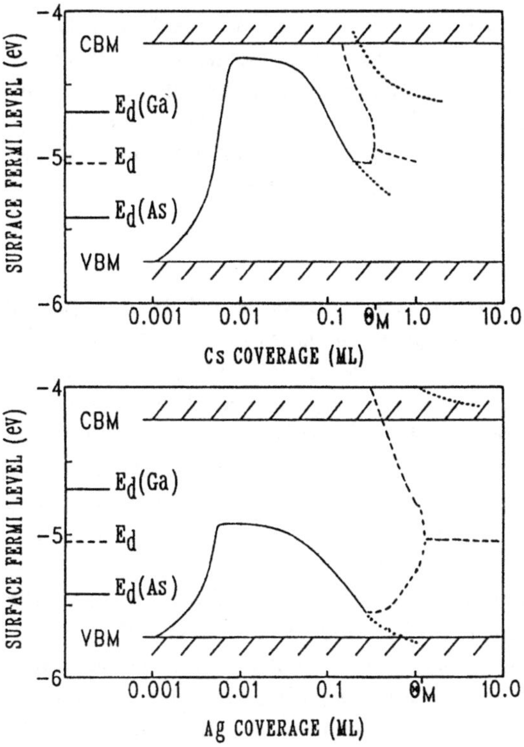

Fig. 6.11. Predicted coverage dependence of the surface Fermi level for Cs and Ag. Full line is the isolated donor limit; dashed lines are the onset of metallization; dotted lines are extrapolations from the isolated donor

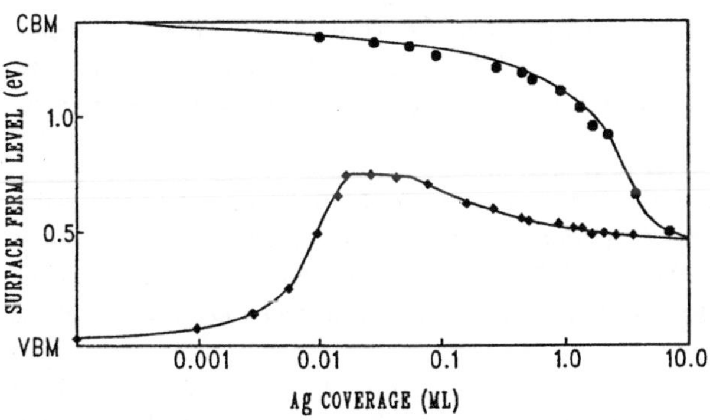

Fig. 6.12. Experimental coverage dependence of the surface Fermi level for Ag, (•) *n*-type substrate, (♦) *p*-type [6.1]

essential physics of the situation. The complete theoretical curves for Cs and Ag are in excellent agreement with experiment (e.g. Fig. 6.12 for Ag except for a dip at $\theta \lesssim \theta_M$ and an abrupt transition at $\theta \gtrsim \theta_M$). The dip is practically nonexistent for Cs but more important for Ag. It results from a delicate balance between electrostatic screening effects (which are essential to explain the pinning of the average dangling bond energy) and the broadening of the bands, which is probably underestimated in our idealized model since it neglects the effects of disorder in the adsorption sites. An important point is that surface diffusion probably still occurs even at 80 K, so that the experimental Fermi level for $\theta \sim \theta_M$ might well be some weighted average of the positions at high and low coverages. This would explain the smooth variation observed experimentally. An interesting test of this interpretation would be to perform experiments at still lower temperatures where diffusion should become negligible.

6.1.5 Discussion

A very interesting property derived from the calculation is that the final pinned value of the Fermi level at higher coverage is always very close (within 0.1 eV) to the average dangling bond energy. This behavior is clear in Fig. 6.11 for the Cs and Ag cases. We shall see in Chap. 7 that this energy indeed plays a key role in the understanding of thick metal–semiconductor interfaces. The physical reason for this is that, in the case of a thick metal layer, the center of gravity of the metallic band (which is given by E_s) will align with the average dangling bond energy in order to maintain local charge neutrality as required by self-consistency. Details of the arguments are given in the next chapter.

The model which we have detailed in the previous sections was originally presented in [6.21]. One point which can be criticized in this model is the assumption of an equal number of metal–Ga and metal–As bonds. The advantage of this hypothesis is that it automatically satisfies surface neutrality for adatoms which have one s-electron. However, there might be situations in which the adatom preferentially couples to one type of dangling bond. In such cases the corresponding gap states will be partly filled. The same is true when the adatom p-shell begins to play a role. Such possibilities have been discussed schematically by *Klepeis* and *Harrison* [6.22] in a model essentially similar to the one developed here.

Note Added in Proof: Since the completion of the book it has been shown that the experimental data at low temperature have to be considered with caution in view of the possibility of a substantial photovoltage effect. This could possibly alter the detailed comparison with experiment but in a way that is not clear at the moment.

6.2 An Ideal Chemisorbed Layer: Arsenic on Si(111) or Ge(111)

An almost ideal case for the study of chemisorption is provided by the arsenic terminated Ge(111) and Si(111) surfaces which exhibit a 1×1 surface periodicity (see [6.23] and references therein). For such surfaces As appears to substitute for the final layer of Si or Ge, resulting in a chemically stable surface. Such surfaces can thus be considered as prototypes for theoretical studies because of their geometrical simplicity. We consider them here from two points of view: i) simple theoretical arguments obtained from considerations given in previous sections, ii) the most complete theory that can be achieved at the moment based upon the calculation of the quasiparticle energies in the GW approximation (Sect. 2.1.1) [6.23].

Let us begin with simple considerations. We start from the unrelaxed ideal (111) surface of the covalent material, Si or Ge. This has been treated in Sect. 4.2 and is characterized by a relatively broad dangling bond band (of width ~ 1 eV) centered on the dangling bond energy which lies just above the top of the valence band. The general shape of this band is represented in Fig. 4.9. We can discuss the As terminated (111) surface on the basis of the following arguments:

i) substitution of the surface Si or Ge atoms by As atoms is equivalent to applying an attractive perturbation potential to the surface atoms of pure Si or pure Ge. Simple first-order perturbation theory shows that the dangling bond surface band will be shifted downwards. Furthermore since there is now one more electron per surface atom this band will be completely filled, resulting in the existence of two non-bonding or lone-pair electrons per As atom.

ii) We have seen in Sect. 5.2.2 that, when the dangling bonds are completely filled, there is a general tendency for the triply coordinated atom to relax away from the plane of its three neighbors. In this process the energy of the dangling bond state is lowered. This downward shift in energy must be added to the previous one.

iii) The fact that the As atoms correspond to an attractive perturbation potential should also lower the energy of the back-bond anti-bonding and bonding states. The energy of the bonding states will lie deeper in the valence band (which has no practical consequence) but some combinations of the anti-bonding states may be stabilized in the forbidden gap at a given $\boldsymbol{k}_{\parallel}$.

Such qualitative arguments can be readily illustrated by a tight-binding calculation. However in the present case there exist more complete calculations, performed not only in the LDA but also in the GW formalism as described briefly in Sect. 2.1.1. Here we summarize the different steps of this calculation performed in [6.23] and show that the results can be understood with the simple arguments presented above. They also serve as a test of the LDA calculations and we also discuss this aspect in the following.

The first step of the treatment reported in [6.23] was to perform a full total energy calculation in the LDA approximation. For this the authors have used a periodic slab geometry which allowed them to use the conventional formalism of band theory. The total energy was minimized with respect to the distance between the surface plane and the first inner plane, resulting in an increase of this distance by about 0.17 Å [this corresponds to the general fact discussed in ii) above that the As atoms tend to move outwards]. Then the LDA eigenvalues have been computed for this relaxed geometry. They are plotted as dashed lines in Fig. 6.13 for Ge(111):As and Fig. 6.14 for Si(111):As. The lower surface band corresponds to the filled dangling bond states while the upper one corresponds

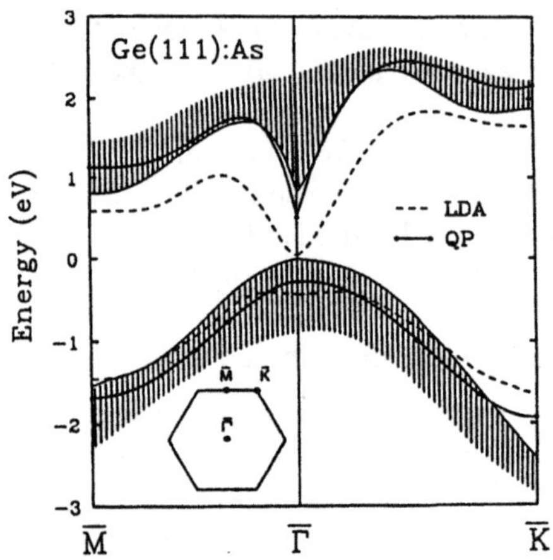

Fig. 6.13. Difference between quasiparticle (QP) and LDA eigenvalues for Ge(111):As. From [6.23]

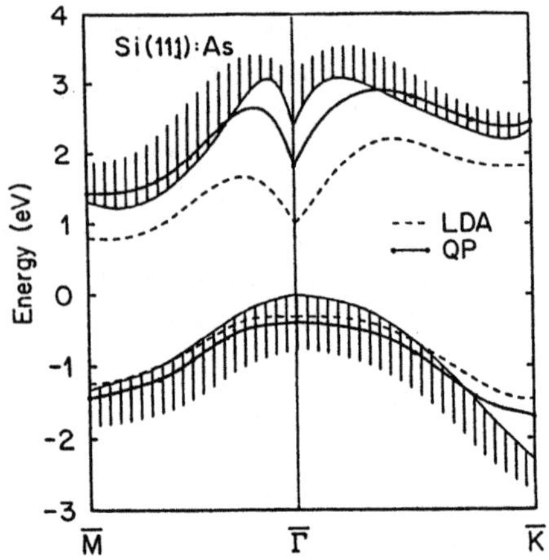

Fig. 6.14. Difference between quasiparticle (QP) and LDA eigenvalues for Si(111):As

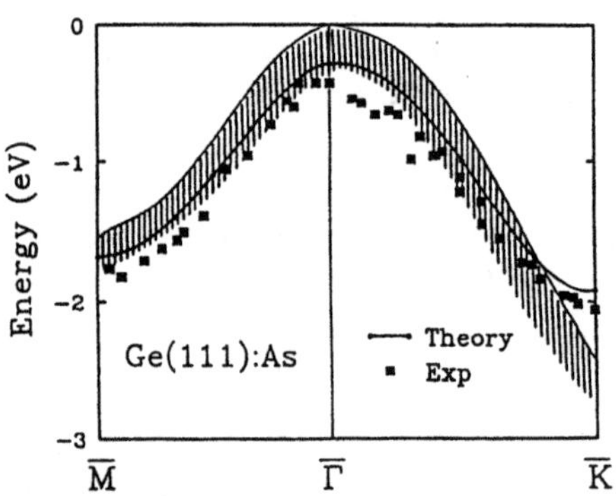

Fig. 6.15. Comparison between theory (quasiparticle) and experiment for the dispersion of the filled dangling bonds of Ge(111):As. Squares are experimental points

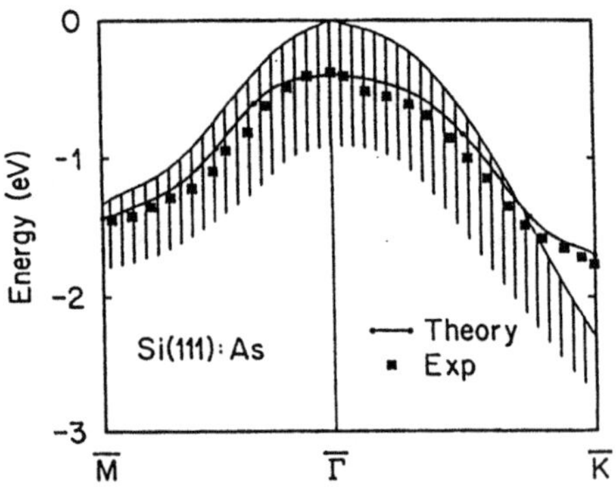

Fig. 6.16. Same comparison as in Fig. 6.15 between theory (quasiparticle) and experiment for Si(111):As

to empty states with anti-bonding character as discussed above. The predicted gap between the two surface bands is much too low (for a detailed discussion see [6.23]) as is the case for the bulk band-gap in LDA.

Hybertsen and *Louie* have improved on this point by calculating the quasiparticle energies in the same way as was done recently for the bulk band structure of these materials, i.e., in the framework of the GW approximation. The advantage of the periodic slab geometry is that the technique can be applied in the same way as for the bulk problem, except that the size of the unit cell is increased. The results are plotted as full lines in Figs. 6.13 and 6.14. The main features of the corresponding corrections are:

i) the quasiparticle and LDA results for the filled dangling bond band are almost identical, the LDA band being slightly narrower;
ii) the quasiparticle empty band is shifted almost rigidly upwards with respect to that of the LDA by an amount which is practically identical to the shift of the conduction band.

The second feature is in agreement with the nature of the empty surface states which are anti-bonding states and should experience the same correction as the other anti-bonding states of the conduction band.

To conclude this section we show a comparison between experiment and theory in Figs. 6.15 and 6.16 taken from [6.23]. The agreement is almost perfect and can be considered as strong support for the validity of the theory.

Appendix

6A. Correction to the Intra-Atomic Coulomb Interaction in a Solid

We consider the atomic energy levels E_s and E_p for an sp shell with the $s^n p^m$ electronic configuration in the Hartree–Fock approximation. For the free atom we can write

$$E_\alpha = E_{\alpha 0} + (n+m-1)U \ , \qquad \text{(A.1)}$$

where the U are terms of the form

$$\frac{\int \psi_i^*(\boldsymbol{r})\psi_j(\boldsymbol{r})\psi_k^*(\boldsymbol{r}')\psi_l(\boldsymbol{r}')d^3\boldsymbol{r}d^3\boldsymbol{r}'}{|\boldsymbol{r}-\boldsymbol{r}'|} = U\delta_{ij}\delta_{kl}$$

and with $\alpha = s$ or p, U being the average Coulomb repulsion between electrons. The term $-U$ in (A.1) is due to the electron self-exchange term. In a solid, however, even if an atom has the same electronic configuration the situation is different. In such a case one must write

$$\begin{aligned} E_s &= E_{s0} + (n+m)U - \frac{nU}{2} \ , \\ E_p &= E_{p0} + (n+m)U - \frac{mU}{6} \ , \end{aligned} \qquad \text{(A.2)}$$

due to the different values of the self-exchange term. Two cases are of interest here. The first concerns the situation $n = 1$, $m = 0$ characteristic of the adatoms. In this case we have

$$E_s = E_{s0} + \frac{U}{2} \ , \qquad \text{(A.3)}$$

which differs from the free-atom value E_{s0} given by (A.1) by an amount equal to

$U/2$. Equation (A.3) is a general property of systems in which the electrons are equally distributed in the orbitals, regardless of the environment [6.24]. Another interesting case is the average sp^3 energy, which from (A.2) is given by

$$\overline{E} = \overline{E_0} + \tfrac{7}{8}(n+m)U \ . \qquad \text{(A.4)}$$

For a covalent system such as Si, $n+m = 4$ and we find an upward shift equal to $7/2\ U$ while the similar quantity in the free atom is only shifted by $3U$. Again the level in the solid is $U/2$ higher than in the free atom. This conclusion still holds for the average sp^3 level in GaAs. This is why we have shifted all our free-atom levels by $U/2$.

Exercises

6.1 Use Table 6.1 and the theoretical treatment of Sect. 6.1.1 to calculate the position of the donor levels in the gap of GaAs for the alkali atoms as well as Ag and Au. Plot the corresponding results against the free atom energy. Discuss the results in a simple way.

6.2 In the manner of exercise 5.3 evaluate the discrete corrections to expression (6.23) for a square lattice. Compare them to the integral approximation.

6.3 *Band bending*: Consider the situation of Sect. 6.1.2 where one has a uniform charge density eN_s^+ on the surface of a p-type semiconductor doped with acceptor concentration N_A. Discuss why a negative space charge region tends to build up in the vicinity of the surface. Calculate the macroscopic band bending ΔV of (6.15) by using the depletion approximation in which the space charge is taken to be $-eN_A$ within a distance X from the surface and zero elsewhere. Discuss the origin of this approximation and also its validity.

6.4 *Onset of metallization*: Analyze the onset of metallization as described in Sect. 6.1.4. To do this use a simplified model where the Ga and As dangling bonds of energies ε_{Ga} and ε_{As} are on the edges of a square lattice and are directed perpendicular to the surface plane. Consider one planar monolayer of adatoms whose coupling with the anion and cation dangling bond is the same and equal to V, while the nearest neighbor adatom-adatom coupling is β.

To these parameters one must add the intra-atomic adatom energy ε_S. Discuss why this is likely to be equal to the average dangling bond energy and take it as the origin of energies.

Use the above model to calculate the broadening of the adsorbate levels. For a given value $\boldsymbol{k}_\parallel$ of the parallel wavevector, write down the 4×4 matrix to be diagonalized. Solve the eigenvalue problem analytically to get the energy bands.

7. Interfaces

This chapter deals with the electronic properties of interfaces, i.e. the abrupt transition from one bulk material to the other one. There has been substantial progress in recent years in the growth of well-defined and reproducible interfaces. One is also beginning to obtain reliable experimental values for the band discontinuities, e.g. Schottky barrier heights, heterojunction band offsets, etc. These values now represent severe tests of the predictive value of the different theories. In this regard the first theoretical models concerning the interface between two materials have always attemped to relate the interface discontinuity to the properties of the material-vacuum interface, i.e. essentially the work function for a metal and the electron affinity for a semiconductor. Such treatments neglect interface dipoles and take the vacuum level as reference. For reasons discussed in the following these treatments usually fail. However, other simple models have been devised based on the use of a neutrality level whose common denominator is shown to be that dangling bond states are involved. The dangling bond states allow one to understand why the Schottky barrier height, the heterojunction band offsets and the transition impurity levels are all correlated. This point of view is confirmed by recent quantitative calculations and the degree af accuracy now obtained for the prediction of band discontinuities is of the order of 0.1 eV. Finally we discuss the Si-SiO_2 interface where again dangling bonds play an important role but this time as interface defects.

7.1 The Metal–Semiconductor Interface

The problem of Schottky barriers is very old but it is only recently that the theory has made any substantial progress. We first review the basic models that have been used in the past, and then discuss the origin of the so-called metal-induced gap states that always exist at such interfaces and show that, in most cases, they can be identified with resonant dangling bonds. This allows us to relate the Schottky barrier height to the average dangling bond energy which is then identified as the semiconductor neutrality level. We then present the results of first principles calculations which confirm the existence and the role of these resonant states. Finally we discuss the different models that describe the pinning of the Fermi level and the conditions under which this pinning is removed.

7.1.1 Evolution of Our Understanding

The key quantity that characterizes the metal–semiconductor interface is the Schottky barrier height Φ_B (i.e., the distance in energy between the conduction band and the Fermi level at the interface, as pictured in Fig. 7.1). The simplest model due to *Schottky* [7.1] attempts to deduce Φ_B from the properties of the metal–vacuum and semiconductor–vacuum interface. This can be done if one neglects the dipole layers at each interface, in which cases one can write

$$\Phi_B = \phi_M - \chi_s \, , \tag{7.1}$$

where Φ_M is the metal work function and χ_s the semiconductor electron affinity, i.e. the ionization energy from the bottom of the conduction band (both quantities are true bulk properties from which dipole effects are excluded). However this simple relation is not generally obeyed although experimental evidence [7.2,3] suggests that there is still a linear relation between Φ_B and Φ_M which can be expressed as

$$\Phi_B = S\Phi_M + \Phi_o \, , \tag{7.2}$$

where the proportionality factor S is characteristic of each semiconductor. Typical values of S are given in Fig. 7.2 as a function of the semiconductor ionicity. A major point is that for common semiconductors like Si and GaAs, S practically vanishes, which means that Φ_B does not depend on the metal, i.e. that the Fermi level E_F is "pinned" at the interface. Such an extreme situation can be explained by *Bardeen*'s model [7.4] which considers that there exists a partially filled narrow band of interface states which is an intrinsic property of the semiconductor and which is able to pin the Fermi level (this is possible if the corresponding interface density of states is large enough). Figure 7.2 tells us that in most cases S takes values between 0 and 1, i.e. the situation is intermediate between the

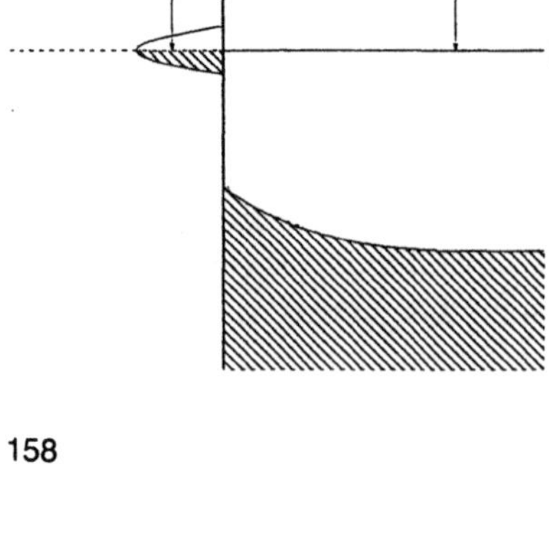

Fig. 7.1. Metal–semiconductor abrupt heterojunction. The interface is represented by the thick vertical line. On the left is the metal, on the right the semiconductor; ϕ_B is the Schottky barrier height, ϕ_M the metal work function and χ_s the semiconductor electron affinity

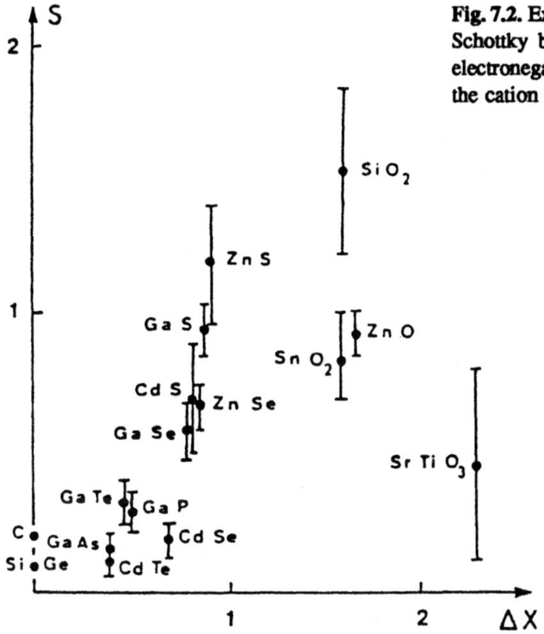

Fig. 7.2. Experimental determination of the empirical Schottky barrier parameter S as a function of the electronegativity difference between the anion and the cation [7.3]

Schottky and Bardeen limits (in some situations it even becomes larger than unity).

To explain the fact that S can take different values one must at least understand Bardeen's limit $S = 0$, i.e. find the microscopic origin of the intrinsic interface states. Some progress in this direction is due to the proposal by *Heine* [7.5] of the existence of metal–induced gap states (MIGS). Their origin lies in the fact that the bulk metal is characterized by a continuous density of states which overlaps with the semiconductor gap. This means that in the interface region there exists a non-vanishing density of states whose wave function corresponds to a propagation in the metal and an exponential decay in the semiconductor. These MIGS should thus correspond to the interface states of Bardeen's picture. However at this level of generality the MIGS should depend upon the metal and it is not easy to see why there should be Fermi level pinning at an energy characteristic of the semiconductor. In the following sections we show that this effect is related to screening and to the properties of the semiconductor dangling bonds.

Before discussing the microscopic origin of these MIGS it is important to notice that another explanation of the Fermi level pinning is the unified defect model (UDM) proposed by *Spicer* at al. [7.6, 7]. This is based on the assumption that the states that pin the Fermi level are due to defects created in the semiconductor during the formation of the interface. The relative merits of the UDM and MIGS models will be discussed later.

7.1.2 The Pinning Energy and the Neutrality Level or the Average Dangling Bond Energy

An interesting step towards an understanding of the nature of MIGS and Fermi level pinning is the concept of a "neutrality level" E_B, introduced by *Tejedor* et al. [7.8] and applied systematically by *Tersoff* [7.9] with impressive success. This neutrality level should play, for the semiconductor, the role analogous to the Fermi level of the metal. The metal–metal interface is characterized by the alignment of the Fermi levels of the two materials. In the same spirit, the Schottky barrier height should be directly obtained by aligning the metal Fermi level E_F and the semiconductor neutrality level E_B. The main problem is to find a reasonable criterion for determining the position of E_B in the gap and also to justify this criterion on physical grounds. To be the equivalent of the Fermi level E_B should represent the natural energy that separates the filled states (with bonding character) from the empty states (with antibonding character). On the other hand, due to the very efficient screening characteristics of many IV, III-V, and II-VI semiconductors like Si, Ge or GaAs (Sect. 5.1.1), E_B should also be chosen in such a way that the criterion of local neutrality be practically fulfilled. This explains the terminology "neutrality level" used for E_B.

The two criteria just discussed do not give a clearcut method for calculating E_B. To achieve this goal *Tersoff* [7.9] has used a more or less ad hoc equation based on the analytic continuation of bulk states with complex wave vectors for energies within the forbidden gap, and corresponding approximately to the transition between bonding and antibonding states. These values of E_B [7.10] are given in Table 7.1. It can be seen in this table that the predicted E_B give good values of ϕ_B for Au on different semiconductors.

Table 7.1. Semiconductor "midgap" energy E_B, and measured Fermi-level positions at metal–semiconductor interfaces, relative to valence-band maxima

	E_B [eV]	E_F (Au) [eV]	E_F (Al) [eV]
Si	0.36	0.32	0.40
Ge	0.18	0.07	0.18
AlP	1.27		
GaP	0.81	0.94	1.17
InP	0.76	0.77	
AlAs	1.05	0.96	
GaAs	0.50	0.52	0.62
InAs	0.50	0.47	
AlSb	0.45	0.55	
GaSb	0.07	0.07	
InSb	0.01	0.00	
ZnSe	1.70	1.34	1.94
MnTe	1.6		
ZnTe	0.84		
CdTe	0.85	0.73	0.68
HgTe	0.34		

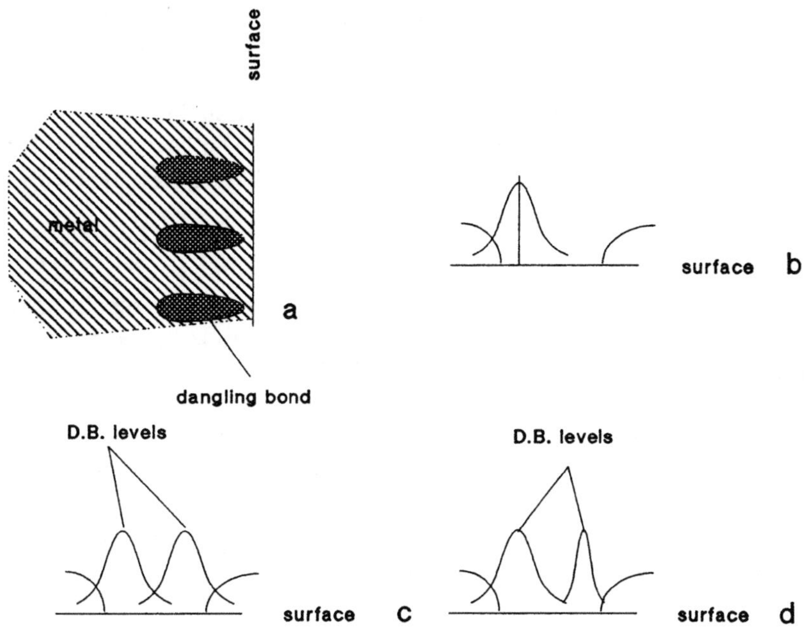

Fig. 7.3. Interaction of surface dangling bonds of the semiconductor with the continuum of states in the metal (a) leading to a resonant state (b) for a simple semiconductor or to two resonant states (c and d) for a compound semiconductor with two different types of atoms at the surface. The resonant states in the compound case can have similar widths (c) or different widths depending on the coupling of anion and cation dangling bonds with the metal states

Further physical insight concerning the meaning of E_B can be gained by considering the coupling between the semiconductor and the metal on a microscopic scale. As discussed in the following, this will lead to the concept of resonant dangling bond states already used by *Mele* and *Joannopoulos* [7.11] to discuss the behavior or S with ionicity. Here we show that this concept leads to a natural definition of E_B and also allows a discussion of the strength of the pinning under different conditions. Let us first consider the interface between a metal and a covalent semiconductor. We assume an idealized situation in which there is no relaxation and no reconstruction in the vicinity of the interface (we thus consider that coupling with the metal inhibits the relaxation or reconstruction that occurs at the free semiconductor surface). In such a situation each dangling bond of the semiconductor surface interacts with the metal as shown schematically in Fig. 7.3a. As the metallic states form a continuum each dangling bond state becomes a resonant state as shown in Fig. 7.3b. If the resonance is symmetrical it will remain centered on the dangling bond level and, as there is one electron per dangling bond, the Fermi level will also be equal to the dangling bond energy E_{db}. In this case we thus find that

$$E_B = E_{db} \,. \qquad (7.3)$$

We can now consider the interface between a metal and the non-polar face of a compound semiconductor like GaAs(110). We now have two dangling bond states at energies E_A for the anion and E_C for the cation. They again become resonant states (as shown in Fig. 7.3c) which overlap if they are broad enough. If this is the case and if the broadening is symmetrical we again have a half-filled situation where E_F is equal to the average dangling bond energy $\bar{E}_{db}$ defined as $(E_A + E_C)/2$. This means that the neutrality level becomes

$$E_B = \bar{E}_{db} \ . \tag{7.4}$$

Of course pinning at these values requires that certain conditions should hold. First there must be a large enough interface density of states at the Fermi level. In other words, as discussed in [7.11], pinning can only occur (i.e., $S \sim 0$) if the width Γ of the resonance is large compared to $(E_C - E_A)$. This means that S should increase with the ionicity as is observed if Γ remains reasonably constant. Another condition for pinning to occur at $\bar{E}_{db}$ is that the coupling between the anion or cation dangling bonds and the metal be reasonably similar. An extreme situation where this is not realized is shown in Fig. 7.3d where E_F is no longer pinned at $\bar{E}_d$ but at the cation dangling bond energy E_C.

7.1.3 Quantitative Comparison Between the Schottky Barrier Height and the Average Dangling Bond Energy

The arguments of the previous section are qualitative and it is necessary to strengthen them by more rigorous mathematical formulation as well as more precise numerical check, as is done in [7.12].

At the interface between a metal and a covalent semiconductor, the dangling bonds of the semiconductor surface are coupled to the continuum of metallic states. The coupling between dangling bonds is small compared to the direct coupling with the metal states (typical bandwidths of surface states are of order 1 eV). Each individual dangling bond state is then treated separately and becomes a resonant state. This can be described by the Hamiltonian

$$H = E_d|d\rangle\langle d| + \sum_k \varepsilon_k|k\rangle\langle k| + \sum_k (V_{dk}|k\rangle\langle d| + \text{c.c.}) \ , \tag{7.5}$$

where $|d\rangle$ is the individual dangling bond state of energy E_d, $|k\rangle$ are the metal eigenstates whose energies ε_k form a pseudo-continuum and V_{dk} are the coupling terms. We proceed as in the treatments of chemisorption on metals [7.12] and define the change $\delta N(E)$ in total number of states of energy lower than E. Using Green's function theory (Sect. 2.3) one has

$$\delta N(E) = -\frac{1}{\pi} \operatorname{Im}\{\ln[\det(I - gV)]\} \ , \tag{7.6}$$

where g is the Green's function of the uncoupled systems and V is the coupling matrix of elements V_{dk}. This leads to

$$\delta N(E) = -\frac{1}{\pi} \tan^{-1} \frac{I(E)}{E - E_d - R(E)} , \tag{7.7}$$

where one has

$$R(E) + iI(E) = \lim_{\eta \to 0} \sum_k \frac{|V_{dk}|^2}{E - \varepsilon_k + i\eta} . \tag{7.8}$$

We now apply the Friedel sum rule which states that $2\delta N(E_F)$ must be equal to the number of chemisorbed electrons, i.e., 1 for a neutral dangling bond at a covalent semiconductor surface. Using (7.7) one finds

$$E_F = E_d + R(E_F) . \tag{7.9}$$

Let us consider the case of a half-filled metallic band which corresponds to the s-band of Au. If this band is symmetric with respect to E_F, then $R(E_F)$ is zero. If it is not, $R(E_F)$ will be a small fraction of V^2/W (where $V^2 = \sum_k |V_{dk}|^2$ and W is the metal band width) and will be negligibly small when W is reasonably large compared to V. Thus, *even if the resonance is broad and the dangling bond states no longer exist, it is their energy E_d which is aligned with the metal Fermi energy E_F*. As the relative position of E_d with respect to the top of the valence band is a constant E_d, the Schottky barrier height is simply equal to this value.

The argument can be extended to the nonpolar face of a compound semiconductor. The total $\delta N(E)$ contains equal contributions of cation and anion dangling bonds. Thus $\delta N(E_F)$ per pair of dangling bonds is a sum of two terms like (7.7) which must equal 1, half the number of electrons per neutral pair. If the coupling of a cation (V_{ck}) and an anion (V_{ak}) dangling bond with the metal states are equal, one easily gets

$$E_F = \frac{E_A + E_C}{2} + R(E_F) , \tag{7.10}$$

meaning that E_F is aligned with the average dangling bond energy $(E_A + E_C)/2$ of the anion and of the cation. If V_{ck} and V_{ak} are not equal, their difference remains small as V_{ck} and V_{ak} mainly depend on the interatomic distances between the semiconductor and the metal and not on the nature of the atoms. In any case the error in (7.10) will be negligible as it is of the order of $(V_{ak} - V_{ck})^2/(V_{ak} + V_{ck})^2$. To check the validity of our conclusion that E_F should be aligned with the average dangling bond energy we have computed the individual dangling bond states of covalent and III-V semiconductors by a charge dependent tight-binding technique discussed in Sect. 5.1.3. We have used two sets of tight-binding parameters, the sp^3s^* nearest-neighbor scheme of [7.14] and the second-nearest-neighbor sp^3 scheme of [7.15].

In Fig. 7.4 the results are compared to the experimental E_F (Au) at the interface. The agreement is impressive and the trend is clearly reproduced for both sets of parameters (the differences between the two sets of results are simply indicative of the error bars inherent in empirical tight-binding theory).

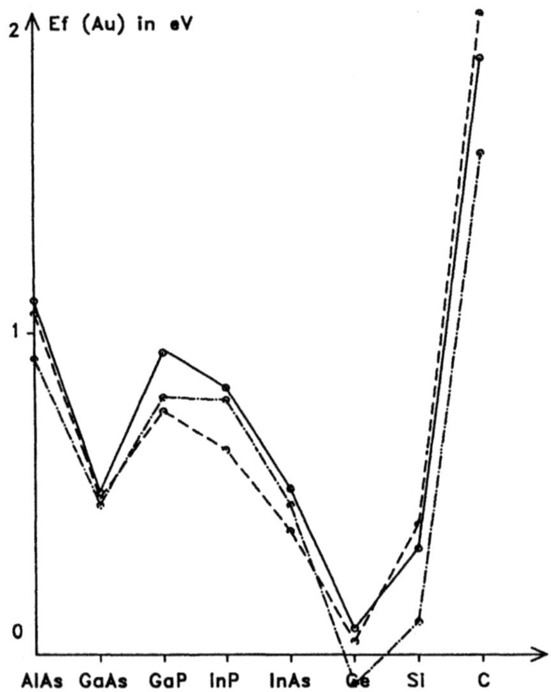

Fig. 7.4. Alignment of E_F (Au) with the (average) dangling bond energies for different III-V and IV semiconductors. Full line: experiment, dashed and dash-dotted lines represent the prediction of two different tight-binding calculations

7.1.4 Discussion

The general model described above does not offer a detailed picture of each individual metal-semiconductor interface. However it shows quite generally why for moderately ionic semiconductors the factor S is small, i.e. why there is pinning of the Fermi level. At first sight it seems that the unified defect model (UDF) of *Spicer* et al. [7.6, 7] explains this pinning equally well. The UDF model is based on the assumption that the energy released in the deposition of the metal atoms allows the creation of point defects in the vicinity of the interface and it is the levels associated with these defects that are supposed to pin the Fermi level. Some support for the UDM picture has been obtained from the calculations of *Sankey* et al. [7.16] of cation dangling bond levels and of III-V antisites which correlate with the barrier height of Au on the corresponding semiconductors. However the basic reason for this correlation is that such defects involve dangling bonds. Thus this agreement with experiment does not discriminate between the UDM and MIGS (or average dangling bond) pictures. Probably the most serious argument is that thick metallic overlayers require a high density of interface states (of order of $10^{14}\,cm^{-2}$ [7.17] for a pinning of the Fermi level to become possible. While such a density (of order 0.1–1 state per interface atom) is naturally obtained in the MIGS model where there is one dangling bond state per atom it is hard to understand how it can be realized in the UDM model. Another point is that the resonant dangling bond states of the MIGS model are always present so that there is no real need of the UDM except in situations where the MIGS are not able to pin the Fermi level by themselves (but in such cases, like the more ionic systems where $S \rightarrow 1$, there seems to be no pinning at all, which again rules out the UDM). Finally, as discussed in Chap. 6, low temperature studies in the submonolayer or monolayer regime again tend to favor the MIGS model over the UDM.

A number of detailed calculations have been performed for thick metal overlayers on semiconductors. They are reviewed in [7.18] but here we mainly comment on the work by *Louie* et al. [7.19] who have studied the interface between jellium (representing Al) and different semiconductors. The importance of this work is that it clearly demonstrates the existence of MIGS having the required interface density of states for Si and GaAs to pin the Fermi level. It also confirms the basic validity of the dangling bond picture and, in particular, the reduction in the density of interface states for more ionic systems, due to reduced overlap between the anion and the cation resonances. This reduction explains the increase in S as shown in detail in [7.11].

The case where there is a chemical reaction has been discussed in [7.20, 21] but as underlined in [7.18] an interlayer should not affect the barrier height too much in most cases in view of the reduction of the dipole layer by screening effects. In the same spirit, let us finally mention two fairly recent LDA calculations [7.22, 23] for the structurally well-defined $CoSi_2$-Si and $NiSi_2$-Si system which represents one of the very few cases where the atomic structure is well known.

7.2 The Heterojunction Between Two Semiconductors

One can now fabricate extremely good and abrupt interfaces between two semiconductors. Most of the interesting properties of these systems come from band discontinuities such as those pictured in Fig. 7.5. In particular a very precise

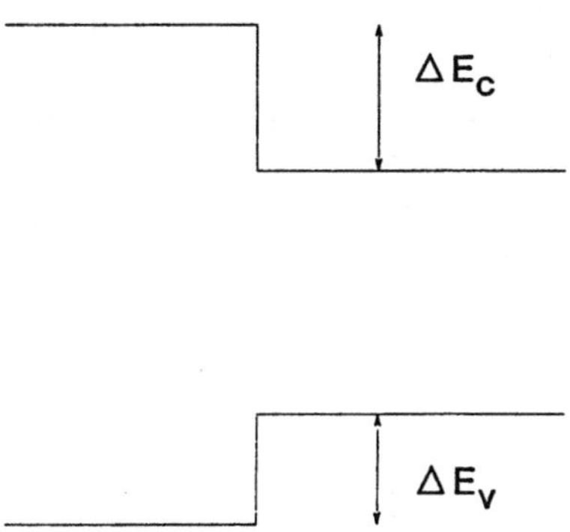

Fig. 7.5. Band offsets between a wide gap and a small gap semiconductor. The interface is ideally abrupt in the figure

knowledge of the band offsets ΔE_c or ΔE_v is of great importance for band structure engineering (accuracy better than 20 meV is required). The first theories of band offsets have been based on the existence of an absolute reference level, such as *Anderson*'s affinity rule [7.24] or *Harrison*'s rule [7.25]. On the other hand there have been many recent calculations, some based on the LDA + pseudopotential methods [7.26, 27] but also semi-empirical tight-binding self-consistent techniques [7.28, 29]. As we shall see, the theoretical predictions now achieve an accuracy of the order of 0.1 eV.

In this section we first consider the relation between band offsets and Schottky barrier heights and justify it in terms of dangling bonds. We then discuss some useful simplified treatments based on zero charge and zero dipole approximations. This is followed by an overview of the full calculations for lattice-matched systems. Finally we summarize recent work on strained heterojunctions.

7.2.1 Relation Between Band Offsets and Schottky Barrier Heights

Tersoff [7.9] pointed out that the differences in Φ_B for Au on various semiconductors give fairly good values of the band offsets for the heterojunction between the corresponding pair of semiconductors. In his approach this is justified by the fact that the neutrality level E_B plays the role of a Fermi level and gives directly the Schottky barrier height while the ΔE_B (with E_B referred to the top of the valence band) gives the valence band offset ΔE_V. However, as for the Schottky barrier, this proposal can be justified microscopically with the help of the dangling bonds.

Let us then consider lattice-matched systems where dangling bond states will couple by pairs across the interface to form bonding and antibonding states as in any diatomic molecule. The contribution of one such bond to charge transfer across the interface is given by

$$\Delta q_i = \frac{\Delta E_{\mathrm{db}i/2}}{\sqrt{(\Delta E_{\mathrm{db}i/2})^2 + V_i^2}} , \tag{7.11}$$

where $\Delta E_{\mathrm{db}i}$ is the difference in dangling bond levels and V_i their covalent coupling. Our general self-consistency requirement will be that the global charge transfer is screened in such a way that a good approximation is to take it equal to zero. This is in line with *Tersoff*'s argument and with local neutrality conditions. This approximation will be discussed in Sect. 7.2.2.

If we apply the zero charge transfer condition to the heterojunction between two covalent semiconductors with all bonds equivalent one immediately finds $\Delta E_{\mathrm{db}i} = 0$, i.e. the dangling bond levels E_{db} must align. For the case of two different bonds one must add two contributions like (7.11) and it is easy to show that the zero charge transfer condition is equivalent to

$$V_2 \Delta E_{\mathrm{db}1} + V_1 \Delta E_{\mathrm{db}2} = 0 . \tag{7.12}$$

As $V_1 \sim V_2$ for the same reasons that $V_{ak} \sim V_{ck}$ at Schottky barriers we again obtain the condition that the two average dangling bond levels must align.

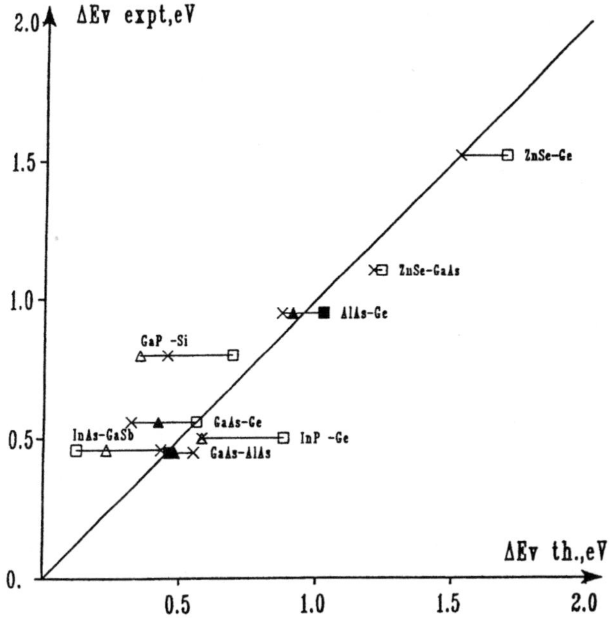

Fig. 7.6. Experimental and theoretical band offsets for different systems. Lattice-matched systems are indicated with filled symbols. Triangles are from [7.14]; squares from second nearest neighbor tight-binding calculations; crosses from [7.9]

This is likely to occur for lattice matched systems where nearest-neighbor bonds have similar lengths. The theoretical offsets determined from the alignment of E_{db} are compared with experiment in Fig. 7.6. Again the agreement is very good for unstrained cases and a little poorer for strained cases.

We thus conclude that the equivalent of the Fermi level for a non-polar semiconductor surface is the average energy of dangling bonds. Such a quantity is an intrinsic characteristic of a given semiconductor and its line-up leads to a direct connection between Schottky barrier heights and band offsets. This formal relation is also confirmed by the charge-dependent empirical tight-binding calculations described previously and in the following.

7.2.2 Zero-Charge and Zero-Dipole Approximations

An accurate evaluation of dipole layers is needed to get very precise values for the band offsets. This requires a self-consistent treatment which leads to complex numerical calculations. For this reason self-consistent semiempirical calculations are of great interest since they are much simpler than LDA calculations but at the same time they can be very accurate since the heterojunction represents a weaker perturbation than a free surface (for this to be realized both semiconductors must be treated at the same level of approximation). As we have discussed earlier, it is even possible to devise approximate versions of the self-consistent treatment. Here we consider two such approximations, i.e. the zero-dipole and zero-charge approximations and test them by comparing the predicted band offset to the full self-consistent value [7.30].

Let us then use the self-consistent tight-binding procedure discussed in Sect. 5.1.2. We consider a polar interface where all atoms of plane i are identical and bear the same charge Q_i. The charges Q_i calculated by the tight-binding approximation are nonlinear functions of the perturbative potentials V_j (on each atom of plane j):

$$Q_i = f_i(V_1, V_2, \dots) . \tag{7.13}$$

Starting from a potential V^0 and charges Q^0 such that

$$Q_i^0 = f_i(V_1^0, V_2^0, \dots) , \tag{7.14}$$

the equations (7.13) are linearized and can be written in matrix form as

$$Q = Q^0 + \chi(V - V^0) , \tag{7.15}$$

where χ is the dielectric susceptibility matrix. On the other hand, the potentials V_j depend linearly on the charges through the Coulomb interactions and we write

$$V_j = \sum_i (C^{-1})_{ij} Q_i , \tag{7.16}$$

or in matrix form

$$V = C^{-1} Q . \tag{7.17}$$

Q can be eliminated from (7.15) and (7.17) which leads to a new approximation V for the potential

$$V^1 = V^0 + (I - C^{-1}\chi)^{-1}(C^{-1}Q^0 - V^0) . \tag{7.18}$$

If V^0 is close enough to the exact solution then the corrective term will be small and at the same time V will be practically the exact value. In practice there are two ways of obtaining a fairly good initial potential V^0: the zero-dipole and the zero-charge approximations, the latter being equivalent to the charge neutrality potential already discussed at some length.

Let us first discuss the zero-dipole potential. The macroscopic dipole layer across the heterojunction is

$$\Delta V = V_{i=+\infty} - V_{i=-\infty} \tag{7.19}$$

and we choose an initial potential such that $\Delta V^0 = 0$. However there is an infinity of charge distributions that satisfy this zero-dipole condition. We then use the fact (discussed in Sect. 5.1.1) that the screening length in these materials is very short (less than the bond length) and define the zero-dipole approximation to be the one that gives the most localized charge disturbance. This is pictured in Fig. 7.7c. It gives rise to a ΔV^0 that we call ΔV_{ZD}, the zero-dipole approxima-

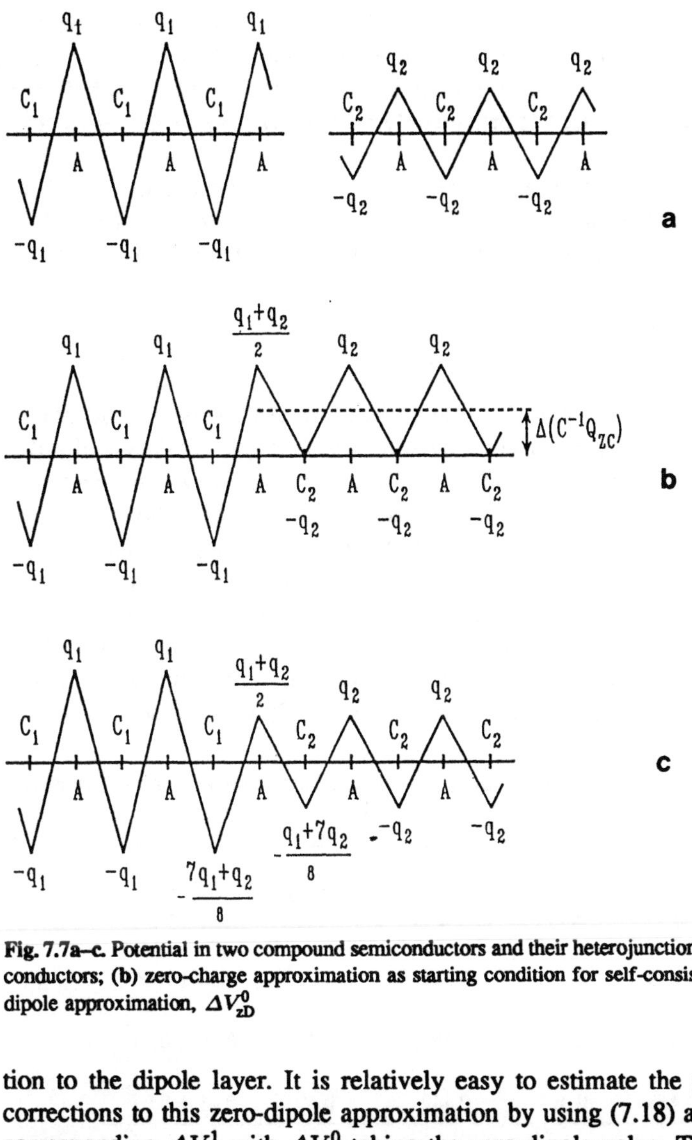

Fig. 7.7a–c. Potential in two compound semiconductors and their heterojunction. (**a**) Decoupled semiconductors; (**b**) zero-charge approximation as starting condition for self-consistency, ΔV^0_{ZC}; (**c**) zero dipole approximation, ΔV^0_{ZD}

tion to the dipole layer. It is relatively easy to estimate the magnitude of the corrections to this zero-dipole approximation by using (7.18) and evaluating the corresponding ΔV^1 with ΔV^0 taking the zero-dipole value. This is done in detail in [7.30] for the case of two materials having similar dielectric properties. However the result is fairly easy to understand on simple physical grounds. One must first realize that $(I - C^{-1}\chi)^{-1}$ is the inverse dielectric matrix while $C^{-1}Q^0$ is the potential created by the initial charge distribution. When evaluating the dipole layer ΔV^1 from (7.18) one obtains directly:

$$\Delta V^1 = \Delta V_{\mathrm{ZD}} \left(1 - \frac{1}{\varepsilon(0)}\right) . \tag{7.20}$$

Since the term $C^{-1}Q^0$ gives no contribution (Q^0 corresponds to no dipole layer),

$\Delta V^0 = \Delta V_{ZD}$ and $(I - C^{-1}\chi)^{-1}$ reduces in this limit to the inverse dielectric constant which is assumed to be the same for the two materials. Of course (7.20) does not hold exactly when the screening properties of the two materials are different but it can still be used approximately with an average inverse dielectric constant.

A second way of obtaining an initial V^0 is from the zero-charge condition, again for the most localized charge disturbance pictured in Fig. 7.7b. Then ΔV^0 becomes the zero-charge approximation and again using (7.18) we can now show that

$$\Delta V^1 = \frac{\Delta(C^{-1}Q_{ZC})}{\varepsilon(0)} + \Delta V_{ZC}\left(1 - \frac{1}{\varepsilon(0)}\right) . \tag{7.21}$$

In principle the two expressions (7.20) and (7.21) for ΔV^1 should be identical. As $\varepsilon(0)$ is usually large this means that the difference $(\Delta V_{ZD} - \Delta V_{ZC})$ is of order $\Delta(C^{-1}Q_{ZC})/\varepsilon(0)$, where Q_{ZC} is the distribution of charge shown, for example, in Fig. 7.7b for the (100) common anion case. In all our calculations we have found this term to be at most of order of 0.1 eV, so that both approximations tend to give fairly accurate band offsets. This is confirmed by the numerical values given in Table 7.2 which compare the valence band offsets obtained in the zero-dipole and zero-charge treatments with the complete self-consistent tight-binding calculation performed along the lines of Sect. 5.1 and [7.29]. In particular the correct zero-dipole expression (7.20) gives an extremely good value.

Table 7.2. Comparison of the zero-charge and zero-diple approximations with the self-consistent result and the experimental values. For more details, see [7.30]

	$\varepsilon(0)$	ΔE_V^{ZD} [eV]	ΔE_V^{ZC} [eV]	Corrected ΔE_V^{ZD} [eV]	ΔE_V [eV]	Exp [eV]
AlAs	10.1					
GaAs	12.7	0.618	0.475	0.546	0.527	0.50 ± 0.05
CdTe	9.6					
HgTe	21.	0.305	0.383	0.355	0.379	0.35 ± 0.05
AlSb	12.0					
GaSb	14.8	0.465	0.407	0.449	0.452	0.45 ± 0.05

7.2.3 Full Calculations for Lattice-Matched Systems

Here we describe briefly the results of band offset calculations performed for unstrained systems. Pseudopotential LDA calculations have been performed by *Martin* and *van de Walle* using a periodic slab geometry (the thickness of the slab corresponding to four atomic layers of each material) [7.26]. Similar calculations, but within the context of the linear-muffin-tin orbital (LMTO) technique, have been carried out by *Christensen* [7.27] on a 7 + 7 periodic slab geometry (note that *Christensen* and *Cardona* [7.31] have also devised a "dielectric-gap" model

Table 7.3. a: self-consistent, ab initio LDA calcualtion [7.26]; b: SC LMTO-Sr 7+7 [7.27]; c: LMTO "dielectric midgap energy model" [7.31]; d: Tight-binding, self-consistent, strong screening [7.29]; e: Values calculated by Tersoff [7.9]. The underlined values indicate an agreement with experiment of better than 0.10 eV

System	Experimental band offset	a	b	c	d	e
AlAs/GaAs	0.45, 0.55	0.37	0.53	0.43	0.50	0.55
InAs/GaSb	0.51	0.38	0.46	0.55		0.43
ZnSe/GaAs	1.10	1.59	1.07	0.99		1.20
ZnSe/Ge	1.52	2.17	1.58	1.46		1.52
GaAs/Ge	0.53	0.63	0.46	0.51		0.32
AlAs/Ge	0.95	1.05	1.03	0.87		0.87
AlSb/GaSb	0.45	0.38	0.45	0.34	0.45	0.38
CdTe/HgTe	0.35		0.22	0.64	0.39	0.51
AlP/Si		1.03		0.92		0.91
GaP/Si	0.80	0.61	0.27	0.53		0.45
GaP/AlP		0.36		0.38		0.46

for predicting band offsets). All results are reproduced in Table 7.3 where it can be seen that the two LDA calculations give results that are quite different from one another in several cases. A reason for this discrepancy might perhaps be the different approximations used in the two techniques. In this particular case, LMTO gives results that are closer to experiment. The fact that the full LDA calculation of [7.26] does not give very good results might certainly be attributed to the failure of this approximation. This has been confirmed recently by the work of *Zhang* et al. [7.32] who have performed a full first principles quasi-particle theory (based on the GW approximation discussed in Sect. 2.1.1) and find, for AlAs-GaAs(001) an offset $\Delta E_v = 0.53 \pm 0.05$ eV which is 0.12 eV larger than their LDA result of [7.26].

Another class of calculations correspond to the self-consistent tight-binding (TB) techniques of *Munoz* et al. [7.28] and of *Priester* et al. [7.29]. These require a good fit to bulk band structures (usually the parametrization of *Vogl* et al. [7.14] is used) and a good description of screening (see the discussion of Chap. 5). The point is that the atomic charges must be known with extreme precision (0.01 eV if one needs offsets with a 0.1 eV accuracy). For this the best method is the decimation procedure for calculating the Green's function (Sect. 2.3.4). In Table 7.3 we give only the TB results for common anion systems for which the prescription of the determination of the interatomic interactions is well defined across the interface. For example, at the AlAs/GaAs polar interface the interactions at the interface correspond to As-Ga or As-Al bonds for which we use the corresponding known bulk values. This is not so at other interfaces (like Ge-GaAs) where the Ge-Ga or Ge-As interactions have no bulk counterpart, leading to some uncertainty in these parameters which can induce errors larger than 0.1 eV for the offsets (see [7.29] for a discussion).

7.2.4 Strained Heterojunctions – Quantum Wells

Up to now there have been only a few published calculations of band offsets for strained systems [7.26]. Here we shall discuss the case of the $In_xGa_{1-x}As/GaAs$ system which has some importance for high speed and optoelectronic devices. For this we summarize the work detailed in [7.33].

For GaAs/InAs quantum well structures (QWS) where the thickness of the strained layers is kept small enough to avoid the generation of misfit dislocations, the lattice mismatch is entirely taken up by elastic strain. Then the effect of the strain in the QWS can be considered in terms of atomic rearrangement: the in-plane parameter is imposed by the substrate or the buffer layer and this determines the biaxial stress in each layer. From a tight-binding point of view, this atomic rearrangement induces a modification of off-diagonal interactions [7.34], and also a modification of atomic level parameters [7.35]. The biaxial stress has to be decomposed as the sum of a purely hydrostatic part plus a uniaxial part. For the sufficiently small deformations we are dealing with, it is convenient to linearize the dependence of the band extrema with respect to the hydrostatic and uniaxial strain. This leads to the definition of the so-called deformation potentials. The deformation potentials of the gap at Γ under hydrostatic strain are usually called "a". Those at the same point of the Brillouin zone under [100] uniaxial strain are written "b".

Using a self-consistent (zero-dipole) tight-binding description one can calculate the band edge alignment in systems made by a well of a few (1 to 8) layers of InAs between two thick (semi-infinite) GaAs crystals. In this study three strain configurations were chosen. The in-plane parameter was strained to:

- the GaAs value,
- the InP value,
- the InAs value,

whereas the perpendicular parameter was obtained from the Poisson ratio.

A first important result is the small dependence of the band offset on the thickness of the wells for two or more InAs atomic layers.

The effect of strain may be seen in Fig. 7.8 where the GaAs/InAs band alignment is plotted for the three different substrates. The offset of the top of the valence bands may be split into two components for a strained heterojunction:

– ΔE_V^{AV} is the discontinuity of the average top of the valence bands, the light hole and heavy hole being split by the strain. It is related to charge transfer at the interface.

Since ΔE_V^{AV} is the quantity that best compares to the valence band offsets of an unstrained heterojunction, we would like to comment on one important result of this study. The ratio of the conduction band minimum offset, ΔE_C, to the valence band maximum offset, ΔE_V for the unstrained case, is the driving parameter for band gap engineering applications. It is the essential parameter for estimating the gain of possible avalanche devices using heterojunctions. One usually represents it as the ratio of percentages of the total band gap difference.

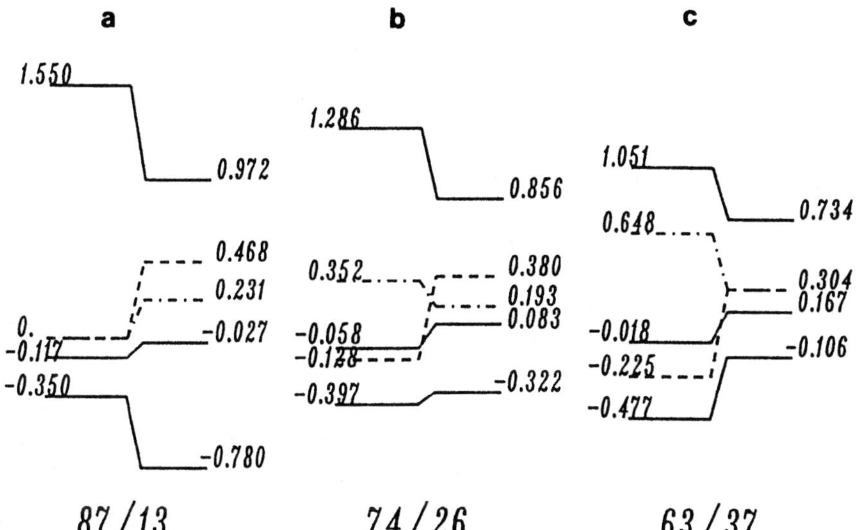

Fig. 7.8a–c. Band alignment in GaAs/InAs heterojunctions strained to (a) the GaAs parameter (5.654 Å); (b) the InP lattice parameter (5.869 Å); and (c) the InAs lattice parameter (6.058 Å). At the top of the valence band the dotted curve corresponds to E_V^{AV}, the dashed curve to heavy holes, the dashed-dotted curve to light holes, and the full curve to the last nearly degenerate band near the valence band maximum at Γ. GaAs is always on the left side. The values of $\Delta E_c/\Delta E_V^{AV}$ are indicated in each case. Energies are given in eV. The valence band splittings are those obtained in the tight-binding approximation, which are different from those obtained with experimental deformation potentials

It has been found experimentally that $\Delta E_C/\Delta E_V$ is constant over the whole range $0 < x < 1$ for GaAs/$Ga_xAl_{1-x}As$ heterojunctions. For these systems $\Delta E_C/\Delta E_V = 60/40$. This means that 60% of the band gap difference between the two semiconductors occurs in the conduction band offset and 40% in the valence band offset.

In the case of GaAs/InAs the calculations stated above show that this ratio varies with the strain conditions: On GaAs, $\Delta E_C/\Delta E_V^{AV} = 87/13$, on InP (intermediate strain) it is 74/26; and on InAs it is 63/37.

Furthermore, due to the strain-induced splitting of the valence bands, one can see in Fig. 7.8 that on a GaAs substrate, both light and heavy holes should be confined in the InAs layer. On the other hand, for an InAs or an InP substrate, the heavy holes should be confined but the light holes spread into the surrounding GaAs.

To compare this theoretical result with the pioneering experimental work of *Kowalczyk* et al. [7.36] obtained by combined core-level and valence-band photoemission on the InAs-GaAs(100) heterojunction, it is not possible to talk in terms of ΔE_v, as [7.36] neglects the effect of strain (in the InAs layer) on the valence band. However, if one assumes that the atomic levels are rigidly bound to the core levels, and uses experimental values of unstrained materials $(E_{Ga3d}^{GaAs} - E_v) = 18.81 \pm 0.02\,eV$ and $(E_{In4d}^{InAs} - E_v) = 17.43 \pm 0.02\,eV$, one can deduce from the calculated value of the unstrained $\Delta E_v = 0.235$ eV, a theoretical value of $\Delta E_{c1} = 1.61 \pm 0.05$ eV, which agrees fairly well with the experimental determination, i.e. 1.55 ± 0.02 eV.

The estimated numerical accuracy of this calculated ΔE_V^{AV} is a few meV. The greatest limitation of the calculation is related to the fact that tight-binding predictions for the uniaxial deformation potentials "b" differ somewhat from experimental values [7.33]. A numerical estimate of this error in ΔE_V^{AV} gives a few tens of meV. The error in the splittings of the top of the valence band is much more important. A more reliable way to deduce offsets in our strained systems would thus be to use the calculated ΔE_V^{AV}, and to add the splittings calculated with the exact experimental values of b. This is done in the following discussion.

From the results corresponding to GaAs/$In_xGa_{1-x}As$ with $x = 1$, it is possible to deduce the band configuration for every $x \neq 1$. In a virtual potential approximation, ΔE_V^{AV} varies linearly with indium concentration. For $In_{0.05}Ga_{0.95}As$/GaAs strained to GaAs, experimentally studied in [7.37], the difference between the calculated (20 meV) and experimentally deduced (36 meV, 41 meV) ΔE_{Vh} is about 20 meV greater than the numerical accuracy for $x = 0$. However, the discrepancy is small enough that one cannot consider that the calculation really contradicts the results of [7.37]. At this point, it is important to notice that a slight shift of the indium concentration (e.g. 7% instead of 5%) will firstly raise the ΔE_V^{AV} by 4 meV and will also increase the difference between the E_{hh} and ΔE_V^{AV} by 6 meV, so that the theoretical ΔE_{Vh} for $Ga_{0.93}In_{0.07}As$ becomes 30 meV. Note that, both in the experimental and theoretical cases, the weak difference between ΔE_{Vh} (20–40 meV), the heavy hole band offset, and ΔE_{lh} (26 meV), the splitting between light and heavy holes, does not allow one to draw a direct conclusion about light hole localization since the well or barrier height is only about 10 meV.

The predictions for $In_{0.15}Ga_{0.85}As$/GaAs strained to GaAs are in excellent agreement with the results of [7.38]; one obtains $\Delta E_{Vh} = 56$ meV and $\Delta E_{Vl} = -15$ meV while the experiment gives +52 meV and -18 meV. It is interesting to note that a superlattice like GaAs/$In_xGa_{1-x}As$ strained to GaAs can form a type I superlattice for heavy holes only when $x < 1$ (as here when $x = 0.15$). For the case of $In_{0.40}Ga_{0.60}As/In_{0.63}Ga_{0.37}As$ strained to InP studied in [7.39], the predicted band configuration is summarized in Fig. 7.9. The photoluminescence experiments [7.39] do not allow one to deduce the band offset precisely, but the configuration obtained corresponds to the assumptions made in the interpretation of the observed excitonic transitions. In conclusion, band lineups strongly depend on the strain distribution in the system (strain confined either in one side of the heterojunction, or shared between both sides). The value of the strain is also very important; for example, for In(Ga)As/GaAs strained to the GaAs lattice parameter, depending on the indium concentration, light and heavy holes can be localized either both in the same layers, or in the two separate layers. Comparison with different experiments has provided a quite good agreement.

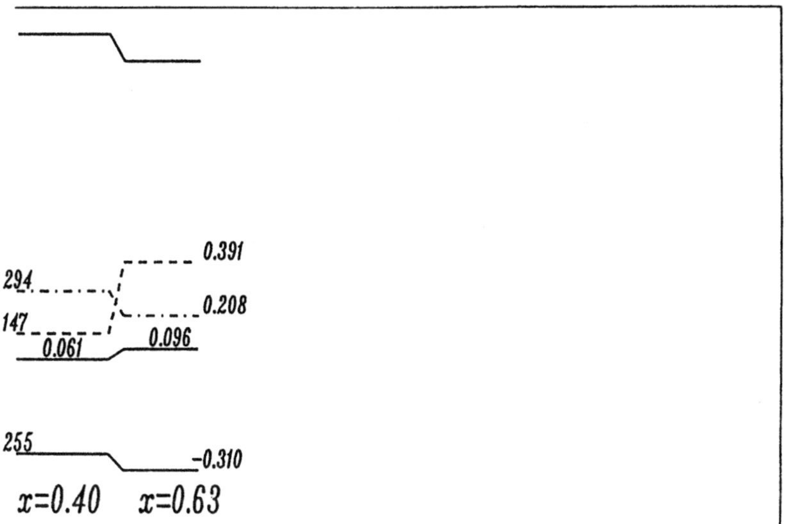

Fig. 7.9. Band alignment at $In_{0.40}Ga_{0.60}As/In_{0.63}Ga_{0.37}As$ heterojunction strained to InP parameter. The notation is the same as in Fig. 7.8. Left side $x = 0.40$, right side $x = 0.63$

7.2.5 Strained Heterojunctions – Superlattices

Recent interest has centered around the strained superlattices of Si and Ge. Experimental results [7.40, 41] have indicated new optical transitions in very thin superlattices with periods of 4 to 12 atomic layers. In particular, in the case of the $Si_4Ge_4/Si(001)$ superlattice (i.e. four layers of Si and four layers of Ge per period on a Si substrate in the (001) direction), the electroreflectance spectra show strong features in the infrared region at roughly 0.8–0.9 eV [7.40]. The features are characteristic of the microscopic structure of the sample and suggest the possibility of structurally induced direct optical transitions. Since Si and Ge are both indirect gap semiconductors (i.e. with the conduction band minimum at a different point of k-space (than the valence band maximum), this result would enable one to build layers on a Si substrate that could be used for optical or optoelectronic applications. Several microscopic calculations [7.42–49] have been made using various approaches. In all cases the band structure shows an indirect gap for the 4+4 case on Si or on Ge. First principles self-energy calculations such as those described in Sect. 2.1.1 unambiguously show that this indirect gap coincides in energy with the lowest electroreflectance feature. The lowest direct dipole-allowed transition is 0.4 eV higher in energy. It is also observed in electroreflectance. Furthermore, the infrared features mentioned above have never been observed by more direct techniques such as photoluminescence or infrared absorption. This confirms the idea that the observed features are probably related to some electric-field induced electron transition observed in the electroreflectance spectra. However, the challenge to obtain light emission from a combination of two indirect gap semiconductors remains.

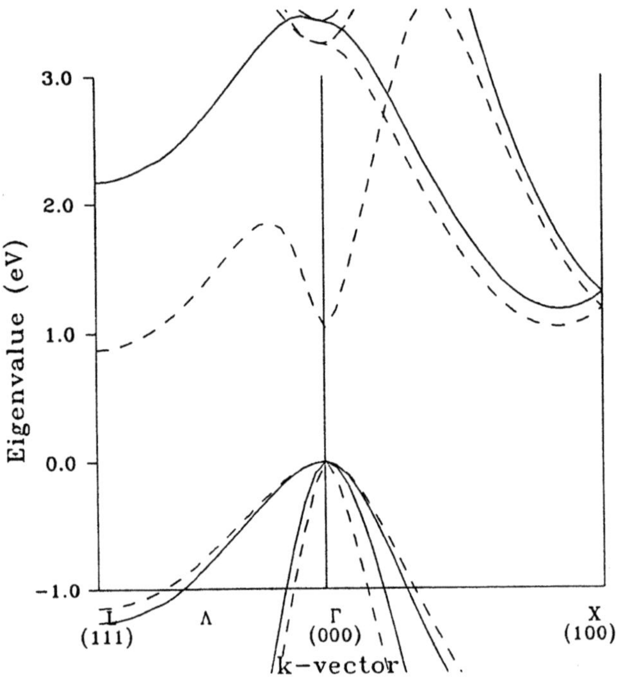

Fig. 7.10. Comparison of Si and Ge band structures. Full lines: Si; dashed lines: Ge

More recently, the growth of different periodicity superlattices under a variety of strain conditions [7.50–52] has shown that one can reproducibly obtain a photoluminescence feature in the infrared region. Let us first discuss the origin of a direct gap in such a system.

The band superlattices of Si and Ge in the region around the gap may be superimposed (Fig. 7.10). They show only very minor differences except in the vicinity of the Γ point where the Γ_2' state of Ge comes down to an energy very close to that of the Δ minimum in the [001] direction. Apart from that, the bands are almost superposable in the [001] direction. The other major difference occurs in the [111] direction where the Ge conduction band has its absolute minimum at the L point. The whole of the lower conduction band in this direction is lower in Ge than in Si. However, the effective mass of electrons in this direction is small. Confinement is thus expected to pull the L minimum high above the band in the Γ-X direction. Thus, simple perturbative arguments for the Γ-X direction may help to predict the essential features of the band in the gap region of the SiGe superlattices.

The average valence band offset between Ge and Si is about 0.5 eV, neglecting any effect of strain as a first step [7.26]. Consequently, although the band gap of Ge is significantly smaller than that of Si, the conduction band minimum will usually lie in the [001] direction. It should contain a large component localized in the Si region.

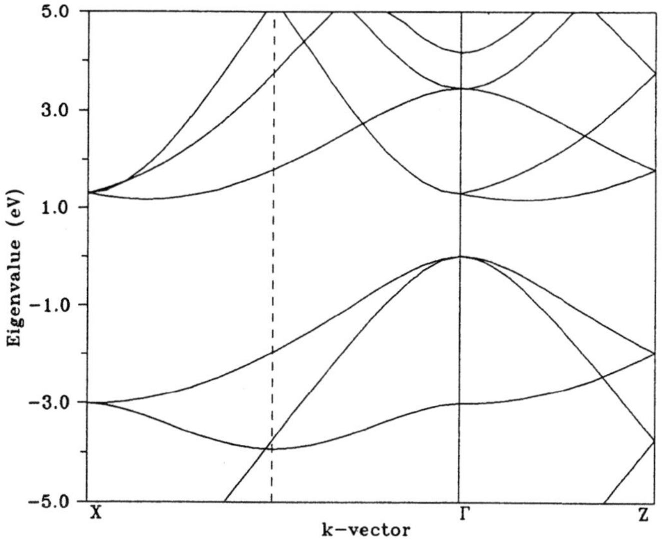

Fig. 7.11. Folding of the bands in the growth direction of a two-fold periodic superlattice grown on (001). In the [100] direction (*left*) one sees bands projected from the $k_z = \pi/2a$ plane of Si. In the [001] direction the folding clearly brings the conduction band minimum close to Γ

In a superlattice grown in the [001] direction on a substrate of Si or Ge, the layers of the same material as the substrate are unstrained, whereas the other partner of the superlattice experiences biaxial strain. The bulk deformation potentials (see [7.26, 48] for values) of Si and Ge are such that the strain strongly affects the splitting of the conduction bands in the [001] direction between valleys that become inequivalent under strain. The corresponding deformation potentials are about 5 times larger than any other deformation potential in these materials. In the case of a compressive biaxial strain (SiGe grown on Si for instance), the conduction band mimima in the two valleys parallel to the growth rise above the four minima in the transverse direction. The inverse is true when the sign of the strain is reversed.

The most obvious effect of a superlattice is to give a larger unit cell in the growth direction. Bulk Si has 2 atoms per unit cell. A 2 + 2 Si-Ge superlattice has 4 atoms in the unit cell. This leads to a folding of the electronic energy bands in reciprocal space. Figure 7.11 shows the effect of a double periodicity on a simplified sketch of the Si (or Ge) bands in the [001] direction. As one can see, the conduction band minimum moves closer to the Γ point. A first attempt to predict a direct gap is thus very easily made by considering an increased period in the growth direction, from which one can deduce the rough position to which the minimum should fold. This is illustrated in Fig 7.12 for unit cells with between 2 and 14 atoms. The horizontal lines indicate the lengths of the Γ-Z segments (where Z is the intersection of the first Brillouin zone with the [001] reciprocal space axis). The diamonds show the conduction band minimum when folding pure Si bands, the circles when folding Ge bands and the crosses are

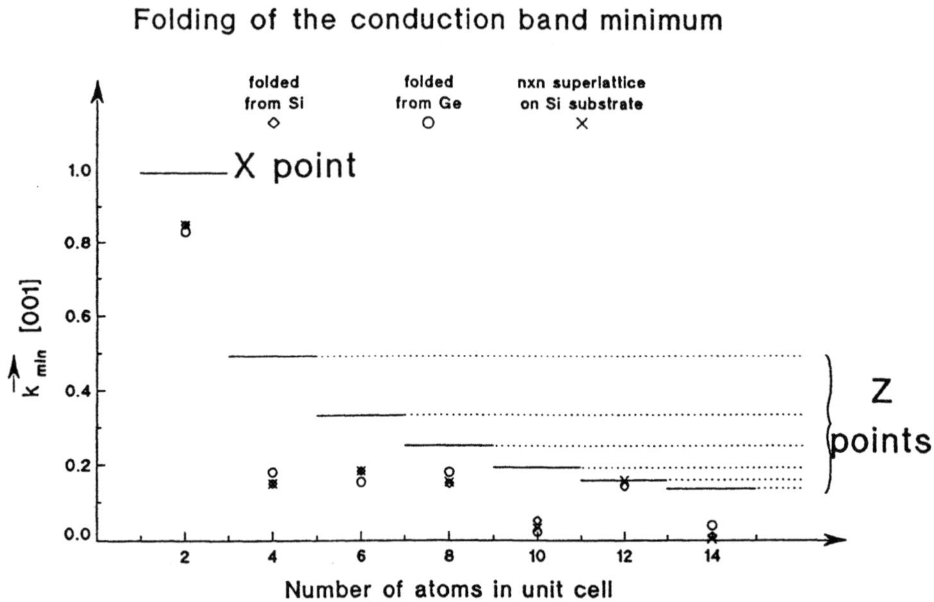

Fig. 7.12. Folding of the conduction band minimum. For superlattices of increasing number of atoms in the unit cell the figure shows the predicted positions of the conduction band minimum (see text)

the result of an empirical pseudopotential calculation described in [7.48]. One can see that the two most promising cases are those of 10 and 14 atomic layer periodicities. Indeed, the conduction band minimum then lies very close to Γ.

However, the effect of the super-periodicity only acts on the Γ-X axis in the growth direction. The two other directions still have their Δ minimum away from the center of the Brillouin zone. This is where the strain should play a role. As mentioned above, the largest deformation potential in Si or Ge splits the inequivalent Δ bands. To push the Δ mimima in the transverse directions above the folded minimum one should thus grow the superlattice so that one of the layers experiences a biaxial extension. This corresponds to a substrate rich in Ge.

Finally, this Δ minimum does not quite fall at the Γ point according to this simple argument. However, as we now intend to show, the chemical difference between Si and Ge induces a perturbative potential sufficient to pull the conduction band at Γ down to an absolute minimum. This may be done in a perturbation approach starting from the folded band situation of an unstrained large unit cell semiconductor. One should choose the starting point carefully. A simple prescription is to take pure Si or pure Ge with a large unit cell corresponding to the superlattice unit cell without the strain. Since we showed above that a Ge substrate offers the best chances to get a direct gap, one would be tempted to start from pure Ge. It should be noted however, that the Δ conduction band minimum in Ge falls very close in energy to the Γ_2 conduction band minimum. If pure Ge is taken as a starting point one may expect that the perturbation approach will

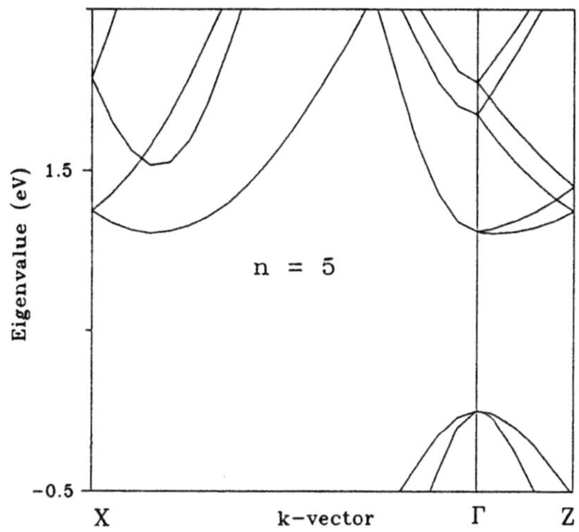

Fig. 7.13. Band structure of the virtual crystal corresponding to a $Si_{0.5}Ge_{0.5}$ composition, with a unit cell artificially extended over 10 periods in the [001] direction

be very sensitive to the exact difference between the Δ minimum energy and the Γ_2 minimum. Large errors could then occur. To avoid this it is wise to choose a starting point that is a virtual crystal $Si_{1-x}Ge_x$ where $0 \leq x < 1$, preferably with x close to 1. The folded band superlattice will then be intermediate between that of pure Si and that of pure Ge. This is depicted in Fig. 7.13 in the region of Γ. As one may observe, the Δ band is very flat in the Γ-Z direction whereas the parabola origination from Γ_2 is very steep due to a small effective mass in Si and Ge. We now can study the effect of adding the perturbation H_χ due to the different atomic potentials of Si and Ge. The first-order perturbation shifts are roughly equal for the Γ_2 and Δ condution band minima. The second-order perturbation terms involve the non-diagonal matrix element $|\langle\Psi_\Gamma|H_\chi|\Psi_\Delta\rangle|$ where $|\Psi_\Gamma\rangle(|\Psi_\Delta\rangle)$ represents the wavefunction of the band that contains the $\Gamma_2'(\Delta_{\min})$ point. This matrix element varies slowly along the Γ-X direction. The variation of the second-order term,

$$\frac{|\langle\Psi_\Gamma|H_\chi|\Psi_\Delta\rangle|^2}{E_\Gamma - E_\Delta} ,$$

with wave vector $\boldsymbol{k}$ is then dominated by the denominator. Since the Γ band is much more rapidly varying than the Δ band, the second-order term decreases very quickly as $\boldsymbol{k}$ goes from Γ to Δ. Thus the repulsion between bands Γ and Δ is maximum at the Γ point of the Brillouin zone. The initial gap between Γ_2 and Δ_{folded} is a decreasing function of the composition x of the virtual crystal Si_{1-x} Ge_x. It is then possible, under the hypothesis that the matrix element is weakly dependent on the actual composition to evaluate the composition x_0 above which the repulsion is sufficient to give a direct gap due to the chemical effect. One can calculate the important parameters of this approach using a simple empirical

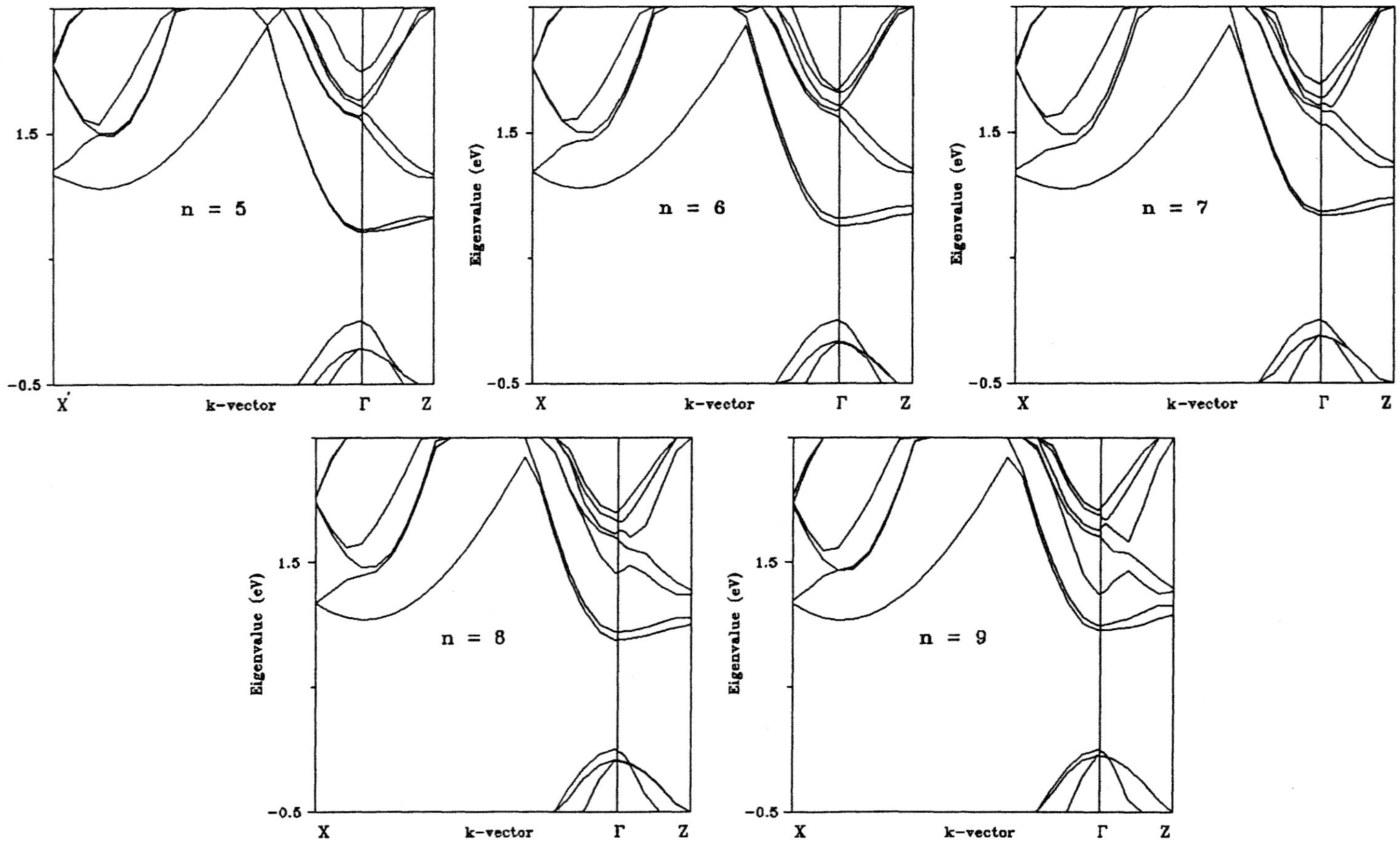

Fig. 7.14. Bands of the Ge_nSi_m superlattice with $n+m = 10$, and $n = 5, 6, 7, 8, 9$ on a Ge substrate in the [001] direction

pseudopotential model [7.48]. Using the notation of Fig. 7.13 and a periodicity of 10 layers of a 50/50 virtual crystal, the relevant parameters are:

- the $\boldsymbol{k}_0$ vector corresponding to the position of the Δ minimum in the Γ-Z direction, $\boldsymbol{k}_0 = 0.2$ a.u.;
- the matrix element $|\langle\Psi_\Gamma|H_\chi|\Psi_\Delta\rangle| \sim 4 \times 10^{-3}$ Ryd;
- the value of the Γ band parabola at $\boldsymbol{k}_0, ak_0^2 \sim 7.5 \times 10^{-3}$ Ryd;
- the value of the Δ band parabola at $\Gamma, bk_0^2 \sim 4.4 \times 10^{-4}$ Ryd.

With these values, it is possible to show that the virtual crystal should have an energy difference between the Δ minimum at $\boldsymbol{k}_0$ and the Γ_2 energy of less than 0.08 Ryd or 1.1 eV. This is the case for $x \gtrsim 0.5$.

The outcome of this study is thus that a Si-Ge superlattice should have a direct gap on a Ge-rich [100] substrate, for a periodicity of 10 layers, and more Ge layers than Si layers. Another possibility is to take a total periodicity of 14 layers. Other growth directions could also be investigated. However, due to the very large offset at the L point in the [111] direction, it seems hopeless to think of bringing the L minimum below the Γ minimum by growing the superlattice in the [111] direction. Other directions such as the [110] could also show interesting features.

These results may be compared with the outcome of theoretical calculations on the one hand and experimental results on the other. Figure 7.14 shows the band structures of $Si_{10-n}Ge_n$ for $n = 5, 6, 7, 8, 9$ on an [001] Ge substrate. As predicted, the folded bands produce a direct gap whose value decreases to the pure Ge value as n increases. To our knowledge, the only structures grown in the conditions specified above are Si_3Ge_7 and Si_4Ge_6 superlattices on a [001] Ge substrate [7.51]. Luminescence was obtained on these samples. More studied have been the symmetrically strained Si_nGe_m superlattices with $n \sim m \sim 4$ [7.50, 52]. By symmetrically strained one means that the superlattice is grown on a buffer layer of $Si_{1-x}Ge_x$ where x is chosen such that the strain in the Ge layers of the superlattice is equal and opposite to that in the Si layers. In all these experiments, however, it was debatable whether the observed photoluminescence lines are intrinsic to the superlattice or defect induced; in particular, dislocations may be the origin. Some preliminary experiments by *Abstreiter* [7.50] and by *van de Walle* [7.52] indicate that this is not the case.

7.3 Relations Between Transition Metal Impurity Levels, Band Offsets and Schottky Barrier Heights

Very recently, a novel empirical approach, offering surprising accuracy in the prediction of band offsets (BO) for semiconductor heterojunctions (HJ), has been proposed [7.53,54]. This approach is based on the assumption that a reference level, which seems to exist for transition metal (TM) impurities in semiconductors, may be used in the band alignment procedure in HJs. It has been observed that the relative ordering of the ionization energies for the whole series of TMs is preserved when the isovalent host is changed. It is therefore possible, by a shift in the valence band edges of two isovalent semiconductors, to align all the respective TM ionization levels. An explanation for this has been proposed by *Tersoff* and *Harrison* [7.55] who have concluded, on the basis of a defect molecule model previously introduced by *Picoli* et al. [7.56], that TM impurity levels are pinned to bulk cation vacancy levels. On the other hand, it was also proposed [7.57] and then further argued [7.58], that the reference level is simply the vacuum level.

The alignment of the TM ionization levels allows one to express the ionization energy of the ith element of the TM series in an AB semiconductor (values are referred to the top of the valence band) as a simple sum

$$E_i(AB) = E_i(\mathrm{TM}) + E(AB)\,, \tag{7.22}$$

where the first term $E_i(TM)$ is characteristic of the TM, and the second term $E(AB)$ of the host AB. With GaP and ZnSe as references [E(GaP) = E(ZnSe) = 0], a global least-squares minimization was performed to extract the averaged $E(AB)$ parameters. For consistency of treatment, all the energies for a given TM impurity charge state were extracted from the same type of measurement, following the same procedure. The results obtained in this way for $E(AB)$ are given in Table 7.4. If one assumes that the TM levels are pinned relative to the vacuum level [7.57], then the quantity $E(AB) - E(CD) = \Delta E$ should be equal to the difference $\Delta\chi$ between the corresponding electron affinities. Table 7.4 presents a comparison of the two quantities (again taking GaP and ZnSe as references) and shows that differences of affinities $\Delta\chi$ produce significantly larger valence band separations than the use of ΔE derived from our TM averaging procedure. This gives some experimental indication that TM vacuum level pinning is not verified on the scale of 0.1 eV. The observed alignment of TM impurity levels has led *Langer* and *Heinrich* to postulate [7.53] that the valence band discontinuity in a HJ is given by the difference in the energy level positions of a TM impurity in

Table 7.4. Comparison of the positions of the top of the valence band (GaP and ZnSe are taken as references in III-V and II-VI compounds respectively) obtained by: a) estimation from the TM averaging procedure (ΔE_{exp} and ΔE_{th} are experimental results and theoretical values, respectively); b) difference between measured photothresholds $\Delta\chi$; c) difference between average dangling bond energy levels $\overline{\Delta E_{db}}$ (see text) (all the energies are in eV). For more details, see [7.59]

	III-V					II-VI			
	ΔE_{exp}	ΔE_{th}	$\Delta\chi$	$\overline{\Delta E_{db}}$		ΔE_{exp}	ΔE_{th}	$\Delta\chi$	$\overline{\Delta E_{db}}$
GaAs	+0.33	+0.39	+0.57	+0.34	HgTe	+1.10			
InP	+0.17	+0.11	+0.37	+0.03	CdTe	+0.80	+0.96	+1.04	+0.89
GaP	0	0	0	0	CdSe	+0.09		+0.20	
AlAs	−0.12		−0.15	−0.11	ZnSe	0	0	0	0
					CdS	−0.37		−0.44	
					ZnS	−0.56		−0.68	

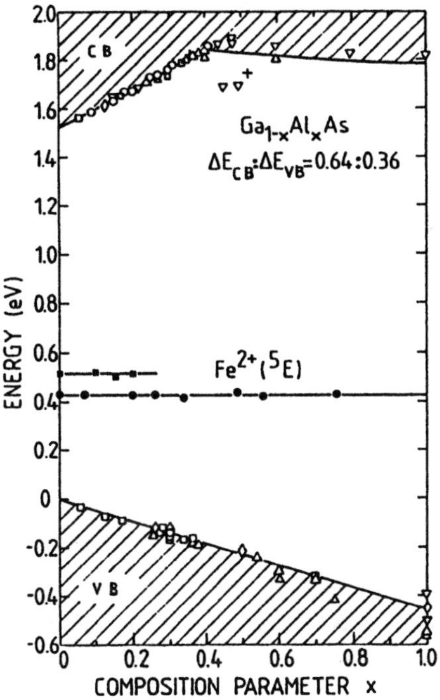

Fig. 7.15. Conduction band and valence band offsets compared to the Fe energy level as functions of x in $Ga_{1-x}Al_xAs$ on GaAs [7.59]

the two compounds forming the HJ, or by the difference in the ΔE if one can use an average procedure. The applicability of their proposal is confirmed in a striking way by Fig. 7.15–17. This agreement is not easy to explain if TM levels are pinned to a vacuum level, since this would imply the existence of a similar vacuum level pinning rule for HJ (i.e. the electron affinity model) which does not seem to work satisfactorily [7.59–61].

The problem of band alignment in HJs has already been analyzed in terms of the existence of a certain internal reference level [7.9, 62], which was identified with the neutrality level E_B by *Tersoff*. Following the conclusions of the previous section, we shall present arguments for the identifying of this level with the average self-energy of dangling bonds $\overline{E_{db}}$. This rule will be confirmed by the charge-dependent tight-binding calculation of $\overline{E_{db}}$ for several compounds.

Some physical insight concerning the relation between TM levels and band offsets can be gained from the defect molecule model recently formulated by *Picoli* et al. [7.56]. *Tersoff* and *Harrison* [7.55] have recently shown, using this model, that the TM impurity levels follow the T_2 cation vacancy level at a distance independent of the bulk material, because of the strong intra-defect Coulomb interaction U_d. In the defect molecule approach, one only considers the interaction between the d-states of the transition metal impurity and the sp^3 dangling bonds of the four nearest neighbor anions (these characterize the cation vacancy). The five d-states transform as the E and T_2 representations of the T_d point group while the sp^3 dangling bonds transform as A_1 and T_2. The

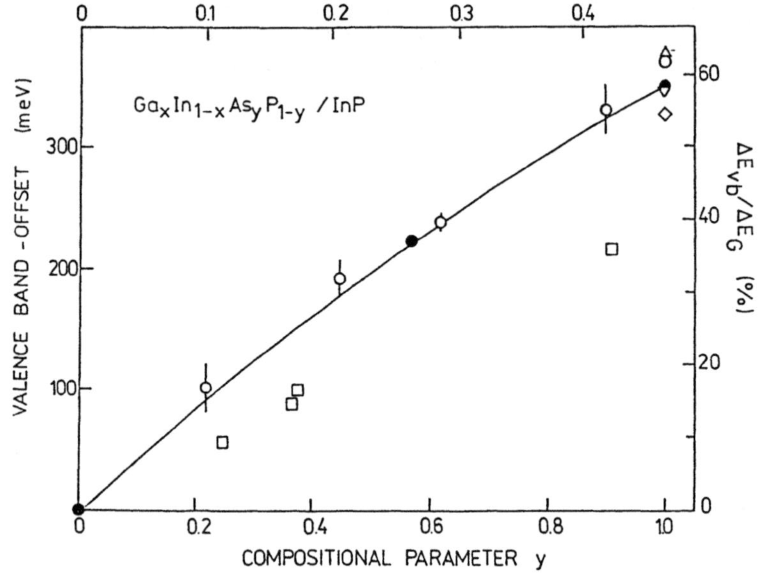

Fig. 7.16. Valence band offset (◇, □, ▽, △, o) and the Fe levels (•) in $Ga_xIn_{1-x}As_yP_{1-y}$/InP (see [7.59] and references therein). The full line is the result of a best fit described in [7.59]

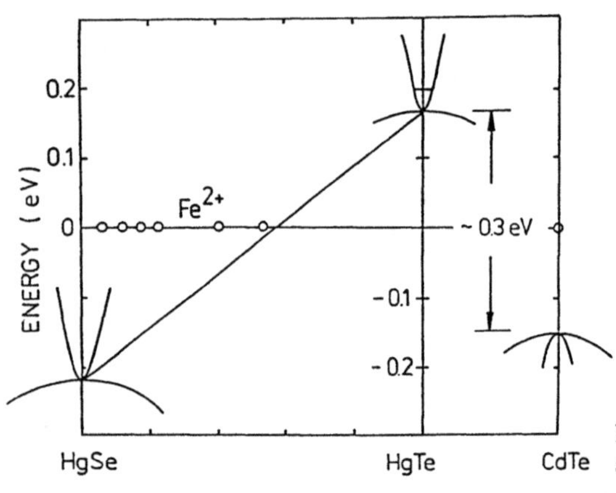

Fig. 7.17. Band structure of HgSe, HgTe and CdTe relative to the Fe level [7.59]

two d-states of E symmetry will remain uncoupled and the three d-states of T_2 symmetry will couple to the T_2 cation vacancy states leading to bonding and antibonding triply degenerate states. These are solutions of the 2 × 2 matrix

$$\begin{pmatrix} E_d & V \\ V & E_v \end{pmatrix}, \tag{7.23}$$

where E_d is the atomic level, E_v the cation vacancy level and V the covalent coupling. It is easy to show that the fraction of "d" character in the bonding and antibonding states is $1/2(1 - \delta/\Delta)$ and $1/2(1 + \delta/\Delta)$ respectively, with

$\delta = 1/2(E_d - E_v)$ and $\Delta = \sqrt{\delta^2 + V^2}$. As there are six electrons in the bonding state, n_E in the E state and n_A in the antibonding state, the total d-electron population n_d is given by

$$n_d = 3\left(1 - \frac{\delta}{\Delta}\right) + n_E + n_A\left(\frac{1}{2} + \frac{\delta}{\Delta}\right) . \tag{7.24}$$

In the limit of strong screening, i.e. when the intra-atomic Coulomb term U_d is large, a very good approximation is provided by the local neutrality condition which states that n_d given by (7.24) must be equal to its value n_{d0} for the neutral atom which is independent of the host. From (7.24) this means that δ is independent of its non-self-consistent value δ_0 and practically takes the same value for all III-V or all II-VI semiconductors. In other words the transition metal impurity level E_d is pinned, within a constant, to the cation vacancy level.

Following this argument, *Tersoff* and *Harrison* have then tried to relate the cation vacancy level to the neutrality level E_B but this is the weakest part of their argument. The problem is that, although the molecular approach offers a transparent explanation, its numerical accuracy on the required scale of 0.1 eV is not sufficient to provide a clear-cut picture of these properties. In fact, a validation of the two-fold TM and HJ connection would require the TM energy level to pin not to the cation vacancy level but to the average dangling bond energy, for reasons explained in the previous sections.

Let us show that this is indeed the case. For this we make use of a Green's function calculation [7.63] based on a tight-binding formalism where the TM atom is represented by its five d orbitals and the bulk atoms by an sp basis. The band structure of a perfect crystal is obtained by the procedure of *Talwar* and *Ting* [7.14] (for II-VI materials, we have derived our own set of parameters). We then remove an atom to replace it by an TM ion, using for this purpose the Dyson equation

$$G = G_0 + G_0VG , \tag{7.25}$$

where G_0 is the perfect crystal Green's function, G the final Green's function, and V the perturbation induced by the defect. For the interatomic terms of the matrix V the first neighbor interaction parameters are deduced from Harrison's rules [7.64], the others are neglected. The energy E_d of the d orbitals of the TM ion is dependent on the electron population n_d on the atom. Proceeding as in [7.63], we write in a spin-unrestricted formalism

$$\begin{aligned} E_{d\uparrow} &= E_d^0 + U_d(n_d - n_{d0}) - \frac{J}{2}(n_{d\uparrow} - n_{d\downarrow}) , \\ E_{d\downarrow} &= E_d^0 + U_d(n_d - n_{d0}) + \frac{J}{2}(n_{d\uparrow} - n_{d\downarrow}) , \end{aligned} \tag{7.26}$$

where $E_{d\uparrow}$ is the d energy for the spin up, J an average exchange interaction, U_d an average Coulomb interaction and E_d^0 the d atomic energy for the configuration with $n_d = n_{d0}$.

As discussed in [7.63], we take $U_d = 8\,\mathrm{eV}$ and $J = \Upsilon(5/2B + C)$, where B and C are the well known Racah parameters and Υ is a reduction factor ($\Upsilon = 0.72$). The final results are found to be insensitive to the exact value of U_d, provided it is large. The reason for this was given first by *Haldane* and *Anderson* [7.65]. It comes from the fact that, when U_d is large (i.e. $U \gg 1\,\mathrm{eV}$ here), the screened potential is small and independent of U_d, as shown analytically in the defect molecule model [7.56]. Finally, self-consistency is achieved by applying a potential U_1 to the first neighbors of the defects. For a defect with charge state q, we impose a net charge q/ε, where ε is the dielectric constant. In the spirit of the empirical tight-binding approximation the atomic energies, i.e. E_d^0 and the average sp^3 energy of the bulk crystal are taken to be the free atom values, deduced from *Herman* and *Skillman* [7.66]. As for any empirical theory its justification lies in the fact that it is successful for predicting a broad range of physical properties as demonstrated in [7.63]. As it stands the model is completely free of adjustable parameters. The ionization energies are calculated with the Slater transition state for maximum spin configuration. We have calculated the relative valence band offsets using the same TM averaging procedure as described above. As shown in Table 7.4 they are in excellent agreement with the values deduced from experiment [7.59].

To clarify the relation between TM levels and band offsets, we have also calculated the energy differences between computed TM energy levels and i) the t_2 cation vacancy level and ii) the average dangling bond energies determined by the same Green's function calculations. The results, depicted in Fig. 7.18 show that the TM impurity level is connected more to E_{db} than to the vacancy level. This behavior is due to the potential U_1 on the first neigbors, introduced to impose global neutrality, and to the delocalization of the impurity states included in our description, factors which increase the influence of the second nearest neighbor cations. Comparison between the valence BOs predicted by the TM average

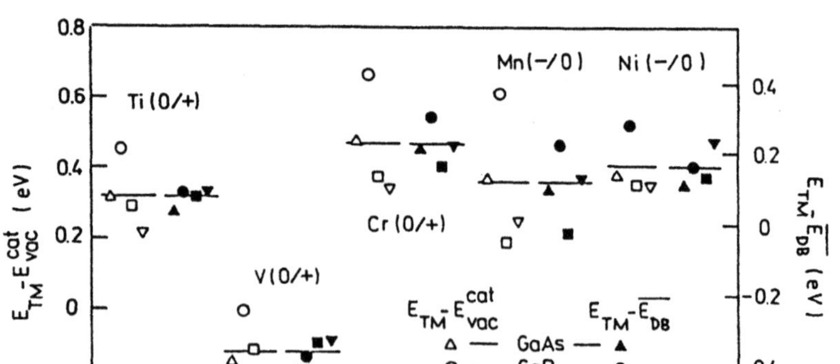

Fig. 7.18. Difference between the calculated transition metal levels and the calculated T_2 cation vacancy level or the average dangling bond level [7.59]

procedure [7.53] and by differences between average dangling bond energies [7.12] is also impressive (see Table 7.4).

The physical origin of TM level alignment is thus transparent: because of strong Coulomb interactions ($U_d \gg 1$ eV) in the $3d$ shell, the TM impurity level position becomes independent of U_d and is pinned to a bulk level, the latter being related to the average dangling bond energy and not just to the cation vacancy level because of long range effects (delocalization, screening). This calculation gives sound theoretical support for the empirical rule connecting TMs and HJs [7.53, 54], and shows again the importance of the concept of a dangling bond state in the description of HJs and defects in semiconductors.

In conclusion the central physical quantity which relates Schottky barrier heights, heterojunction band lineups and transition metal impurity levels is the average dangling bond energy which also plays the role of Tersoff's neutrality level. We have demonstrated this by using defect molecule models and charge dependent tight-binding calculations.

7.4 The Semiconductor–Oxide Interface: Si-SiO_2

Although there is a wealth of experimental information concerning the Si-SiO_2 interface, its atomic and electronic structures are not yet completely understood. There is now good evidence that the interface is structurally abrupt with low roughness as shown by high resolution transmission electron microscopy [7.67]. There is then a transition layer of intermediate structure SiO_x and of thickness equivalent to two monolayers, i.e. ~ 5 Å [7.68, 69]. It is also likely that, at least in some cases, there is a strained layer of about 20 Å as discussed by *Grunthaner* et al. [7.69]. However the general situation is more complex and the detailed atomic structure of the interface depends sensitively on the oxide growth conditions.

In this context the aim of the present section is to discuss some key features of the electronic structure of the Si-SiO_2 interface. This will be based on a very simple tight-binding analysis, starting from molecular models and using them to understand the results of more complex numerical calculations. This is first tested in the case of bulk Si and SiO_2 and then applied to the study of the intermediate SiO_x composition where we analyze the results of cluster Bethe lattice calculations and compare with experimental data essentially obtained by photoemission [7.70]. In a second part we concentrate on recent photoemission measurements [7.71] on thin SiO_2 films and show that the relative amplitude of the peaks can only be understood on the basis of the existence of valence alternation pairs (VAP). We perform a cluster tight-binding total energy calculation of such VAP and show that their stability is close to that of a normal SiO_4 unit, explaining why they are likely to be formed under conditions of interface formation. In a final part we discuss the best known interface defect: the P_b center and show that theory is able to account for most of its properties as determined from electron paramagnetic resonance (EPR), optical spectroscopy and deep level transient spectroscopy (DLTS). We also comment on the possibility that this defect could behave as a negative U center.

7.4.1 Electronic Structure of Si, SiO_2 and SiO_x

Our analysis makes use of a tight-binding description of these materials where the electron wave function is expressed as a linear combination of atomic orbitals. These are assumed to form an orthonormal basis set (i.e. interatomic overlaps are neglected) and furthermore we limit ourselves to a "minimal" basis set consisting of one s- and three p-orbitals on each Si and O atom. The electronic states are then obtained by diagonalization of the Hamiltonian matrix $H_{i\alpha,j\beta}$ where i and j specify the atoms, α and β the orbitals. Such a diagonalization must be performed numerically but here we proceed in a different way, making use of molecular models to discuss the formation of the band structure. For this we make a basis change chosen in such a way that the Hamiltonian matrix becomes dominated by some interatomic terms, the other elements being much smaller. When this is possible the neglect of the smaller matrix elements corresponds to a molecular description (introduced in Sect. 4.1.1), intimately connected to the nature of the chemical bonds, where each band reduces to a highly degenerate set of molecular levels. Inclusion of the smaller matrix elements will then broaden these levels into bands (first order effect) and eventually slightly shift these bands (second order).

The simplest illustration is for bulk Si. We recall the sp^3 basis orbitals that were introduced in Sect. 4.1.1 [7.72]:

$$\varphi_{ij} = \tfrac{1}{2}(s_i + \sqrt{3}p_{ij}) \tag{7.27}$$

with

$$\langle\varphi_{ij}|H|\varphi_{ji}\rangle = -\beta \quad (\beta > 0)\,.$$

In this very crude description where all other matrix elements of the Hamiltonian are neglected we saw that the system was equivalent to a set of diatomic molecules. The diagonalization of the Hamiltonian leads to a set of two highly degenerate states. The degeneracy is equal to the number of bonds, N. With two electrons per bond the bonding state was completely filled, and the anti-bonding state was completely empty.

It will be instructive for the following to discuss now in more detail the broadening of these states into bands to get an idea of the shape of the density of states. For this we consider the bonding states (only valence bands will be studied in the following in relation to photoemission data) and call $-\Delta(\Delta > 0)$ the Hamiltonian matrix element between two adjacent bonding states, which must be the next important interaction to include. To discuss the broadening effect we write the valence band wave function Ψ_{v} as a combination of the bonding states:

$$B_{ij} = \frac{\varphi_{ij} + \varphi_{ji}}{\sqrt{2}}\,, \qquad \Psi_{\mathrm{v}} = \sum_{\text{pairs } ij} \alpha_{ij} B_{ij}\,. \tag{7.28}$$

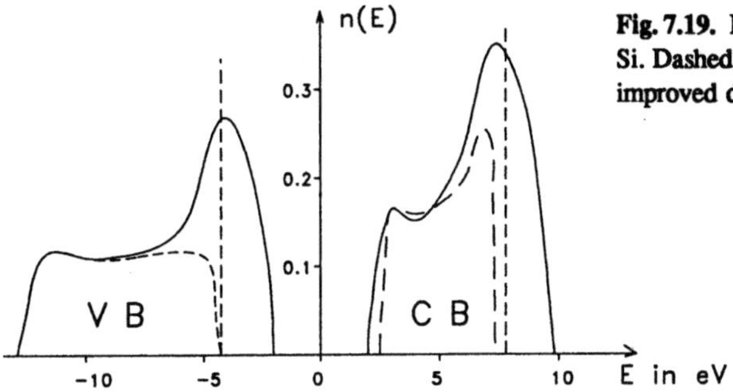

Fig. 7.19. Broadening of the flat bands in Si. Dashed line: simplified model; full line: improved density of states [7.78]

Projecting Schrödinger's equation $(E - H)\Psi_v = 0$ onto one basis state B_{ij} gives

$$(E - \bar{E} + \beta)\alpha_{ij} + \Delta \sum_{\substack{k \neq j \\ l \neq i}} (\alpha_{ik} + \alpha_{jl}) = 0 \tag{7.29}$$

where k and l are nearest neighbors of i and j. Introducing $S_i = \sum_k \alpha_{ik}$ on each atom we get

$$(E - \bar{E} + \beta - 2\Delta)\alpha_{ij} + \Delta(S_i + S_j) = 0 \ , \tag{7.30}$$

which, when summed over the four neighbors j, gives

$$(E - \bar{E} + \beta + 2\Delta)S_i + \Delta \sum_j S_j = 0 \ . \tag{7.31}$$

Equation (7.31) is formally equivalent to the nearest neighbor tight-binding problem of a s-band on the silicon lattice. It gives rise to a broad band whose extrema correspond to all S_i equal or to alternant S_i leading to $\bar{E} - \beta - 6\Delta < E < \bar{E} - \beta + 2\Delta$. This band contains one state per atom. Other non-trivial solutions correspond to all $S_i = 0$, i.e., from (7.30), to $E = \bar{E} - \beta + 2\Delta$. These states have pure p character; their number is one per atom and they correspond to a flat band. The resulting density of states is pictured schematically in Fig. 7.19. Inclusion of further interactions will mainly broaden the upper flat band as shown again in Fig. 7.14. Thus with no calculations we have obtained correct information about the valence band density of states of Si.

Let us now apply similar reasoning to the case of SiO_2. The molecular model is essentially the one developed by *Harrison* and *Pantelides* [7.73]. The Si-O-Si building unit is shown in Fig. 7.20. Again one builds sp^3 hybrids on the Si atoms while on the oxygen atom one keeps the natural sp basis. The oxygen s-state is by far the lowest in energy and, to first order, its coupling with other states can be neglected. It will remain atomic like at its atomic value $E_s(0)$. On the other hand the oxygen p energy $E_p(0)$ is closer to the silicon sp^3 energy $\bar{E}$ and the interaction of the corresponding states must be taken into account. The molecular states of the Si-O-Si unit of Fig. 7.20 are then constructed from the two sp^3 hybrids a and b pointing towards the oxygen atom and the p_x, p_y and p_z oxygen p-states. It is clear that p_z, being perpendicular to the Si-O-Si

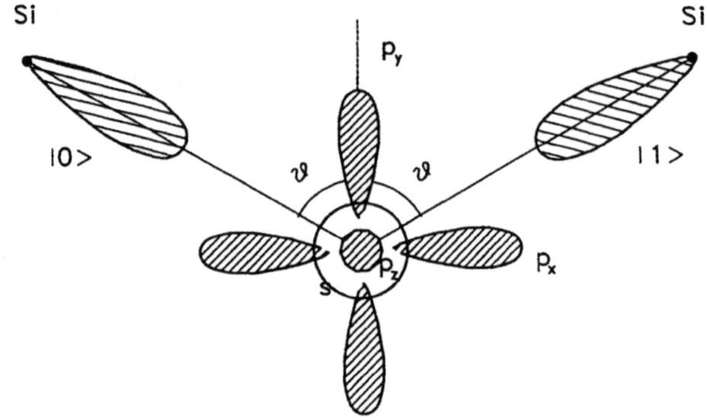

Fig. 7.20. Basis for the SiO_2 molecular model

plane, will remain uncoupled at this level of approximation giving one state at the atomic value $E_p(0)$. Thus the sp^3 states a and b will only couple to p_x and p_y via the projection of these states along the axis of the corresponding nearest neighbor direction. All these interactions reduce to only one parameter β_p defined as the interaction between an sp^3-orbital and the p-orbital along the corresponding bond. By symmetry p_x only interacts with $(a-b)/\sqrt{2}$ giving rise to strong bonding and anti-bonding states at energies

$$E_{S_A^B} = \frac{\bar{E}+E_p(0)}{2} \mp \sqrt{\left(\frac{(\bar{E}-E_p(0))}{2}\right)^2 + \beta_p^2 \sin^2\theta}\ , \tag{7.32}$$

while p_y and $(a+b)/\sqrt{2}$ lead to weak bonding and anti-bonding states:

$$E_{W_A^B} = \frac{\bar{E}+E_p(0)}{2} \mp \sqrt{\left(\frac{\bar{E}-E_p(0)}{2}\right)^2 + \beta_p^2 \cos^2\theta}\ , \tag{7.33}$$

where 2θ is the Si-O-Si bond angle.

The resulting valence band density of states per Si-O-Si unit is displayed in Fig. 7.21. It consists of delta functions at energies $E_s(0)$, E_{SB}, E_{WB} and $E_p(0)$, the weight of each state per Si-O-Si being equal to unity. The influence of further interactions can now be analyzed as for pure Si. If we call Δ_s and Δ_w the interaction between strong and weak bonding states belonging to adjacent Si-O-Si units then we can repeat the treatment previously applied to Si simply by replacing Δ by Δ_s or Δ_w. This means that we get densities of states in the strong and weak bonding bands that have exactly the same shape as for Si, consisting of the superposition of a broad and a narrow almost flat band. This behavior is apparent from the calculated density of states of Fig. 7.21a. These results are in good qualitative agreement with photoemission data [7.70, 74, 75] and more sophisticated numerical calculations [7.76, 77]. The essential information provided by (7.32) and (7.33) is that the splitting between the strong and weak bonding bands is a very sensitive function of the Si-O-Si angle 2θ. Any cause

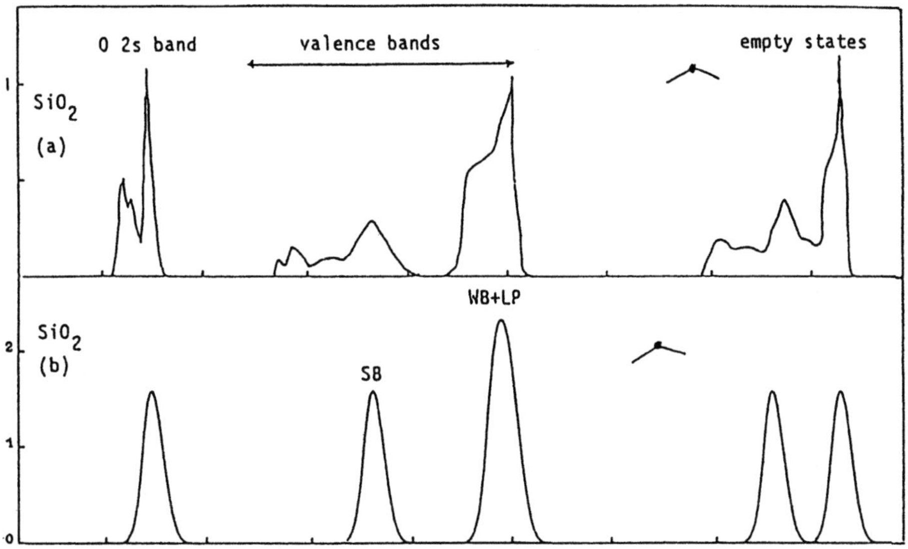

Fig. 7.21. Comparison between the computed density of states (top) and the broadened molecular model (*bottom*) for SiO_2

of randomness in θ such as the existence of a strained SiO_2 layer is then likely to induce some broadening of these two bands and partially fill the gap between them. Note that in an extreme situation where $\theta \sim 90°$ as in $GeSe_2$ the strong and weak bonding energies become identical, leading to qualitative changes in the shape of the density of states.

We are now in position to discuss qualitatively the behavior of the SiO_x layer as a function of composition. Numerical calculations have been performed that are all based on a tight-binding treatment combined with a more or less refined cluster Bethe lattice approximation [7.78, 79]. Again the molecular model is confirmed by the full calculation. For example let us consider the results of [7.78] which are pictured in Fig. 7.22. In the molecular model each Si-O-Si bond corresponds to a density of states as given in Fig. 7.21b while each Si-Si bond corresponds to one bonding state which, for the parameters corresponding to Fig. 7.22 falls at an energy slightly higher than $E_p(0)$, the energy of the SiO_2 non-bonding state. If we start from the SiO_2 limit each Si-Si bond acts like an isolated defect giving rise to a defect level at the energy of the bonding state of Si, i.e. just above the SiO_2 valence band. When the concentration of Si-Si bonds increases this defect state will begin to broaden into a band with no defined structure at low concentrations. This corresponds to $SiO_{4/3}$ in Fig. 7.22. It is only when the concentration is high enough for Si chains to appear that the situation changes qualitatively. If, as before, Δ characterizes the interaction between two adjacent Si-Si bonding states, then the DOS of a Si-Si chain exhibits two divergences at $\pm 2\Delta$ from the bonding energy. At still higher Si concentrations there will be three and then four Si-Si bonds connected to each Si atom. For reasons discussed

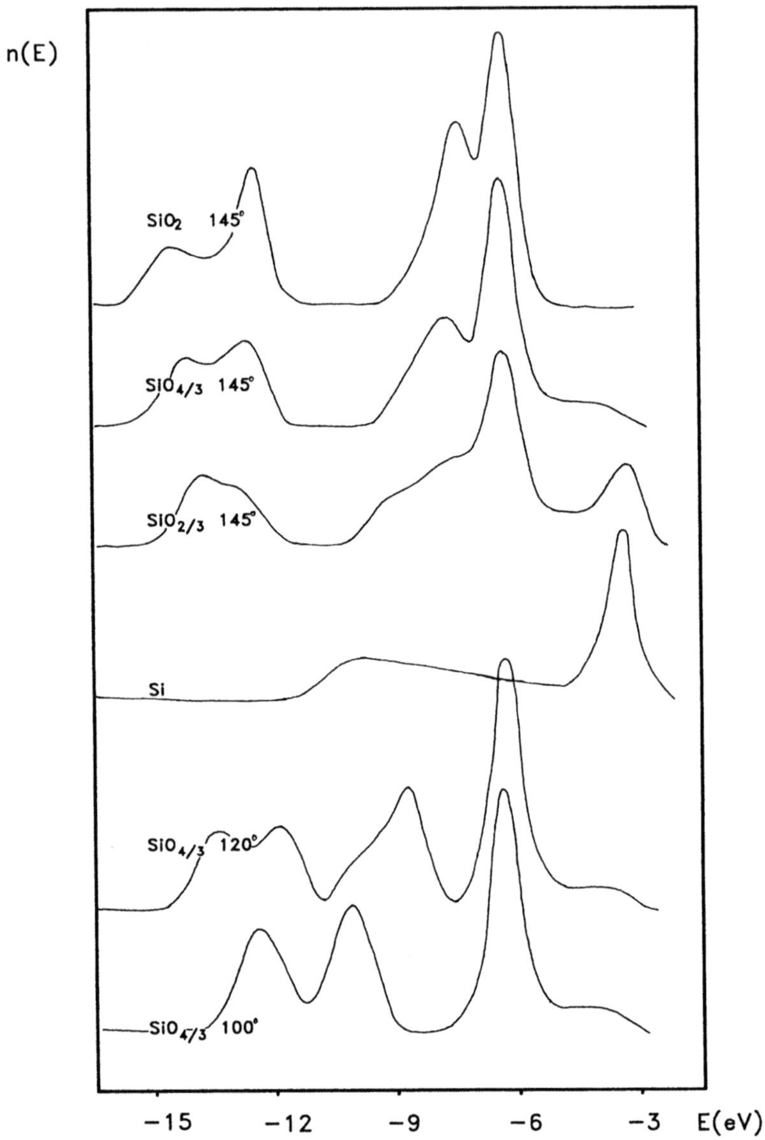

Fig. 7.22. Theoretical predictions for the density of states of different SiO_x compounds [7.78]

above such situations are characterized by a flat band at $+2\Delta$ and a broad band at lower energies. This regime will thus exhibit a peak of increasing height at the energy of the top of the pure Si valence band. This is exactly what happens in Fig. 7.22. Note also that the strong bonding band has exactly the reverse behavior: the height of its peak and its width both decrease. The last point that is clearly seen from the figure is that, for smaller θ, the separation between the strong bonding states and the weak non-bonding states is smaller.

All these conclusions are in qualitative agreement with experimental observations of the valence band of the SiO_x systems and of the transition layer at the Si-SiO_2 interface [7.70].

7.4.2 Very Thin SiO_2 Films on Silicon

The case of thin SiO_2 films thermally grown at low pressure (10^{-6} Torr) and low temperature (700°C) reveals very interesting features. Figure 7.23 shows the UPS valence band spectra coresponding to cases where the oxide layer is 11, 20 and about 100 Å thick. The last two layers are sufficiently thick for the spectra to correspond only to the SiO_2 layer. This is not true for the 11 Å layer where interface and bulk components have been substrated by a procedure described in [7.71]. The thickness of the residual top SiO_2 layer is estimated to be $\sim$ 6 Å. As for bulk SiO_2 the valence band spectra of this layer are characterized by three structures labelled I, II and III in Fig. 7.23. The most striking feature of these data is that the ratio R between the intensities of bands II + III and of band I varies significantly between 0.42 for the thick layers and 0.74 for the thin one in the (100) orientation, and between 0.46 and 0.78 for the (111) orientation. From the discussion of the last section, bands II and III correspond to the strong bonding band of SiO_2 (one state per Si-O-Si unit) and band I to the sum of the weak and non-bonding bands (two states per such unit). Thus the value characteristic of SiO_2, i.e. of 4:2 coordination, is $R = 0.5$ which is not far from what is obtained for the thick layer (the actual value of R cannot be directly compared to 0.5 since the experimental bands overlap and the dip between I and II is arbitrarily chosen

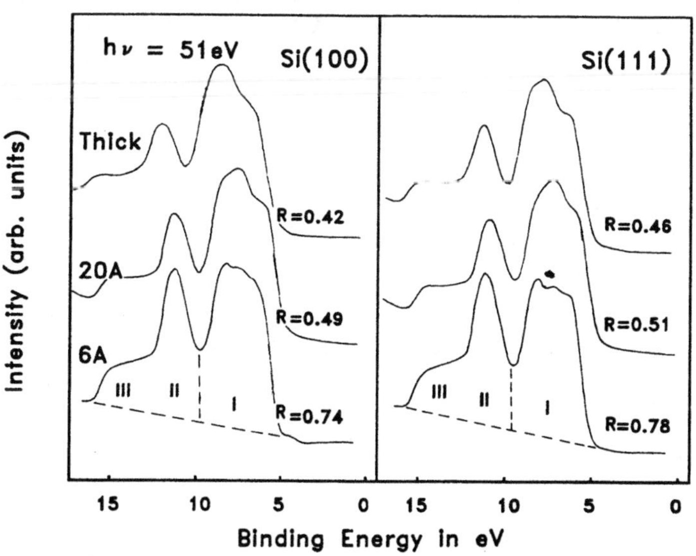

Fig. 7.23. Photoemission measurements of the valence bands of layers of varying thickness of SiO_2 on Si(100) and Si(111) [7.71]

as their separation). Higher values of R cannot be explained by bond-angle or bond-length variations since, as shown in Sect. 7.4.1, these only modify the band positions and spacings but not the value of R. We therefore have to assume that there are deviations from the classical 4:2 coordination.

One constraint that such deviations must respect is that the valence bands should remain similar to the SiO_2 case, with simply an increase in R. From this one can rule out the possibility of 6:3 coordinated SiO_2 stishovite which does not display the dip between bands I and II [7.81]. This is also the case for 3:3 coordinated Si-O whose band diagram is completely different [7.82]. The only hypothesis left seems to correspond to a high concentration of defects in a 4:2 SiO_2. Many types of defects have been proposed in crystalline and vitreous SiO_2 [7.77]. Among these one can reject peroxyl bridges or radicals which give oxygen states in the gap. Other defects such as oxygen vacancies or Si-Si bonds cannot be easily detected in UPS and should not strongly modify the spectra as observed.

We now discuss the defect model which we believe to be the most appropriate to explain these experimental observations. It corresponds to the valence alternation pair (VAP) [7.71, 83 and references therein] which consists of one three-fold overcoordinated oxygen O_3^+ and a one-fold undercoordinated oxygen O_1^-. In this model the four-fold coordination of Si, the charge neutrality, the number of Si-O bonds and therefore the SiO_2 composition are all preserved. To check quantitatively the validity of the VAP model, we have calculated the electronic structure of O_3^+ and O_1^- components using the simple molecular models of Fig. 7.24, with the tight-binding (TB) approximation. We have used the TB parameters given by *O'Reilly* and *Robertson* [7.77] that reproduce well the UPS spectrum of SiO_2. We have also included the relaxation of the distance between the O_3^+ or O_1^- atom and its nearest neighbors (in the O_3^+ case we have also varied

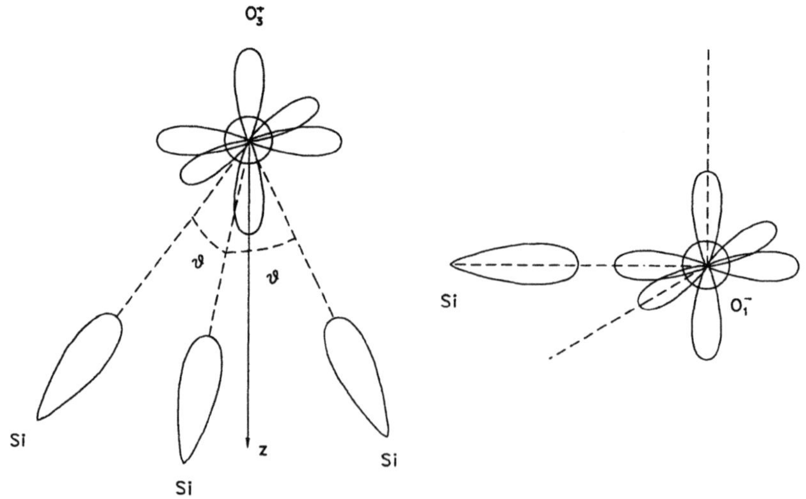

Fig. 7.24. Valence alternation pairs in the molecular models for oxide layers on Si [7.83]

the angle θ) in making a total energy calculation. We write the total energy for each molecular model as

$$E(R) = 2 \sum_{i}^{\substack{\text{occupied}\\ \text{states}}} \varepsilon_i(R) + nf(R) \,, \tag{7.34}$$

where R is the Si-O bonding distance, $\varepsilon_i(R)$ are the single-particle energies, $f(R)$ is the interatomic repulsion energy and n the coordination number. As in [7.84] we consider that the TB parameters vary exponentially with distance as $\exp(-qR)$ while $f(R)$ varies as $\exp(-pR)$, with $qR_0 = 2$, $pR_0 = 4$ for the normal Si-O distance $R_0 = 1.61$ Å of α quartz. The prefactor of $f(R)$ has been adjusted to get the correct R_0 for the molecular model of SiO_2. Applying the method to the two units of Fig. 7.24 the essential results of the approach are:

i) The total energie of a VAP is within 2% of the energy of two Si-O-Si units.
ii) We find a reduction in Si-O bond length and angle for the VAP which is in agreement with the increase in atomic density expected at an abrupt interface.
iii) The densities of states of our simple models which are plotted in Fig. 7.25 show that the VAP produces the right transformation of weak or non-bonding states into strong bonding states. Each VAP corresponds to a ratio $R = 1$.

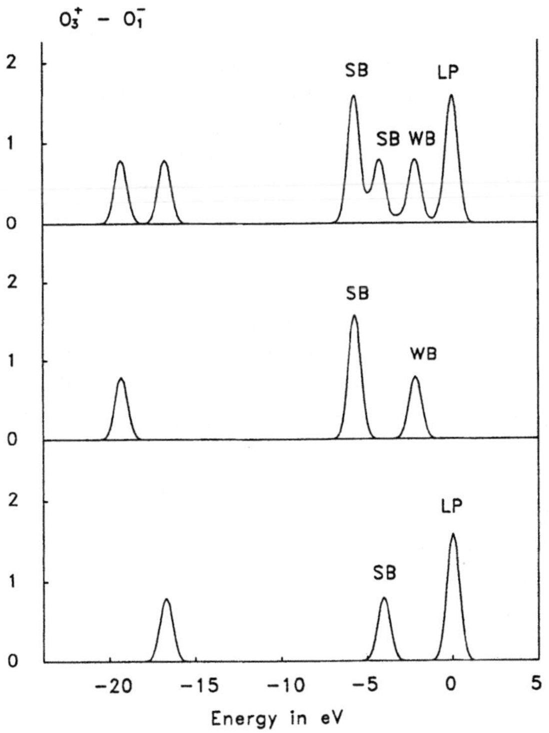

Fig. 7.25. Schematic densities of states of the valence alternation pairs obtained from the broadened levels of the molecular model [7.83]

(Note that for SiO_2 there is good agreement with the full calculation of [7.77].)

From these results it is likely that thin SiO_2 oxide films grown on Si(100) or (111) contain a large number of VAPs. Their energy is comparable to that of two normal SiO_2 units and they could be stabilized near the interface to release the strain. Furthermore they can explain the qualitative change observed in the UPS spectra.

7.4.3 The P_b Center at the Si-SiO_2 Interface

Trivalent silicon atoms (P_b centers) occur not only at the Si-SiO_2 interface [7.85–89], but also in amorphous silicon, grain boundaries in silicon and perhaps at some dislocations too. Each trivalent Si is characterized by an isolated dangling bond which has been identified mainly through electron spin resonance (ESR) [7.85], deep level transient spectroscopy (DLTS) [7.87, 90], capacitance measurements versus frequency and, more recently, through optical experiments [7.91, 92]. The qualitative picture which emerges is the following: i) the isolated dangling bond can exist in three charge states: positive D^+, neutral D^0 and negative D^-; ii) the effective Coulomb term U, i.e. the difference in energy between the acceptor and donor levels ranges from 0.4 eV in a-Si [7.91] to about 0.6 eV at the Si-SiO_2 interface [7.87]; iii) in the paramagnetic state D^0 (one electron in the dangling bond state), the spin density is 80% localized on the trivalent atom with 15% s-character.

There have been several calculations of the electronic properties of dangling bonds. Here we present the results of a Green's function treatment [7.93] that gives a quantitative description of the following experimental data: i) the order of magnitude and shape of the optical absorption versus frequency; ii) the magnitude of the effective Coulomb term (often termed the correlation energy); and iii) the localization and character of the defect wave function as given by ESR experiments.

To calculate the optical properties of a trivalent silicon atom at the Si-SiO_2 interface and especially for transitions between the dangling bond level and the silicon valence band we need a model which correctly describes the top of the valence band. This eliminates the possibility of using a cluster Bethe lattice approximation which gives an incorrect density of states in this energy region. The calculation is based on the Green's function technique in the tight-binding approximation. We calculate the corresponding Green's functions by numerical integration over the Brillouin zone. We then remove an atom and add a perturbative potential which suppresses the interaction between the four dangling bonds of the resulting vacancy. This is done using Dyson's equation.

The first result is that we get a non-self-consistent localized dangling bond level at +0.14 eV from the top of the valence band. We find that the localization of this bound state on the trivalent atom is about 60% with 16% s-character. This is in good agreement with the ESR data.

The determination of the Coulomb (or correlation) energy U requires a self-consistent treatment. The simplest way to do this is to use the zero charge approximation (discussed in Sect. 7.2.2 for the heterojunction) on the trivalent Si atom. This procedure is quite accurate for defects in Si. It can be done for any occupancy N of the dangling bond level ($N = 0, 1, 2$). The observable quantities are the electrical levels $\varepsilon(0,+)$ and $\varepsilon(-,0)$ [7.94] which can be calculated using Slater's transition state as the one-electron eigenvalues for $N = 1/2$ and $N = 3/2$ respectively. In this way we get $\varepsilon(0,+) = 0.05$ eV and $\varepsilon(-,0) = 0.7$ eV. The effective Coulomb energy U is the difference between these two values, so that we get $U = 0.65$ eV.

The value of U is in quite good agreement with the experimental value (0.6 eV at the Si-SiO_2 interface, 0.4 eV in a-Si). Strictly speaking the local neutrality argument should be more valid for bulk material (a-Si for instance) where screening is more efficient than at the Si-SiO_2 interface. However the effective U can be reduced by relaxation effects [7.94], explaining the difference of 0.25 eV between our predicted value and the a-Si one. On the other hand the almost perfect agreement with the value at the Si-SiO_2 interface is the result of a cancellation between the decrease due to relaxation and the increase due to less efficient screening.

We now consider optical transitions between the valence band and the level $\varepsilon(0,+)$, for which there is experimental information at the Si-SiO_2 interface. We first calculate the electronic contribution to the optical cross section $\sigma_e(h\nu)$ versus frequency ν by using the Fermi golden rule. Our calculated results are compared to the experimental curve of [7.92] in Fig. 7.26 for the same polarization direction

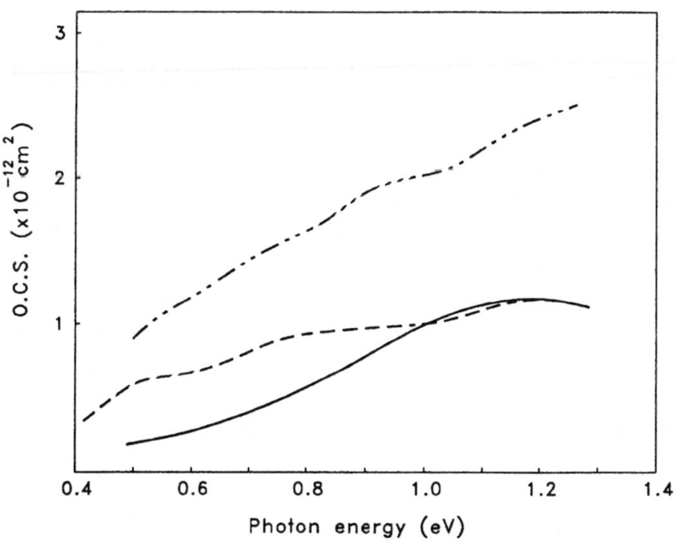

Fig. 7.26. Optical cross section of the P_b center: (- · · · -) theory for perpendicular polarization; (- - - -) theory for parallel polarization; (——) experiment

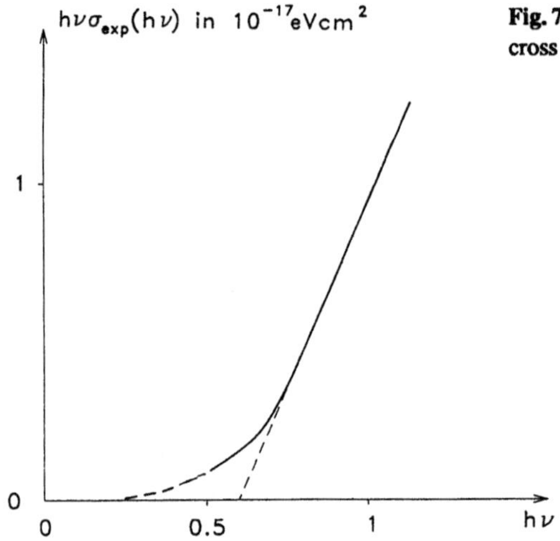

Fig. 7.27. Gaussian tail of the experimental optical cross section for the P_b center

and one perpendicular to it. The order of magnitude is clearly obtained but the shape near the threshold is different. This represents evidence for the existence of electron–lattice coupling, leading to lattice relaxation. Such coupling causes a broadening of the lineshape. In this case $\sigma(h\nu)$ is obtained as the convolution of the purely electronic part $\sigma_e(h\nu)$ by a vibrational lineshape function [7.95]. This can be approximated for relatively strong coupling by a Gaussian [7.94] whose position and width are functions of one parameter, the Franck–Condon shift d_{FC}. An interesting point is that the plot against $h\nu$ of the theoretical $h\nu\sigma_e(h\nu)$ near threshold is a straight line. If this is correct it is easy to show that convolution with a Gaussian also leads to a straight line with a Gaussian tail near threshold. This is precisely what is obtained experimentally in Fig. 7.27 which can be fitted by the theoretical curve for $d_{FC} = 0.27$ eV which is a manifestation of substantial electron–lattice interaction.

This conclusion is quite important and is confirmed by theoretical calculations of this parameter [7.96] which leads directly to $d_{FC} \sim 0.3$ eV. This however has the important consequence that the distance in energy between the two dangling bond levels $\varepsilon(-,0) - \varepsilon(0,+)$ is no longer U but becomes equal to $U^* = U - 2d_{FC}$. If U is equal to 0.6 eV or less then one finds a negative U center ($U^* < 0$) for which the charge state 0 (with one unpaired electron) is not stable [7.97]. This contradicts experimental ESR information. The explanation of this apparent contradiction is that the value found for the purely electronic contribution, $U \sim 0.6$ eV, is an understimate. Indeed it has been calculated with bulk silicon screening whereas the screening should be reduced at the Si-SiO_2 interface. Simple reasoning based on the replacement of ε by $(\varepsilon + 1)/2$ leads to an enhancement by a factor of 2, i.e. $U \sim 1.2$ eV leading to the final result $U^* \sim 0.6$ eV which is consistent with the experimental observations.

Exercises

7.1 Simulate what happens in the theory of the Schottky barrier derived in Sect. 7.1.3 when the metallic band is not symmetric with respect to the Fermi level E_F. Do this by replacing expression (7.8) by a single Lorentzian curve

$$R(E) + iI(E) = \frac{V^2}{E - E_F - \delta + iW} .$$

Determine $E_F - E_d$ in such a case. Under what conditions is this correction negligible? Estimate the order of magnitude of V from the theory of Chap. 6, the metal bandwidth from the free electron gas theory and δ from the energy difference between anion and cation dangling bonds. Use these values to determine the correction numerically.

7.2 Give details of the extension of the proof leading from (7.9) to (7.10).

7.3 Starting from (7.5) detail how one can express $\delta N(E)$ as given in (7.6) in terms of $R(E)$ and $I(E)$. Show that the result is equivalent to (7.7).
Using the logarithmic form of $\delta N(E)$ express the change in density of states as the difference between two quantities. Discuss their significance. Show that in the weak coupling situation ($V_{dk} \to 0$) one gets a well-defined resonance in a narrow energy range near E_d.

7.4 *Magnitude of dipole layers*: Consider a heterojunction interface and assume that the charge transfer is qe per atom between the two planes across the interface. Calculate the potential barrier induced by this dipole layer. Apply this numerically to (111), (110), and (100) interfaces of zinc-blende materials with common nearest neighbor distances equal to 2.3 Å.

7.5 *Charge neutrality*: Use the theory of Sect. 7.3 to deduce the extent to which the charge neutrality argument can be applied to transition metal impurities. Assume that E_d varies linearly with the excess electron population on the impurity

$$E_d = E_{d0} + U(n_d - n_{d0}) ,$$

where U is the intra-atomic Coulomb term on the impurity. Derive a self-consistent equation for δ. Show that the limit $U \to \infty$ corresponds to the criterion of charge neutrality on the impurity. Estimate the order of magnitude of the error introduced by this approximation.

7.6 *Bulk and interface dangling bonds*: State in more detail the arguments of the last paragraph in Sect. 7.4.3 supporting a large difference between the electronic U of a trivalent silicon atom in amorphous Si and that of one at the Si-SiO_2 interface.

8. Surface Phonons

The influence of the perturbation of a bulk crystal by a surface is very important. First of all the breaking of bonds induces forces on the surface atoms which tend to reach an equilibrium configuration different from that of the bulk. We have discussed this effect before in the general context of relaxation and reconstruction of surfaces. This modification in the local bonding induces in turn a change in the vibrational properties which can be observed experimentally. It is this effect that we want to analyze here. For this we first recall some basic properties of phonons in solids with a more detailed discussion of phonon dispersion curves in semiconductors. We then consider surface phonons and try to derive some general results by using simple analytical arguments. We discuss as examples recent results obtained on Si(111) 2×1 and Si(100). In the last part we analyze phonon-related properties paying special attention to the asymptotic decrease of static displacements away from surfaces which can be directly deduced from the bulk vibrational properties.

8.1 General Vibrational Properties

This section can be considered as a general introduction to phonons since the considerations that are developed here apply equally well to bulk, surface or other situations. We begin by recalling how one can solve the Schrödinger equation for lattice displacements. This involves the dynamical matrix and we present two methods that are used for determining its elements: empirical force constants and total energy calculations. We finally discuss the phonon dispersion curves in zinc-blende semiconductors on the basis of a very simple and powerful model based on a valence force field expression for the lattice energy.

8.1.1 The Dynamical Matrix

The possibility of independently analyzing the vibrational properties of a system relies on the use of the Born Oppenheimer approximation, which allows us to separate electronic and nuclear motion. The general technique is to calculate first the total energy $E_{el}(\{\boldsymbol{R}\})$ of the electron system for a fixed set of lattice positions $\{\boldsymbol{R}\}$. This energy then plays the role of a potential energy for nuclear motion. The corresponding Schrödinger equation is then given by

$$\left[-\frac{\hbar^2}{2}\sum_i \frac{\Delta_i}{M_i} + E_{\rm el}(\{\boldsymbol{R}\})\right] \chi(\{\boldsymbol{R}\}) = E\chi(\{\boldsymbol{R}\}) . \tag{8.1}$$

In this expression each atom i has a mass M_i, $\boldsymbol{R}_i$ being its position vector and Δ_i the corresponding Laplacian operator. To obtain general solutions of this equation one considers that $E_{\rm el}$ has an absolute minimum corresponding to the stable configuration. One then assumes that, for most properties, it is sufficient to expand $E_{\rm el}$ to second order in the atomic displacements with respect to this stable configuration. Calling $u_{i\alpha}$ the αth component of the displacement of atom i one can then write

$$\left(-\frac{\hbar^2}{2}\sum_{i\alpha} \frac{1}{M_i}\frac{\partial^2}{\partial u_{i\alpha}^2} + \frac{1}{2}\sum_{\substack{i,j\\ \alpha,\beta}} A_{i\alpha,j\beta} u_{i\alpha} u_{j\beta}\right) \chi = \varepsilon\chi , \tag{8.2}$$

where $A_{i\alpha,j\beta}$ is the force constant matrix whose elements are the second-order derivatives of $E_{\rm el}$ and ε is the energy relative to that of the stable configuration. As the atomic masses M_i are in general different we introduce a change in notation and define

$$v_{i\alpha} = \sqrt{M_i} u_{i\alpha} \tag{8.3}$$

in such a way that we get:

$$\left(-\frac{\hbar^2}{2}\sum_{i\alpha} \frac{\partial^2}{\partial v_{i\alpha}^2} + \frac{1}{2}\sum_{i\alpha,j\beta} D_{i\alpha,j\beta} v_{i\alpha} v_{j\beta}\right) \chi = \varepsilon\chi , \tag{8.4}$$

where

$$D_{i\alpha,j\beta} = \frac{A_{i\alpha,j\beta}}{\sqrt{M_i M_j}} \tag{8.5}$$

is one possible definition of the dynamical matrix D. We can consider the $v_{i\alpha}$ as the components of a column vector $\boldsymbol{v}$ in an orthonormal basis set $\boldsymbol{e}_{i\alpha}$ and the $\partial/\partial v_{i\alpha}$ as the components of a generalized gradient $\boldsymbol{\nabla}_v$ in the same basis. Then (8.4) can be rewritten

$$\left[-\frac{\hbar^2}{2}\boldsymbol{\nabla}_v^+ \cdot \boldsymbol{\nabla}_v + \frac{1}{2}\boldsymbol{v}^+ D\boldsymbol{v}\right] \chi = \varepsilon\chi . \tag{8.6}$$

The square matrix D is symmetric and can thus be diagonalized by a unitary transformation. We write the corresponding eigenvalue equation

$$\left(D - \omega_k^2 I\right) \boldsymbol{v}_k = 0 \tag{8.7}$$

where $\boldsymbol{v}_k$ is the eigenvector of eigenvalue ω_k^2. Calling v_k the components of $\boldsymbol{v}$ in this new basis set we get directly from (8.6):

$$\sum_k \left(-\frac{\hbar^2}{2} \frac{\partial^2}{\partial v_k^2} + \frac{\omega_k^2}{2} v_k^2 \right) \chi = \varepsilon \chi \, . \tag{8.8}$$

This is the Schrödinger equation for a set of independent harmonic oscillators, whose vibration frequencies are the ω_k. The main problem is then to diagonalize the dynamical matrix. This is relatively simple for a bulk crystal where one can use translational periodicity by introducing solutions that obey Bloch's theorem. It is more involved for surfaces where this periodicity is lost in directions perpendicular to the surface.

For the original force constant matrix, equation (8.7) combined with (8.3) gives

$$\left(A - \omega_k^2 M \right) \boldsymbol{u}_k = 0 \, , \tag{8.9}$$

which can also be transformed to

$$\left(M^{-1} A - \omega_k^2 I \right) \boldsymbol{u}_k = 0 \, . \tag{8.10}$$

These last two forms are often convenient to use since they directly involve the lattice displacement vector $\boldsymbol{u}$ and the force constant matrix A.

A final property of A which is very important for the consistency of any physical model comes from translational invariance. If we determine the αth component $F_{i\alpha}$ of the force experienced by atom i we get

$$F_{i\alpha} = - \sum_{j\beta} A_{i\alpha,j\beta} u_{j\beta} \, . \tag{8.11}$$

We now investigate a displacement where all $u_{j\beta}$ for a given β are equal to a common quantity T, all others being zero. This is just a translation T of the whole system along the direction β which should not modify the forces $F_{i\alpha}$. This can only occur if the following sum rule is obeyed:

$$\sum_{j\beta} A_{i\alpha,j\beta} = 0 \, . \tag{8.12}$$

A similar condition can be obtained by imposing rotational invariance.

8.1.2 Empirical Force Constants

The normal procedure for determining the phonon frequencies is to calculate an expression for the total electronic energy and then take its second derivatives to obtain the force constant matrix. This can be done either through a first principles calculation (such as Hartree–Fock or LDA) or by using more empirical methods to get the lattice energy $E_{\text{el}}(\boldsymbol{R})$. We develop here some empirical models in order of increasing refinement.

The most simple-minded idea is to directly parametrize the force constant matrix with the minimum number of nearest-neighbor elements as in empirical tight-binding methods for the electronic properties. The simplest case is then obtained by assuming isotropic nearest-neighbor interactions, i.e.,

$$A_{i\alpha,j\beta} = -k\delta_{\alpha\beta} \ , \tag{8.13}$$

where i and j are nearest neighbors. Translational invariance (8.12) immediately yields

$$A_{i\alpha,i\beta} = Z_i k\delta_{\alpha,\beta} \ , \tag{8.14}$$

where Z_i is the coordination number of atom i. Thus, as in tight-binding theory, the model reduces to one parameter. However isotropy is generally an unreasonable assumption and it is often better to assume full anisotropy (where the axial force constant along the bond dominates) as a starting point. This is particularly true in covalent semiconductors as is shown below.

A more physical empirical determination of the dynamical matrix can be obtained by assuming a given analytical form for the lattice energy $E_{\rm el}(\boldsymbol{R})$. An appropriate expression for covalently bonded systems is the valence force field model derived a long time ago for molecules [8.1]. For a homogeneous bulk system the valence force field expansion takes the form

$$\begin{aligned} \Delta E_{\rm el}(\{\boldsymbol{R}\}) &= \tfrac{1}{2}k_{rr}\sum_{ij}(dr_{ij})^2 + \tfrac{1}{2}k_\theta\sum_\alpha(R_1\,d\theta_\alpha)^2 \\ &\quad + \tfrac{1}{2}k_{r\theta}\sum_{ij\alpha}(dr_{ij}R_1\,d\theta_\alpha) + \dots \ , \end{aligned} \tag{8.15}$$

where $\Delta E_{\rm el}$ is the change in electronic energy with respect to the bulk situation, dr_{ij} the change in bond length between atoms i and j, R_1 the bond length, and $d\theta_\alpha$ the change in bond angle α. Some values of the parameters are given in [8.2] for the case of silicon. We shall return later to this model when we discuss phonon dispersion curves.

A particular version of the valence force field model which has become very popular is *Keating*'s model [8.3], in which $\Delta E_{\rm el}$ is expressed as

$$\Delta E_{\rm el} = \frac{3}{4d^2}\left\{\frac{\alpha}{2}\sum_{\substack{\text{pairs}\\ \text{ij}}}[\Delta(\boldsymbol{r}_{ij}\cdot\boldsymbol{r}_{ij})]^2 + \frac{\beta}{2}\sum_{\substack{i\\ j>k}}[\Delta(\boldsymbol{r}_{ij}\cdot\boldsymbol{r}_{ik})]^2\right\} \ , \tag{8.16}$$

where $\boldsymbol{r}_{ij}$ is the vector connecting two nearest neighbors in the lattice, α is the force constant describing bond stretching and β is related to bond bending.

The two force constants α and β have been tabulated in [8.4] both for covalent and compound semiconductors from the condition that they provide correct values for the elastic constants. These values are reproduced in Table 8.1. We shall see later that although they give reasonable phonon dispersion curves, they cannot reproduce at the same time the elastic constants and the transverse acoustic frequencies which are especially low in this kind of material.

Table 8.1. Values of α and β taken from [8.4]

Material	α [Nm^{-1}]	β [Nm^{-1}]
C	129.33	84.76
Si	48.50	13.81
Ge	38.67	11.35
AlSb	35.35	6.77
GaP	47.32	10.44
GaAs	41.19	8.95
GaSb	33.16	7.22
InP	43.04	6.24
InAs	35.18	5.50
InSb	29.61	4.77
ZnS	44.92	4.78
ZnSe	35.24	4.23
ZnTe	31.35	4.45
CdTe	29.02	2.43
CuCl	12.60	1.00

8.1.3 Total Energy Determination

The total electronic energy $E_{el}(\{\boldsymbol{R}\})$ can be calculated from first principles for varying atomic positions. This is an enormous task but has been done for most semiconductors using the LDA approximation to calculate the elastic constants and also some particular vibrational frequencies. This method can in principle be applied to a determination of the phonon spectrum but the complexity of the calculation becomes prohibitive at surfaces. For this reason empirical tight-binding calculations of E_{el} have been attempted and it is the basis of such calculations that we review here briefly. Such calculations have essentially the same physical basis as the TB total energy calculations for transition metals discussed in Sect. 3.1. We review here what has been done more recently for semiconductors along the same lines.

The first attempts to find an expression for the lattice energy were performed on the basis of the molecular model introduced in Sect. 4.1.1 [8.5]. As discussed previously, the total energy is expressed as the sum of the energies of the occupied one-electron state plus short-range repulsive terms. The tight-binding one-electron energies are obtained in a nearest-neighbor approximation where the usual two-center interactions $H_{\alpha\beta}$ defined in Sect. 2.1.2 are taken to vary exponentially with distance

$$H_{\alpha\beta} = H^0_{\alpha\beta} \exp(-p\Delta r) , \qquad (8.17)$$

where Δr is the change in distance. Similarly, the nearest-neighbor short-range repulsive terms are taken to have the form

$$E_R = E^0_R \exp(-q\Delta r) . \qquad (8.18)$$

In this model the diagonal elements $H_{\alpha\alpha}$, which take two values, the s and p atomic energies E_s and E_p, are assumed to be independent of distance. The molecular model of Sect. 4.1.1 assumes that all interactions between sp^3 hybrids in the tetrahedral structure vanish except for the pair involved in the same bond. This condition implies the following relations between the two-center tight-binding parameters:

$$
\begin{aligned}
&E_s = E_p \qquad H_{ss} = \frac{\beta_0}{4} \\
&H_{s\sigma} = \frac{\beta_0\sqrt{3}}{4} \quad H_{\sigma\sigma} = \frac{3\beta_0}{4} \quad H_{\pi\pi} = 0
\end{aligned}
\tag{8.19}
$$

where β_0 is the interaction between two sp^3 hybrids belonging to the same bond. The quadratic expansion of the lattice energy can be performed analytically in this model [8.5]. It is possible to show that it strictly reduces to the simplest form of the valence force field model with only two force constants, the radial one k_r and the angular one k_θ between adjacent bonds. This forms a bridge between the most simple meaningful models of electronic and vibrational properties. The expressions for k_r and k_θ are:

$$
\begin{aligned}
&k_r = 2pq\left(1 - \frac{p}{q}\right)\beta_0 \,, \\
&k_\theta = \frac{\beta_0}{2R^2} \,.
\end{aligned}
\tag{8.20}
$$

The parameters p and q have been evaluated in [8.5] for Si from the condition that they give the correct equilibrium lattice parameter, bulk modulus and cohesive energy. Then k_θ is fully determined. In the next section we analyze the quality of the corresponding predicted phonon dispersion curves.

Numerical calculations along similar lines have been reported and applied to the calculation of equilibrium geometries near the surfaces [8.6] as well as phonon frequencies for Si(111) and Si(100) [8.7, 8]. These calculations involve slightly different expressions for the parameters. First of all the nearest-neighbor interactions are assumed to follow Harrison's rule and to vary as $(d+\Delta r)^{-2}$ where d is the bulk equilibrium distance. Then the bulk parameters E_s, E_p, H^0_{ss}, $H^0_{s\sigma}$ and $H^0_{\sigma\sigma}$ are deduced from a fit to the band structure. Finally the repulsive terms are not taken to follow an exponential law as in (8.18) but instead are expressed by a quadratic form $U_1\Delta r + U_2\Delta r^2$ which also contains two parameters that have to be fitted. The model, as it stands, is no longer analytic, even for the bulk crystal, but can be applied numerically to various situations as discussed later. The corresponding phonon dispersion curves are discussed in the next section.

8.1.4 Bulk Dispersion Curves in Covalent Semiconductors

We shall analyze these dispersion curves within the framework of the valence force field model which allows a physical understanding of their basic features. We begin with the simplest version which retains only the radial force constant k_r between nearest neighbors, analyzing later the influence of smaller force constants such as k_θ. The lattice energy is then equal to $1/2k_r\sum_{ij}(dr_{ij})^2$ summed over nearest-neighbor pairs i and j. In this model, the matrix of force constants $\boldsymbol{A}$ is given by

$$\begin{aligned} A_{i\alpha,i\beta} &= k_{\mathrm{T}} \sum_j \alpha_{ij}\beta_{ij} , \\ A_{i\alpha,j\beta} &= -k_{\mathrm{T}}\alpha_{ij}\beta_{ij} , \end{aligned} \tag{8.21}$$

where α_{ij}, β_{ij} are cartesian components of the unit vector $\boldsymbol{n}_{ij}$ pointing from atom i to its nearest neighbor j. The eigenvalue equation (8.9), projected on to the local basis vectors gives

$$M\omega^2 u_{i\alpha} = k_{\mathrm{T}} \sum_j \alpha_{ij} dr_{ij} . \tag{8.22}$$

One can write a simliar equation for $u_{k\alpha}$ where k is the neighbor of i, take the difference $(u_{i\alpha} - u_{k\alpha})$, multiply it by α_{ik} and sum over α to obtain

$$\frac{M\omega^2}{k_{\mathrm{T}}} dr_{ik} = \sum_j n_{ik} n_{ij} dr_{ij} - \sum_l n_{ik} n_{kl} dr_{kl} , \tag{8.23}$$

where l are nearest neighbors of k. All scalar products are either 1 or $-1/3$ so that one gets

$$\left(\frac{M\omega^2}{k_{\mathrm{T}}} - \frac{8}{3}\right) dr_{ik} = -\frac{1}{3}\left(\sum_j dr_{ij} + \sum_l dr_{kl}\right) . \tag{8.24}$$

We now label the sum $\sum_j dr_{ij}$ by S_i. Summing (8.24) over k we get the following equation:

$$f\, S_i = \sum_k S_k , \tag{8.25}$$

with f defined as

$$f = 4 - 3\,\frac{M\omega^2}{k_{\mathrm{T}}} . \tag{8.26}$$

The advantage of this formulation is that it is valid not only for the crystalline situation but also for amorphous systems. Furthermore, equations (8.25) are strictly equivalent to the corresponding s-band tight-binding problem with nearest-neighbor interactions equal to unity. The corresponding matrix to diagonalize is the connectivity matrix and, for tetrahedrally bonded systems, general theorems [8.9] show that f lies in the interval $[-4,+4]$. From (8.26) this means that there exists for ω^2 a broad band extending between 0 and $8k_{\mathrm{T}}/3M$. This band corresponds to solutions with $S_i \neq 0$. However there do exist non-trivial solutions for all $S_i = 0$, and these are given either by (8.24), if $dr_{ik} \neq 0$, leading to $\omega^2 = 8k_{\mathrm{T}}/3M$ or by (8.22), if all $dr_{ij} = 0$, giving $\omega^2 = 0$. The model thus predicts, in addition to the broad band of (8.26), two flat bands containing one state per atom, the lowest one being somewhat unphysical at zero frequency and

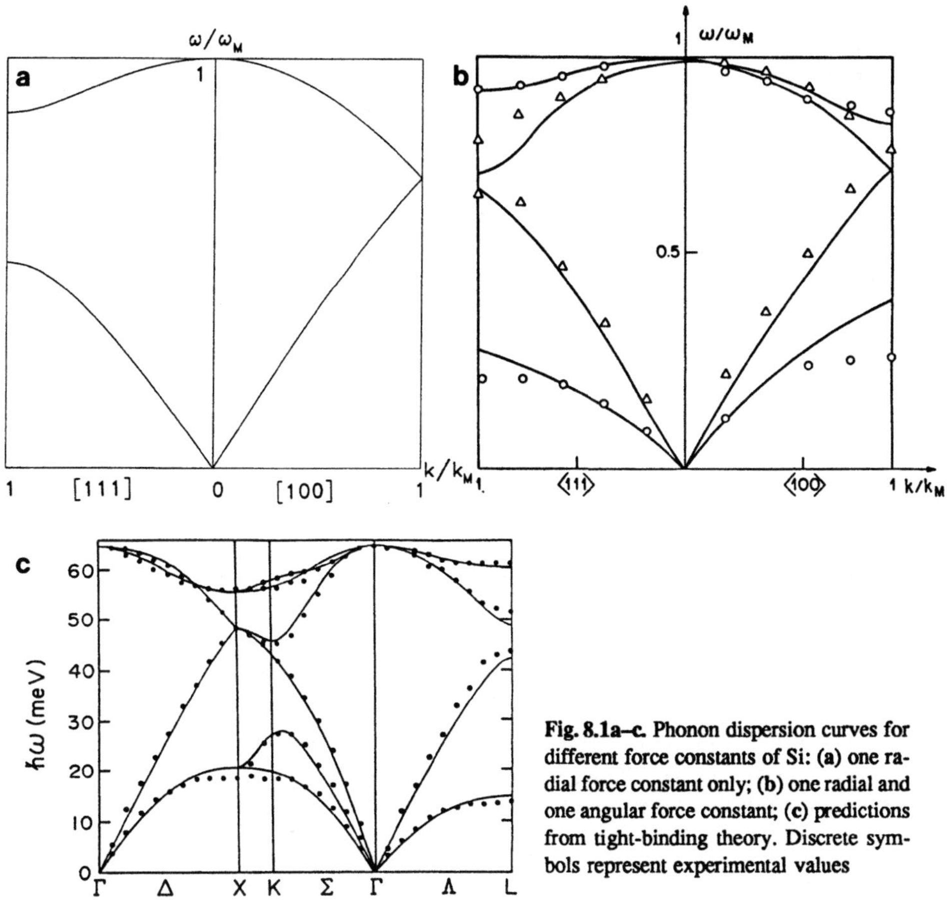

Fig. 8.1a–c. Phonon dispersion curves for different force constants of Si: (**a**) one radial force constant only; (**b**) one radial and one angular force constant; (**c**) predictions from tight-binding theory. Discrete symbols represent experimental values

consisting of pure angular modes (since all $dr_{ij} = 0$). An obvious improvement to this basic description comes from the inclusion of the contribution of distortions in angles between adjacent bonds whose force constant is k_θ. One can even retain the simplicity of the previous analytic description by incorporating only the diagonal terms in k_θ/M in the dynamical matrix. The corresponding contribution can be calculated to be equal to $28k_\theta/3M$ and, when added to the dynamical matrix, it simply shifts the whole phonon band structure by this amount so that there are no more states at zero frequency. The phonon dispersion curve obtained in this model for the zinc-blende structure is depicted in Fig. 8.1a, using the fact that, in this case, the quantity f of (8.26) is wavevector dependent. Its square is given by

$$f^2(\boldsymbol{k}) = 4\left(1 + \cos\frac{k_x a}{2}\cos\frac{k_y a}{2} + \cos\frac{k_y a}{2}\cos\frac{k_z a}{2} + \cos\frac{k_x a}{2}\cos\frac{k_z a}{2}\right). \quad (8.27)$$

It is also possible to calculate the full dispersion curves including the off-diagonal terms in k_θ [8.5]. These are pictured in Fig. 8.1b. The first point to note is the striking similarity with Fig. 8.1a showing directly the validity of the analytical model. The second point is that with the parameters k_r and k_θ used for this figure, and chosen to correctly reproduce the elastic constants (i.e. the long-wavelength limit), one overestimates the transverse acoustic frequencies by a factor of about 40%. This is a major failure of all relatively simple descriptions of the vibrational properties of these systems. In fact the elastic constants as well as the maximum optical frequency ω_{M} and both transverse acoustic frequencies $\omega_{\mathrm{TA}}(X)$ and $\omega_{\mathrm{TA}}(L)$ are given by the following expressions [8.5]:

$$\frac{c_{11}+2c_{12}}{3} = \frac{k_r}{4\sqrt{3}} \qquad M\omega_{\mathrm{M}}^2 = \frac{8}{3}\{k_r + 8k_\theta\}$$
$$M\omega_{\mathrm{TA}}^2(X) = 12\,k_\theta \tag{8.28}$$
$$c_{11} - c_{12} = \frac{3\sqrt{3}}{2R}k_\theta \qquad M\omega_{\mathrm{TA}}^2(L) = \frac{20}{3}\,k_\theta \,.$$

This shows clearly that any improvement in ω_{TA} will destroy the agreement with the elastic constants and vice-versa.

Finally, in Fig. 8.1c, the phonon dispersion curves obtained numerically in the refined tight-binding model of [8.7, 8] are given for comparison. They apparently do a much better job for $\omega_{\mathrm{TA}}(X)$ which is overestimated by about 10%, while the elastic constants are underestimated (for instance the bulk modulus is too small by about 20%). On the whole this is in line with equation (8.28) of the simpler model showing that no significant improvement has been introduced.

8.1.5 Bulk Dispersion Curves in Compound Semiconductors

A similar analysis can be extended to the compound semiconductors where the masses M_A and M_B of the two atoms in the unit cell are different. There are now two types of equation (8.24), one with M_A and one with M_B. Thus (8.25) has to be replaced by a set of two equations, with obvious notation

$$f_A S_{iA} = \sum_k S_{kB} \,,$$
$$f_B S_{kB} = \sum_l S_{lA} \,. \tag{8.29}$$

Eliminating S_{kB} for all k from these equations gives

$$f_A f_B S_{iA} = \sum_{k,l} S_{lA} \,, \tag{8.30}$$

where the sum over l is over the nearest neighbors of k. The same procedure applied to the covalent situation would lead to exactly the same equations but with $f_A f_B$ replaced by f^2. This means that one has

$$f_A f_B = f^2 \,, \tag{8.31}$$

or in other words that the broad bands are solutions of:

$$\left(4-\frac{3M_A\omega^2}{k_r}\right)\left(4-\frac{3M_B\omega^2}{k_r}\right)=f^2\,. \tag{8.32}$$

As before f^2 lies in the interval $[0,16]$. When $f=0$ (i.e. at the X point) there are now two solutions of (8.32) given by $4k_r/3M_A$ and $4k_r/3M_B$ instead of one doubly degenerate solution. This means that a gap opens in the phonon band structure. To complete the picture one can add, as before, the diagonal terms $28k_{\theta_A}/3$ on the A atoms and $28k_{\theta_B}/3$ on the B atoms, still retaining an analytical approach.

8.2 Surface Phonons

In this section we discuss some characteristic properties of surface phonons. We first perform a general moment analysis of the surface phonon density of states along the lines of Sect. 2.3.3. We then apply these considerations to the determination of the mean square displacements of surface atoms and of the surface entropy. We also consider the long-wavelength limit for which the existence of Rayleigh waves is directly obtained by classical elasticity theory. Finally we discuss recent theoretical results obtained for Si(111) 2 × 1 and Si(100).

8.2.1 Moments of the Local Vibrational Density of States

We have seen before that the vibrational frequencies are obtained by diagonalization of the dynamical matrix. This problem, when expressed in terms of the cartesian components of the atomic displacements ($u_{i\alpha}$ or $v_{i\alpha}$ in Sect. 8.1.1) i.e. in terms of three basis vectors $e_{i\alpha}$ per atom, is strictly equivalent to a tight-binding electronic problem with three p-states per atom. Then all considerations developed in Sect. 2.3.3 in the TB context can be directly applied with a slight change in notation. For instance, one can define a resolvent matrix G as in (2.22):

$$G=(\omega^2 I-D)^{-1}\,, \tag{8.33}$$

and a local phonon density of states $n_{i\alpha}(\omega^2)$ given, as in (2.47), by

$$n_{i\alpha}(\omega^2)=-\frac{1}{\pi}\,\mathrm{Im}\left\{G_{i\alpha,i\alpha}(\omega^2)\right\}\,. \tag{8.34}$$

It is also quite instructive to perform a moment analysis from the general expression (2.45) transposed to the phonon case. In particular, the first moment $\mu_{1,i\alpha}$ is simply given by

$$(\mu_1)_{i\alpha}=D_{i\alpha,i\alpha}=\frac{A_{i\alpha,i\alpha}}{M_i}\,, \tag{8.35}$$

and the second moment is

$$(\mu_2)_{i\alpha} = \sum_{j,\beta} \frac{|A_{i\alpha,j\beta}|^2}{M_i M_j} \,. \tag{8.36}$$

Some general conclusions about $n_{i\alpha}(\omega^2)$ near surfaces can be drawn from these expressions. First of all, the center of gravity of the local phonon density of states on surface atoms is generally lower than that for bulk atoms. This is also true of the second-order moment $\mu_2 - \mu_1^2$ which is a measure of the width of this density of states. This can be easily verified using simple force constant models such as the nearest-neighbor isotropic model defined by (8.13) and (8.14). In such a case one obtains for a given atom i

$$\begin{aligned} \mu_{1,i\alpha} &= \frac{Z_i k}{M} \,, \\ \mu_{2,i\alpha} - \mu_{1,i\alpha}^2 &= \frac{Z_i k^2}{M^2} \,, \end{aligned} \tag{8.37}$$

where we have assumed equal masses for all atoms and Z_i is the coordination number of atom i. This quantity is smaller for surface atoms than in the bulk so that both moments are smaller. It is thus likely that, on this basis there should be no localized surface modes at high frequencies above the maximum one. On the other hand localized modes at lower frequencies are more probable. We shall see (Sect. 8.2.4) that a general related property is the existence of Rayleigh waves at acoustic frequencies.

8.2.2 Mean Square Displacement of Surface Atoms

To obtain the general expression of this quantity let us derive it for a single harmonic oscillator of mass M, eigenfrequency ω and lattice displacement u. The mean square displacement $\langle u^2 \rangle$ is the thermal average of this quantity over the eigenstates. A general property of the harmonic oscillator is that the average of the potential energy $(1/2)M\omega^2\langle u^2\rangle$ is one-half of the average of the total Hamiltonian, which gives

$$\tfrac{1}{2} M\omega^2 \langle u^2 \rangle = \tfrac{1}{2}\left(\bar{n} + \tfrac{1}{2}\right)\hbar\omega \,. \tag{8.38}$$

where $\bar{n}$ is the average occupation number

$$\bar{n} = \frac{1}{\exp(\hbar\omega/kT) - 1} \,. \tag{8.39}$$

This leads directly to the following expression for $\langle u^2 \rangle$:

$$\langle u^2 \rangle = \frac{\hbar}{2M\omega} \coth \frac{\hbar\omega}{2kT} \,. \tag{8.40}$$

Such an expression can be generalized to any system involving several normal modes of vibration. We start from the definition (8.3) of $v_{i\alpha} = \boldsymbol{v} \cdot \boldsymbol{e}_{i\alpha}$, expand $\boldsymbol{v}$ in terms of normal modes $\boldsymbol{v}_k$, with components v_k, and write the generalization of (8.38) as

$$\frac{1}{2}\langle v_{i\alpha}^2\rangle = \frac{1}{2}\sum_k (e_{i\alpha}\cdot v_k)^2\langle v_k^2\rangle$$
$$= \frac{1}{2}\sum_k (e_{i\alpha}\cdot v_k)^2\left(n_k+\frac{1}{2}\right)\frac{\hbar}{\omega_k}\,. \qquad (8.41)$$

Using the relation (8.3) we then express $\langle u_{i\alpha}^2\rangle$ as

$$\langle u_{i\alpha}^2\rangle = \frac{\hbar}{2M_i}\sum_k \frac{(e_{i\alpha}\cdot v_k)^2}{\omega_k}\coth\frac{\hbar\omega_k}{2kT}\,. \qquad (8.42)$$

From Sect. 2.3.3 and the notation defined above, the local phonon density of states $n_{i\alpha}(\omega^2)$ may be defined as

$$n_{i\alpha}(\omega^2) = \sum_k (e_{i\alpha}\cdot v_k)^2\delta(\omega^2-\omega_k^2)\,, \qquad (8.43)$$

so that the mean square displacement can be expressed in the form

$$\langle u_{i\alpha}^2\rangle = \frac{\hbar}{2M_i}\int n_{i\alpha}(\omega^2)\coth\frac{\hbar\omega}{2kT}\,\frac{d\omega^2}{\omega}\,. \qquad (8.44)$$

This expression is strictly equivalent to the one derived in [8.10]. In principle it requires a full knowledge of the local phonon density of states $n_{i\alpha}(\omega^2)$. It can however be estimated quite efficiently by making use of an Einstein model approximation. This consists in replacing $n_{i\alpha}(\omega^2)$ by a delta function located at a value of ω^2 equal to the first moment $(\mu_1)_{i\alpha}$ of the exact curve. Defining an average frequency $\bar{\omega}_{i\alpha}$ by

$$\bar{\omega}_{i\alpha} = \sqrt{(\mu_1)_{i\alpha}}\,, \qquad (8.45)$$

we obtain for the Einstein model

$$\langle u_{i\alpha}^2\rangle = \frac{\hbar}{2M_i\bar{\omega}_{i\alpha}}\coth\frac{\hbar\bar{\omega}_{i\alpha}}{2kT}\,, \qquad (8.46)$$

as for a single oscillator in (8.40). Such a straightforward expression gives a very simple and useful answer as discussed some time ago in [8.11]. Let us derive it explicitly for tetrahedral semiconductors in the valence force field model discussed above. Dropping first the terms in k_θ and keeping only k_r, the diagonal term of the dynamical matrix is given in this model by

$$D_{i\alpha,i\alpha} = \frac{k_r}{M_i}\sum_j \alpha_{ij}^2\,, \qquad (8.47)$$

in the notation given above. For a bulk atom with tetrahedral coordination $\sum_j \alpha_{ij}^2$ is equal to 4/3 for all α. Thus $\langle u_{i\alpha}^2\rangle$ is isotropic and given by (8.46) with $\bar{\omega}_{i\alpha}$

replaced by ω_b. The latter is given by

$$\omega_b = \sqrt{\frac{4}{3}\frac{k_r}{M_i}} . \tag{8.48}$$

For atoms at the unreconstructed (111) surface, for instance, one can classify the atomic displacements by symmetry: $u_\perp$ perpendicular to the plane and $u_\parallel$ parallel to it (one has isotropy in the plane). The $\langle u_\perp^2 \rangle$ and $\langle u_\parallel^2 \rangle$ are again given by (8.46) with frequencies $\bar{\omega}_{i\alpha}$ of

$$\begin{aligned} \omega_\perp &= \frac{\omega_b}{2} , \\ \omega_\parallel &= \omega_b . \end{aligned} \tag{8.49}$$

Thus $\langle u_\parallel^2 \rangle$ is practically unchanged with respect to the bulk value, while $\langle u_\perp^2 \rangle$ is about twice as large at low temperature, and about four times as large at higher temperatures where $\coth(\hbar\omega/2kT) \sim kT/2\hbar\omega$. This is in line with the findings of [8.11]. This behavior should remain qualitatively valid along the dangling bond direction on reconstructed (111), (100) or (110) surfaces.

8.2.3 Surface Entropy

A fundamental quantity in the derivation of equilibrium configurations as a function of temperature T is the entropy S, which is defined as

$$S = -\left(\frac{\partial F}{\partial T}\right)_V , \tag{8.50}$$

where V is the volume and F the free energy:

$$F = -kT \ln Z . \tag{8.51}$$

For a simple harmonic oscillator the partition function takes the form

$$Z = \frac{\exp(-\hbar\omega/kT)}{1 - \exp(-\hbar\omega/kT)} . \tag{8.52}$$

From these definitions one obtains directly the vibrational contribution to the entropy

$$\frac{S}{k} = \frac{\hbar\omega/kT}{\exp(\hbar\omega/kT) - 1} - \ln\left[1 - \exp\left(-\frac{\hbar\omega}{kT}\right)\right] . \tag{8.53}$$

This can easily be generalized to the case of several vibrational modes. As the partition function is a product of individual partition functions, the entropy will be a sum over eigenfrequencies which, for our system, can be replaced by an integral:

$$\frac{S}{k} = \int \left[\frac{\hbar\omega/kT}{\exp(\hbar\omega/kT) - 1} - \ln\left(1 - \exp\left(-\frac{\hbar\omega}{kT}\right)\right)\right] n(\omega^2) d\omega^2 , \tag{8.54}$$

where $n(\omega^2)$ is the total vibrational density of states. This total entropy can be expressed as a sum of local contributions,

$$S = \sum_{i,\alpha} S_{i\alpha} \,, \tag{8.55}$$

where each $S_{i\alpha}$ takes the form (8.54) but with the total density of states $n(\omega^2)$ replaced by the local one $n_{i\alpha}(\omega^2)$. Once again these local contributions can be estimated rather simply with the help of the Einstein model. With the same notation as in the previous section, and working at high temperatures we find

$$\frac{S_{i\alpha}}{k} \sim 1 - \ln \frac{\hbar \bar{\omega}_{i\alpha}}{kT} \,. \tag{8.56}$$

The meaningful quantity to calculate here is the change in entropy ΔS_i between a surface and a bulk atom. It is the sum of contributions from the three cartesian coordinates. If there are symmetry axes it is better to choose the cartesian coordinates along these axes so that they will correspond to truly uncoupled displacements. For instance, in the case of the unrelaxed Si(111) surface, it is better to choose α either perpendicular ($\perp$) or parallel ($\|$) to the surface. Thus ΔS per surface atom will be the sum of two contributions,

$$\Delta S = \Delta S_{\perp} + 2\Delta S_{\|} \,. \tag{8.57}$$

When using the high temperature expression (8.56) for calculating ΔS all terms cancel except the term in $\ln \bar{\omega}_{i\alpha}$. Making use again of the valence force field model discussed previously ($k_r \neq 0$, $k_\theta = 0$) we find, since $\omega_{\perp} = \omega_b/2$ and $\omega_{\|} = \omega_b$,

$$\Delta S_{\perp} = k \ln 2 \sim 0.7\, k \,, \tag{8.58}$$

which can be considered as the contribution of one broken (or dangling) bond to the entropy. This is in good agreement with what has been found for vacancies in similar conditions ($\sim 2.8\, k$ for 4 dangling bonds) using more refined Green's function treatments [8.2].

8.2.4 Long-Wavelength Acoustic Modes: Rayleigh Waves

The general tendency, discussed in Sect. 8.2.1, for a lowering of vibrational frequencies near surfaces is confirmed by the existence of localized surface waves at frequencies below the bulk acoustic branch. In the long-wavelength limit these have been known for a very long time as "Rayleigh waves" [8.13]. They can in principle be obtained by direct diagonalization of the dynamical matrix taking the $\omega^2 \to 0$ limit, which also corresponds to the long-wavelength case. This is completely analogous to the use of the effective mass equation in the electronic problem near the extremum of a band.

When treating the bulk problem in the long-wavelength (or elastic) limit one is led to consider the displacement $\boldsymbol{u}_i$ of an atom i as a slowly varying function (essentially constant over the crystal unit cell) of the atomic position $\boldsymbol{R}_i$ so that one can treat it as a continuous function of that position. This is equivalent to searching for solutions of the kind

$$u_i = u \exp(\mathrm{i} q \cdot R_i) . \tag{8.59}$$

In such a case the dynamical matrix at fixed q becomes a 3×3 matrix in the general case. Its eigenvalues ω^2, when expanded to second order in q, give the three acoustic branches. This procedure is strictly equivalent to starting from elasticity theory and gives the relation between the elastic constants and the force constant matrix.

Similar considerations also hold in the case of the surface. However the boundary conditions obviously become different: vibrational waves can still propagate in directions parallel to the surface (with a wavevector $q_{\parallel}$) but not in the perpendicular direction where they are either reflected at the surface or decay exponentially away from it. We are interested in the latter case which represents truly localized states at fixed $q_{\parallel}$. However there is a more convenient procedure in the long-wavelength limit which is to make direct use of elasticity theory. If one assumes that the semi-infinite system extends in the positive z-direction then, away from the surface, one has to search for solutions of the bulk equations that decrease exponentially. These must take the form

$$u_i = u \exp(\mathrm{i} q_{\parallel} \cdot R_i - Kz) , \tag{8.60}$$

or, in other words, they are bulk solutions of the form (8.59) but with complex wavevector $q = (q_{\parallel}, \mathrm{i}K)$. Thus the localized state is, in general, a linear combination of the three possible solutions at fixed $q_{\parallel}$ and the same value of ω^2 (corresponding to different decay constants K). The values of the unknown coefficients and of ω^2 are then obtained by imposing the condition that the solution obeys the correct boundary conditions. These boundary conditions can be obtained either from the detailed form of the dynamical matrix near the surface or, as is much simpler but equivalent, from the general conditions of elasticity theory. This procedure is detailed in [8.14] for the case of isotropic systems, whose dispersion relations are

$$\begin{aligned} \omega_l(q) &= c_l |q| , \\ \omega_t(q) &= c_t |q| , \end{aligned} \tag{8.61}$$

which correspond to pure longitudinal and pure transverse waves respectively, the latter being two-fold degenerate. Writing the frequency of the localized state as

$$\omega_R = C_t |q_{\parallel}| \xi , \tag{8.62}$$

the derivation of [8.14] shows that ξ must be the solution of the following equation:

$$\xi^6 - 8\xi^4 + 8\xi^2 \left(3 - \frac{C_t^2}{C_l^2}\right) - 16 \left(\frac{C_t^2}{C_l^2}\right) = 0 \tag{8.63}$$

satisfying $\xi < 1$ if one wants to get a truly localized state. It is found that ξ varies between 0.86 and 0.96 in the normal range of values for $(C_t/C_l)^2$, which means that there always exist Rayleigh waves in the isotropic systems of interest.

8.2.5 Recent Results for Si(111) and Si(100)

We summarize here recent studies on these two surfaces of silicon [8.7, 8] which have been performed by total energy tight-binding expansions as discussed in Sect. 8.1.3. Phonon frequencies are calculated for a ten layer slab. This thickness is usually enough to prevent interactions between the two opposite surfaces except for the long-wavelength Rayleigh modes which cannot be obtained correctly in the $q_{\parallel} \to 0$ limit, since their decay becomes infinitely slow in the z-direction.

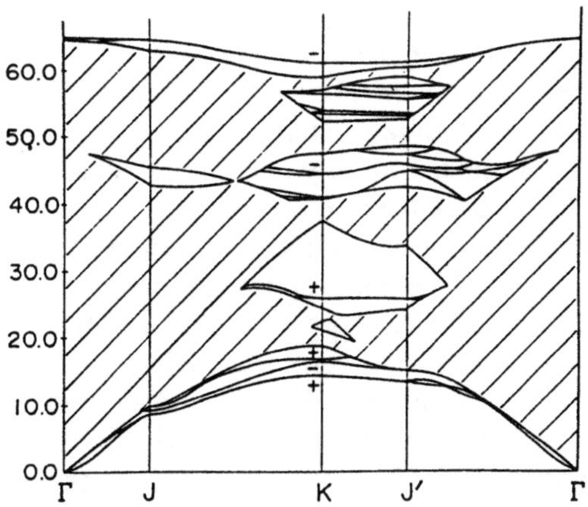

Fig. 8.2. Surface phonons for the asymmetric dimer model of the Si(001) 2 × 1 surface (*solid curves*) and projected bulk vibrations (*hatched*) [8.7]

We reproduce the results of [8.7] for Si(100) in Fig. 8.2. One of the most important features is the existence of several branches of localized modes below the bulk acoustic modes at $q_{\parallel}$. At relatively short wavelength these are mostly localized on the surface atoms and the lowering in vibrational frequency corresponds to the reduced coordination of the surface atoms, for the reasons discussed in Sect. 8.2.1. At higher wavelength these branches should tend towards the Rayleigh wave predicted by classical elasticity which however are absent in the slab model. There are other localized states in the different gaps that open at $q_{\parallel}$. However the most surprising result is the existence of localized states at frequencies higher than the bulk continuum, which seems to contradict the fact that the average frequency near the surface is lowered. However *Allan* and *Mele* [8.7] have analyzed the nature of this mode in detail and have shown that it has the largest amplitude on fully tetrahedrally coordinated atoms belonging to the first and second subsurface layers.

The phonon dispersion curves for Si(111) 2 × 1, in the π-bonding chain model discussed in Sect. 4.3.2, have also been calculated [8.8]. In this case an on-site electron-repulsion term has been added to the tight-binding Hamiltonian

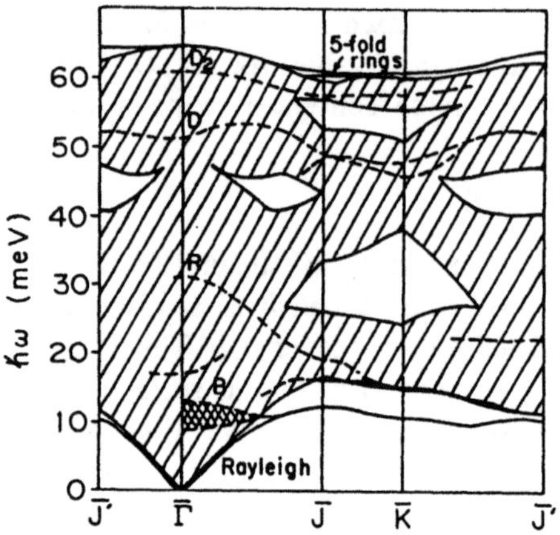

Fig. 8.3. Phonon dispersion curves for the π-bonded chain on Si(111) 2×1. Solid lines are surface phonons and dashed lines surface resonances. The hatched regions represent the projected bulk modes, and the dashed area near the zone center corresponds to a broad surface resonance [8.8]

described previously, which was adjusted to give the measured value for the gap between electronic surface states. We reproduce the phonon dispersion curves in Fig. 8.3. As for the (100) surface one gets localized modes at high frequencies above the continuum. Again they are attributed to fully coordinated atoms but are also connected to the stretching of the back-bonds that participate in the subsurface five-fold rings. In fact this was also the case for the (100) surface and can thus be considered as an intrinsic feature of five-fold rings which should probably characterize them in other situations as well. Apart from the identification of dimer modes, the next important characteristic concerns the localized acoustic branches which involve the vibrations of surface atoms. However, away from the center of the zone, this localized mode becomes essentially dispersionless. This feature is attributed to the strong polarization effects occurring near the surface and is in good agreement with the results of scattering experiments [8.15] which were previously attributed to an Einstein oscillator.

8.3 Phonon-Related Properties

We now consider different physical properties that are more or less directly connected to phonons. In the first section we examine the asymptotic form of the displacement field away from surfaces and show that one quite generally expects oscillatory relaxation or reconstruction. We then give an estimate of the effective force constants that have been used in Sect. 4.3.2. Finally we discuss the electron–lattice interaction at the Si(111) surface.

8.3.1 Oscillatory Relaxation or Reconstruction Near Surfaces

Relaxation or reconstruction does not mean that only the surface atoms are displaced; the displacement field also extends away from the surface. In particular, the existence of oscillatory perpendicular relaxation was established theoretically a long time ago for transition metals [8.16] and was then demonstrated experimentally for many metal surfaces [8.17–21]. There have also been detailed calculations on specific metal surfaces [8.21–24] but it is interesting to give a general and systematic analysis of this phenomenon. For this let us consider a simple case of perpendicular relaxation consisting of a semi-infinite system of equally spaced planes. Its elastic energy per atom is written in the form

$$E = \alpha d_{1,2} + \frac{\beta}{2}\sum_{i=1}^{\infty} d_{i,i+1}^2 + \gamma \sum_{i=1}^{\infty} d_{i,i+1} d_{i+1,i+2} , \tag{8.64}$$

with $\beta, \gamma > 0$ and where the $d_{i,i+1}$ are the changes in distance between planes i and $i+1$ with respect to the bulk values. The term linear in $d_{1,2}$ corresponds to forces due to the breaking of bonds at the surface. The bulk phonon dispersion curves associated with (8.64) in directions perpendicular to the planes are given by

$$M\omega^2(k) = (1 - \cos ka)(\beta + 2\gamma \cos ka) , \tag{8.65}$$

where M is the atomic mass, a the eqilibrium interplanar spacing and k the wave number. From (8.65) the stability of the system requires $\beta > 2\gamma$. The stable positions are then obtained by minimizing (8.64). One finds

$$\begin{aligned} 0 &= \alpha + \beta d_{1,2} + \gamma d_{2,3} \\ 0 &= \beta d_{n,n+1} + \gamma(d_{n-1,n} + d_{n+1,n+2}) \quad \text{for} \quad n > 1 . \end{aligned} \tag{8.66}$$

The second set of equations (8.66) are bulk equations whose general solutions are

$$d_{n,n+1} = d_{1,2}\, e^{ik(n-1)a} , \tag{8.67}$$

$$\cos ka = -\frac{\beta}{2\gamma} . \tag{8.68}$$

As $\beta > 2\gamma$, the only solution of (8.68) corresponds to a complex number

$$k = \frac{\pi}{a} + i \cosh^{-1}\left(\frac{\beta}{2\gamma}\right) . \tag{8.69}$$

This shows that the asymptotic relaxation is exponentially decreasing and oscillatory. The complete solution can be found by inserting (8.67) together with (8.69) into the first of equations (8.66). The solution given by (8.67) and (8.68) corresponds to the bulk dispersion relation (8.65) for complex wave vectors (other than $\mathbf{0}$) such that $\omega^2(k) = 0$. This is completely general and can be extended to study the asymptotic behavior not only of the relaxation but also of the reconstruction that frequently occurs at surfaces. To show this let us start from the general expansion of the energy in terms of the atomic displacements u_i:

$$E = -\sum_i F_i u_i + \tfrac{1}{2}\sum_{i,j} A_{ij} u_i u_j , \tag{8.70}$$

where the F_i are forces located in the vicinity of the surface and A_{ij} is the matrix of force constants. The stable positions are given by

$$F_i = \sum_j A_{ij} u_j \; . \tag{8.71}$$

Away from the surface the F_i vanish and the A_{ij} tend to A^0_{ij}, the bulk force constants. On the other hand, the bulk dispersion relation is obtained from

$$M_i \omega^2 u_i = \sum_j A^0_{ij} u_j \; . \tag{8.72}$$

Comparison of the asymptotic form of (8.71), $0 = \sum_j A^0_{ij} u_j$, and the bulk dispersion relation leads to the conclusion that this asymptotic behavior of the u_i must correspond to a combination of bulk solutions of zero frequency. If we write the phonon dispersion curves as $\omega(k_\perp, \boldsymbol{k}_\parallel)$, where $\boldsymbol{k}_\parallel$ is the wavevector along the surface and $k_\perp$ its component in the perpendicular direction, any type of reconstruction plus relaxation corresponds to a displacement field that is periodic along the surface (i.e. corresponds to a fixed value of $\boldsymbol{k}_\parallel$) and exponentially decaying in the perpendicular direction. From the discussion above we thus have to find the solutions of

$$\omega^2(k_\perp + \mathrm{i}K, \boldsymbol{k}_\parallel) = 0 \; , \tag{8.73}$$

obtained by working out the phonon dispersion relation at complex perpendicular wavevectors. The general asymptotic displacements will be a linear combination of the independent solutions of (8.73) in which the unknown coefficients will be obtained by matching to the short-range solution in the perturbed region.

The practical procedure is to determine the real solutions $\omega^2(k_\perp + \mathrm{i}K, \boldsymbol{k}_\parallel)$ in the complex plane as a function of K (note that this automatically fixes $k_\perp(K)$ by the condition $\mathrm{Im}\{\omega^2\} = 0$). There is a general property that the branching point $k_\perp(K = 0)$ onto the real band structure must be an extremum of $\omega^2(k_\perp, k_\parallel)$ with respect to $k_\perp$ [8.25]. From a branching point one can follow the real branch of ω^2 for complex wavevectors to find solutions of (8.73) (an example of this is given in Fig. 8.4). These solutions are, in general oscillatory ($k_\perp \neq 0$) and exponentially damped away from the surface. In Table 8.2 we give the allowed values of the quantity $\exp \mathrm{i}(k_\perp + \mathrm{i}K)$ for some low index surfaces of Ni and Fe. For Ni(110) the multilayer relaxation has recently been measured [8.27] and the ratio of the changes in the first two interlayer spacings is -0.38 (Table 8.2). Another case is the bcc Fe(111) surface where the first four interlayer spacings have been determined from LEED data [8.28]. Figure 8.5 shows a least squares fit of the asymptotic expression $u_0 \exp(-nKd) \cos(nk_\perp d + \varphi)$ where $k_\perp$ and K are taken from Table 8.2. The agreement is quite good.

The conclusion of this section is that multilayer relaxation is related to solutions of zero frequency in the complex phonon structure. Surprisingly, such an

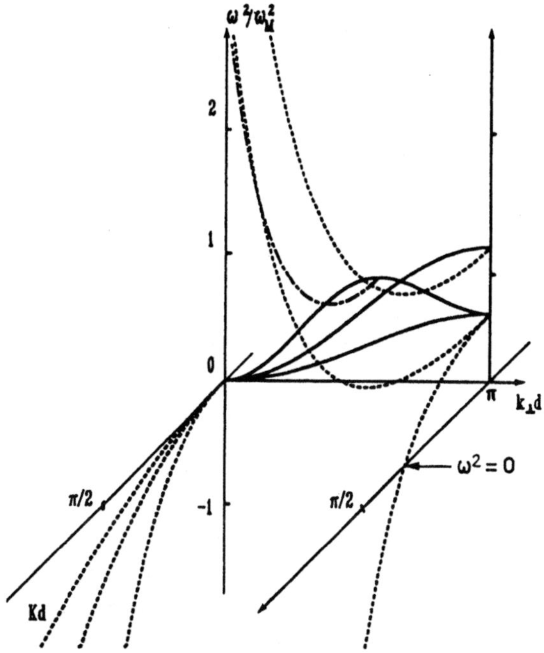

Fig. 8.4. Extension of nickel phonon dispersion curves in the complex k-plane. Full lines are in the real k space and dashed lines in the complex k space. $k_\perp$ is normal to the (110) plane. ω_M is the maximum frequency in bulk nickel

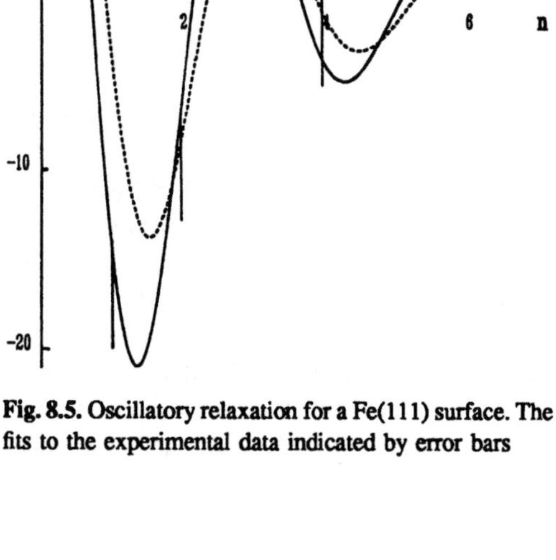

Fig. 8.5. Oscillatory relaxation for a Fe(111) surface. The solid and dashed lines represent two possible fits to the experimental data indicated by error bars

Table 8.2. Allowed values of $y = \exp i(k_\perp + iK)d(k_\parallel = 0)$ for low index surfaces of fcc Ni and bcc Fe as obtained from the force constants of [8.26] ($u_\parallel$ gives the direction of atomic displacements)

Surface	$u^1_\parallel$	y	$u^2_\parallel$	y	$u^3_\parallel$	y
(100) Ni	(010)	0.007	(001)	0.007	(100)	0.013
(100) Fe		0.026		0.026		−0.187
(110) Ni	(110)	0.031	(001)	0.007	(110)	−0.376
(110) Fe		0		0		0
(111) Ni	$(1\bar{1}0)$	0	$(11\bar{2})$	0	(111)	0
(111) Fe		−0.095 ±i0.089		−0.095 ±i0.089		−0.322 ±i0.527

asymptotic expression is often valid close to the surface and could be very useful in the analysis of LEED data.

8.3.2 Effective Force Constants

To introduce this notion let us consider an expression of the lattice energy taking the form of (8.70) but with only one non-vanishing force F_1 corresponding to $i = 1$. The equilibrium situation is obtained by minimizing (8.70) with respect to all u_i which gives directly

$$u_i = (A^{-1})_{i1} F_1 \tag{8.74}$$

so that the total minimum energy is given by

$$E_m = -\tfrac{1}{2}(A^{-1})_{11} F_1^2 . \tag{8.75}$$

For u_1 and E_m this is exactly the same answer as if all u_i were kept equal to zero with the exception of u_1, in which case one would get

$$E = -F_1 u_1 + \tfrac{1}{2} A_{11} u_1^2 , \tag{8.76}$$

leading, after minimization, to

$$u_1 = \frac{F_1}{A_{11}} ; \quad E_m = -\frac{1}{2}\frac{F_1^2}{A_{11}} . \tag{8.77}$$

The only difference is that the direct force constant $k = A_{11}$ is replaced by an effective force constant $k^* = (A^{-1})_{11}$.

Let us now consider the case of Si(111) which was discussed in Sect. 4.3.2. The general problem is to determine the effective force constant corresponding to the axial displacement of a trivalent silicon atom, as pictured in Fig. 8.6. We use the same notation as in Sect. 4.3.1 and first calculate the direct force constant, i.e. we keep all atoms fixed except the trivalent one which experiences an axial displacement u. We make use of the valence force field model of (4.36). We have already seen in Sect. 8.1.4 that the radial contribution to such an axial displacement is equal to $k_r/3$. The angular contribution is the same as in the bulk,

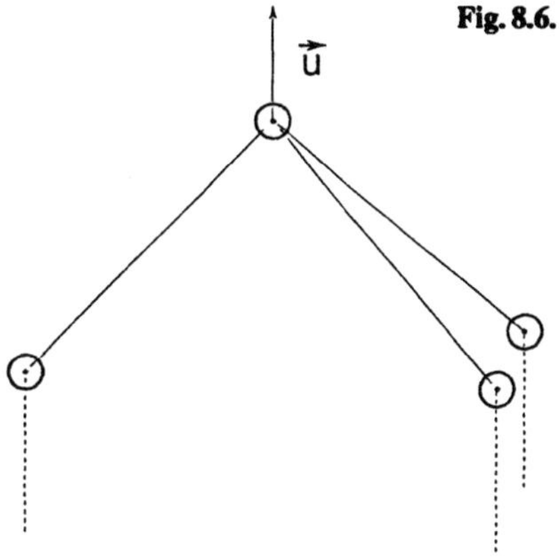

Fig. 8.6. Outward displacement of a tri-coordinated Si atom

i.e. $28\,k_\theta/3$ if one assumes that the angular force constant between the dangling bond and its back-bonds is unchanged with respect to the bulk one between two adjacent bonds:

$$k = \frac{k_r + 28\,k_\theta}{3}\,. \tag{8.78}$$

Taking for k_r and k_θ the values from [8.2], which fit the maximum optical frequency and the transverse acoustic frequency, one gets $k \sim 5.1\,\mathrm{eV/Å^2}$. This is in principle the value for one isolated trivalent atom with the back-bond atoms fixed. If relaxation of further neighbors is allowed for the isolated trivalent atom, then one would get an effective force constant k^* which is much smaller (by more than a factor of 2 as shown in [8.2]). This is due to the fact that the long-range displacement field allows the system to relax by almost pure angular distortions which cost much less energy than bond stretching. However this is not the case at the Si(111) surface where all surface atoms are trivalent and where there is collective motion of these atoms. For instance, in a pure relaxation where only surface atoms are allowed to vary, the direct force constant should be the same as given by (8.78) except for some interference effects coming from angular terms which add $8\,k_\theta/3$ to k. This would increase the direct value of k to 5.6 eV, close to the 6 eV used in Sect. 4.3.2. The displacement field decays rapidly since the lattice energy takes the form (8.64) with γ only involving angular terms. These are much weaker than the radial terms, so that the exponential decrease is very fast. Thus the effective force constant k^* in that case should not be much smaller than k. Similar arguments hold for a motion of the surface atom leading to a buckled 2×1 reconstruction. However, interference effects will compensate and only slightly reduce k with respect to the individual case, giving $k \sim 4.9\,\mathrm{eV}$. Again k^* will be close to this value.

8.3.3 The Electron–Lattice Interaction

This section is motivated by the optical absorption data for the Si(111) 2×1 surface [8.29]. Figure 8.7 shows the experimental curve for different temper-

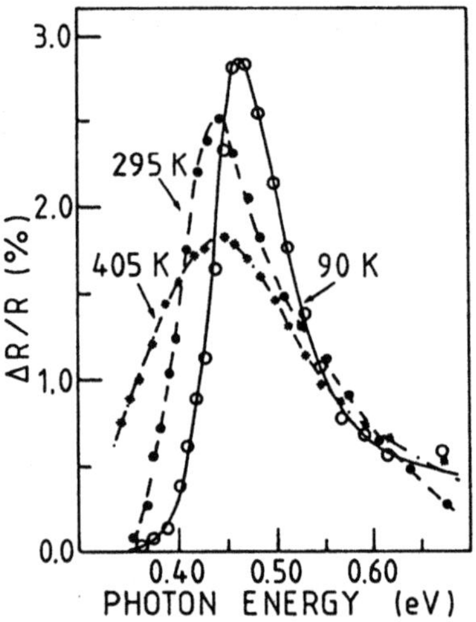

Fig. 8.7. Differential reflectivity as a function of photon energy in Si(111) 2×1 at three different sample temperatures

atures. This absorption occurs between the surface states and the evolution of its shape is similar to what happens for point defects where the electron–lattice interaction broadens the optical absorption curves. A detailed discussion of this effect is given in [8.30] and references therein. We shall only recall here the basic features of this phenomenon.

We consider a localized electronic state containing one electron. We suppose that there is an electron–lattice coupling to one local coordinate Q. Near the stable position $Q = 0$ the electronic energy takes a parabolic form. Upon excitation of the electron into an upper excited state the electronic energy becomes another parabola, eventually centered on a different position Q_e. The corresponding configuration coordinate diagram is pictured in Fig. 8.8. The vibrational solutions in the two parabolic wells are harmonic oscillator states of energies $(n + 1/2)\hbar\omega$ where ω is the phonon frequency. The eigenstates of the total Hamiltonian are Born–Oppenheimer products of the electronic part and the harmonic oscillator wave functions. The optical transition, instead of occurring between two electronic levels is now possible between all states from the lower potential well to the upper one. Each such transition has to be weighted by the appropriate probability. This gives rise to the optical lineshape function discussed in [8.30]. One can readily obtain important characteristics of this function by the method of moments. The moment of order k of the normalized curve can be written as

$$\mu_k = \sum_n p_n \langle n|(H' - H)^k|n\rangle \quad \text{only if} \quad k = 1, 2\,. \tag{8.79}$$

If the two parabolas have the same curvature, $k = d^2E/dQ^2$, the difference $H' - H$ is equal to

$$H' - H = h\bar{\nu} - FQ\,, \tag{8.80}$$

where $h\bar{\nu}$ is the energy of the vertical optical transition at $Q = 0$ and F is the

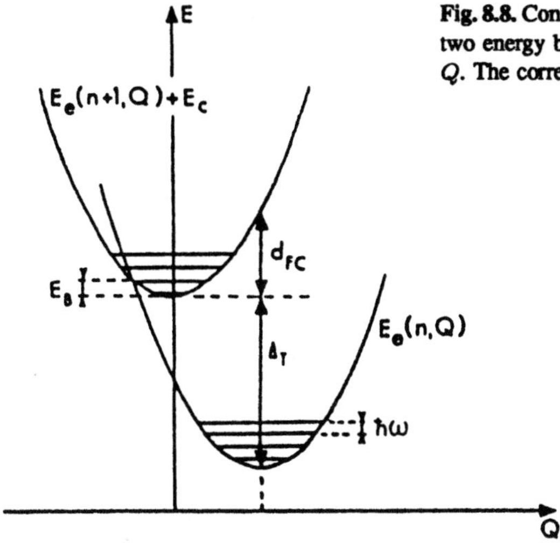

Fig. 8.8. Configuration coordinate diagram representing the two energy branches versus the atomic motion coordinate Q. The corresponding phonon energy is $\hbar\omega$

force on the displacement mode Q in the excited state. One then gets from (8.79)

$$\begin{aligned} \mu_1 &= h\bar{\nu} \,, \\ \mu_2 - \mu_1^2 &= F^2\overline{Q^2} \,, \end{aligned} \tag{8.81}$$

where $\overline{Q^2}$ is the mean square displacement. This can be easily estimated from (8.38) and we find:

$$F^2\overline{Q^2} = (2\bar{n}+1)\frac{F^2}{2k}\hbar\omega \,. \tag{8.82}$$

Now $F^2/2k$ is the relaxation energy in the excited state which is often written $S\hbar\omega$, where S is the Huang-Rhys factor. One thus obtains:

$$\mu_2 - \mu_1^2 = (2\bar{n}+1)S(\hbar\omega)^2 = S(\hbar\omega)^2 \cot\frac{\hbar\omega}{2kT} \,. \tag{8.83}$$

Of course the model discussed above is oversimplified since in general there is a continuous distribution of modes but it still remains valid if ω is given the meaning of a suitable average frequency. This allows a straightforward analysis of the experimental data since $\sqrt{\mu_2 - \mu_1^2}$ is a measure of the width of the spectrum. From (8.83) this quantity behaves as $\sqrt{2S\hbar\omega kT}$ at high temperatures. The plot of the experimentally measured linewidth [8.29] shown in Fig. 8.9, shows that this is indeed the case. Furthermore assuming that the relation between $\mu_2 - \mu_1^2$ and the width is the one given by a Gaussian one gets $S\hbar\omega \sim 0.09\,\mathrm{eV}$. This corresponds to a relatively strong electron–phonon interaction.

These results are evidence that the optical excitation must be localized in real space. As discussed in [8.29] this may occur through polaronic effects.

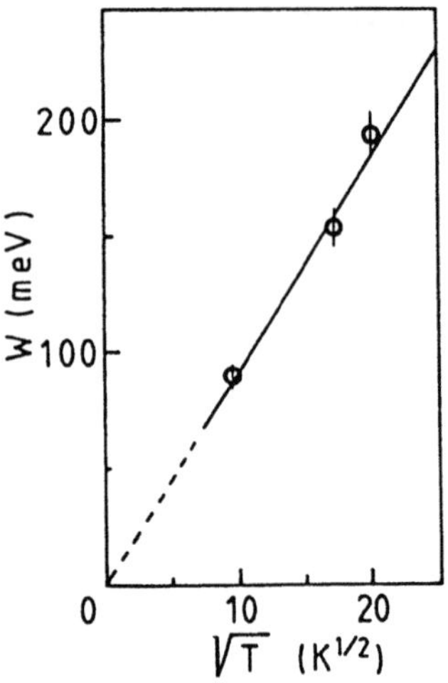

Fig. 8.9. The full width at half maximum of the curves of Fig. 8.7 vs the square root of the absolute temperature [8.29]

Exercises

8.1 *Phonons on a linear chain*: Consider a linear chain of identical atoms of mass M with interatomic separation a and atomic displacements along the axis of the chain. Show that the use of a nearest neighbor approximation with condition (8.13) allows one to express the force-constant matrix in terms of one parameter. Deduce the phonon dispersion curves in this model

8.2 *Rayleigh surface waves*: Start from classical elasticity theory, i.e., the equation $(\partial^2 u/\partial t^2) - c^2 u = 0$, where u is any component of the longitudinal or transverse displacement vectors $\boldsymbol{u}_l$, $\boldsymbol{u}_t$ of wave velocity c_l or c_t. Write down the general expression for a surface wave, propagating along the surface and decaying exponentially away from it. Apply the boundary conditions at the surface to deduce equation (8.63).

8.3 *Surface relaxation*: Use the valence force field model with two parameters k_{rr} and k_θ to calculate the parameters β and γ in expression (8.64) for the (100) face of silicon. From the numerical values given in Chap. 4 for k_{rr} and k_θ discuss the characteristics of the long-range relaxation.

8.4 *Effective force constant*: Using the valence force field model of equation (4.36) calculate the effective force constant k of Sect. 4.3.1 for a trivalent silicon atom in terms of k_r and k_θ. Consider two situations: (i) an isolated trivalent atom in amorphous Si for which only the trivalent atom is allowed to move; (ii) the (111) Si surface for which only the surface plane is allowed to move.
Compare the two values of k. What would be the effect of allowing further neighbors to move?

8.5 *Optical line shape*: Starting from the Fermi Golden Rule describe how one can obtain the expression (8.79) for the first two moments of the absorption curve.

Solutions to Exercises

Chapter 1

1.1 $$-\beta = \frac{\beta_{ss} + 2\sqrt{3}\beta_{s\sigma} + 3\beta_{\sigma\sigma}}{4} = \frac{-24.5\,\mathrm{eV}}{(R\,[\text{Å}])^2}$$

For c and d the arguments are given in [1.37].

1.2 The overall description of the valence band is satisfactory except for the existence of two flat bands at the top instead of two weakly dispersing bands in the full calculation. The conduction band is much worse showing little (if any) resemblence with the more refined treatments. A detailed comparison is given in [1.38].

1.3 *2D surface lattice*
Unit cell and basis vectors: If a is the lattice parameter

$$\begin{aligned}
&(111)\text{face}: \boldsymbol{a}_1 = \frac{a}{2}(1,\bar{1},0) \quad \boldsymbol{a}_2 = \frac{a}{2}(1,0,\bar{1}) \quad 1\ \text{atom/unit cell}\\
&(100)\text{face}: \boldsymbol{a}_1 = \frac{a}{2}(0,1,\bar{1}) \quad \boldsymbol{a}_2 = \frac{a}{2}(0,1,1) \quad 1\ \text{atom/unit cell}\\
&(110)\text{face}: \boldsymbol{a}_1 = a(001) \quad \boldsymbol{a}_2 = \frac{a}{2}(\bar{1},1,0) \quad 2\ \text{atoms/unit cell}\,.
\end{aligned}$$

Reciprocal lattice

$$\begin{aligned}
&(111)\text{face}: \boldsymbol{a}_1^* = \frac{4\pi}{3a}(1\bar{2}1) \quad \boldsymbol{a}_2^* = \frac{4\pi}{3a}(11\bar{2})\\
&(100)\text{face}: \boldsymbol{a}_1^* = \frac{2\pi}{a}(01\bar{1}) \quad \boldsymbol{a}_2^* = \frac{2\pi}{a}(011)\\
&(110)\text{face}: \boldsymbol{a}_1^* = \frac{2\pi}{a}(001) \quad \boldsymbol{a}_2^* = \frac{2\pi}{a}(\bar{1}10)\,.
\end{aligned}$$

1.4 Since $E_p - E_s$ is large, the s and p bands can be treated separately. To lowest order one neglects H_{ss} and the s band reduces to its center of gravity E_s. On each atom one prepares one p_z state perpendicular to the plane of the two bonds and two p_x and p_y states which are oriented symmetrically with respect to the two bonding directions. As the angle between p_x and p_y (90°) is close to θ each of them approximately points towards one nearest neighbor. The molecular model thus only includes the interaction between the pairs of p orbitals, which is practically equal to $H_{\sigma\sigma}$ as defined in Fig. 4.5. In this description the p_z states remain uncoupled. The molecular levels are then E_s, $E_p - |H_{\sigma\sigma}|$ (σ bonding states), E_p, (p_z states) and $E_p + |H_{\sigma\sigma}|$ (σ^* anti-bonding states). As there are 6 electrons per atom only the σ^* states are empty, leading to semiconducting behavior.

1.5 The molecular model of SiO_2 neglects the interactions between the silicon sp^3 hybrids of the same atom and allows one to treat each Si-O-Si unit separately. In view of their low energies, the s states on oxygen remain uncoupled at their free atom value. Each Si-O-Si unit is symmetrical and one prepares the oxygen p states with suitably chosen symmetry: p_z perpendicular to the Si-O-Si plane, p_y parallel to the Si-Si direction and p_x perpendicular to it. Thus p_x will couple with the sum of the two Si sp^3 hybrids, p_y with their difference. The covalent coupling between p_x and the sum of sp^3 hybrids is equal to $[H_{s\sigma} + \sqrt{3}H_{\sigma\sigma}\sin(\theta/2)]/\sqrt{2}$ and between p_y and their difference $[H_{s\sigma} + \sqrt{3}H_{\sigma\sigma}\cos(\theta/2)]/\sqrt{2}$. The level ordering, in increasing order is: oxygen s states, strong bonding states (σ_y, involving p_y), weak bonding states (σ_x), non-bonding p_z states, weak anti-bonding states (σ_x^*) and strong anti-bonding states (σ_y^*).

1.6 – Allowed k values $\dfrac{kL}{2} + \eta(k) = n\dfrac{\pi}{2} \quad (n = 1, 2 \dots N)$

– Density of states $n(k) = \dfrac{2}{\pi}\left(\dfrac{L}{2} + \dfrac{d\eta}{dk}\right)$

– Electron density $\varrho(x) = \displaystyle\int_0^{k_F} |\psi(x)|^2 n(k) dk$

– For $k_F \ll k_0$

$$\varrho(x) \simeq \frac{2}{3}\frac{k_F^3}{\pi k_0^2}\exp(2k_0x) \qquad \text{for} \quad x \leq 0$$

$$\varrho(x) \simeq \frac{k_F}{\pi}\left(1 - \frac{\sin\left[2k_F\left(x + \frac{1}{k_0}\right)\right]}{2k_F\left(x + \frac{1}{k_0}\right)}\right) \qquad \text{for} \quad 0 \leq x \ll \frac{L}{2}$$

– When x increases $\varrho(x)$ tends to its bulk value k_F/π after a few oscillations (Friedel oscillations) of pseudo period π/k_F. This perturbation becomes negligible at a distance equal to a few $1/2k_F$.

1.7 If a is the interplanar distance,

– for the bulk the energy per atom of each plane can be written

$$E_B = f_1(a) + f_2(2a) ,$$

and the stability condition leads to $0 = f_1'(a) + 2f_2'(2a)$ requiring that $f_1'(a)$ and $f_2'(2a)$ be of opposite signs [i.e., $f_1'(a) < 0$ and $f_2'(2a) > 0$].

- At the surface, the energy per atom of the surface plane is $E_s = f_1(\Delta) + f_2(\Delta + a)$ and the derivative of E_s with respect to Δ evaluated at $\Delta = a$ becomes $E'_s = f'_1(a)+f'_2(2a)$. Taking the bulk stability condition into account this quantity is necessarily negative so that the surface plane will tend to relax away from the first subsurface plane.

1.8 The dispersion relation is $R(k) = -2\beta \cos ka$ and one can write $(L/\pi)dk = n(E)dE$ where L is the length of the chain. From this one readily obtains the density of states per atom

$$n(e) = \frac{1}{\pi\sqrt{4\beta^2 - E^2}} .$$

Chapter 2

2.1 The total energy, in terms of the interatomic distance R, is

$$E(R) = 2E_0 - \beta_0 \exp(-qR) + V_0 \exp(-pR) .$$

The stability condition is

$$0 = q\beta_0 \exp(-qR_e) - pV_0 \exp(-pR_e) ,$$

where R_e is the equilibrium distance. The second-order derivative at equilibrium is given by $d^2E/dR^2 = -q^2\beta_0 \exp(-qR_e) + p^2V_0 \exp(-pR_e)$ in which one can express V_0 in terms of β_0 by using the stability condition. This gives $d^2E/dR^2 = pq[1-(q/p)]\beta_0 \exp(-qR_e)$. On the other hand, the cohesive energy $E_c = 2E_0 - E(R_e)$ and can be written $E_c = [1-(q/p)]\beta_0 \exp(-qR_e)$, leading to the relation $d^2E/dR^2 = pqE_c$.
The molecular model of silicon, in which the promotion energy $E_p - E_s$ is neglected, gives exactly the same expression.

2.2 *General expressions*: If we denote the nearest neighbor tight-binding interaction by β and take the origin of energies at the atomic level then, if $|i\rangle$ are the basis states, we have

$$\mu_1 = \langle i|H|i\rangle = 0$$

$$\mu_2 = \sum_j |\langle i|H|j\rangle|^2 = N\beta^2$$

$$N = 4 \text{ (diamond)}, \ 6 \text{ (sc)}, \ 8 \text{ (bcc)}, \ 12 \text{ (fcc)}$$

$$\mu_3 = \sum_{j,k} \langle i|H|j\rangle\langle j|H|k\rangle\langle k|H|i\rangle = M\beta^3$$

$$M = 0 \text{ (sc, diamond, bcc)} \ M = 48\beta^3 \text{ (fcc)}$$

2.3 $a_1 = 0 \quad b_1 = 6\beta^2 \quad a_2 = 0 \quad b_2 = 5\beta^2$

2.4 $G(E) = R(E) + \mathrm{i}I(E)$ with

$$I(E) = -\pi \sum_k |k\rangle\langle k|\delta(E - E_k)$$

$$R(E) = \mathcal{P} \sum_k \frac{|k\rangle\langle k|}{E - E_k}$$

Hilbert transform $R(E) = -\frac{\mathcal{P}}{\pi} \int \frac{I(E')}{E - E'} dE'$.

2.5 From the definitions of exercise 2.4 the real part R_{00} of G_{00} can be written as

$$R_{00}(E) = \mathcal{P} \int \frac{n(E')}{E - E'} dE'$$

with $n(E)$ given from exercise 1.8. It is easy to show that

$$R_{00}(E) = \begin{cases} 0 & |E| < 2|\beta| \\ \dfrac{1}{\sqrt{E^2 - 4\beta^2}} & E > 2|\beta| \\ -\dfrac{1}{\sqrt{E^2 - 4\beta^2}} & E < -2|\beta| . \end{cases}$$

2.6 The two steps correspond to (i) decoupling the sp^3 dangling bond $|0\rangle$ from the three other sp^3 hybrids $|i'\rangle$ in each back-bond; (ii) the repeated application of Dyson's equation giving

$$G_{00} = \cfrac{1}{E - \cfrac{3\Delta^2}{E - 2\Delta - \beta^2 g_{ii}}} .$$

After these two steps each $|i\rangle$ is the sp^3 dangling bond of a new trivalent atom which means that $g_{ii} = G_{00}$. This leads to a quadratic equation for G_{00}. Details of its solution are given in [2.34].

Chapter 3

3.1 The number of d states per atom is 5. If we simulate the corresponding density of states $n(E)$ by a Gaussian normalized to 5 we get

$$n(E) = \frac{5}{\sqrt{2\pi\mu_2}} \exp\left[-\frac{(E - \mu_1)^2}{2\mu_2}\right] .$$

Taking $\mu_1 = 0$ we easily get the equivalent of (3.3)

$$E_{bs} = -10\sqrt{\frac{\mu_2}{2\pi}} \exp\left(-\frac{E_F^2}{2\mu_2}\right) ,$$

while N_d takes the integral form

$$N_d = \frac{10}{\sqrt{2\pi\mu_2}} \int_{-\infty}^{E_F} \exp\left(-\frac{E^2}{2\mu_2}\right) dE \,.$$

It is thus only possible to relate E_{bs} to N_d via E_F and this in a numerical way. However, it is easy to show that the curve $E_{bs}(N_d)$ is fairly close to the analytic form (3.3).

3.2 Application of (3.35) directly gives

$$\delta N_{i0} = \frac{10\varepsilon_F}{\sqrt{12\beta}} \left(\frac{1}{\sqrt{Z_i}} - \frac{1}{\sqrt{Z_b}}\right) ,$$

$$\chi = \frac{10}{\sqrt{12\beta}} \frac{1}{\sqrt{Z_i}} \,.$$

Writing $\delta N_i = \delta N_{i0} - \chi U_i$ and $U_i = U\delta N_i$ one readily finds

$$U_i = \frac{U\delta N_{i0}}{1+\chi U} \,.$$

The limit $\chi U \gg 1$ leads to the approximate result $U_i \simeq \delta N_{i0}/\chi$, which corresponds to directly imposing $\delta N_i \simeq 0$, i.e. the criterion of local charge neutrality leading to (3.36). The relative error made in U_i is

$$\frac{\delta U_i}{U_i} = -\frac{1}{1+\chi U} \,.$$

3.3 The shift in core level $\delta\varepsilon_c(n_c - 1/2)$ due to a change δN_i in electron population on atom i can be written

$$\delta\varepsilon_c = U_{cv}\delta N_i \,,$$

where U_{cv} is the average Coulomb repulsion between core and valence electrons. This can be rewritten as

$$\delta\varepsilon_c = \frac{U_{cv}}{U} U_i \,.$$

Calculation of the ratio U_{cv}/U shows that it is of order 1.1 (see Ref. [3.22]) so that a fair order of magnitude is given by $\delta\varepsilon_c \sim U_i$ which leads to (3.43).

3.4 *Stoner condition*: The first relation relates U to ΔN and is $U = J\Delta N$. The second results from the calculation of the $\Delta N(U)$ due to a given applied perturbation U. It is linear for small U with slope $-\chi$ (where χ is the susceptibility) but must saturate at large U. The solution can be obtained graphically by plotting the two curves ΔN versus $-U$. It is easy to show that a solution (i.e. an intersection) only exists if $\chi > 1/J$, which is Stoner condition.

Chapter 4

4.1 One can solve the matrix (4.8) exactly by building symmetric and antisymmetric combinations of the atomic states, i.e. $S_A \pm S_B/\sqrt{2}$, $\sigma_A \pm \sigma_B/\sqrt{2}$, $\pi_A \pm \pi_B/\sqrt{2}$, $\pi'_A \pm \pi'_B/\sqrt{2}$. For the π states, the solutions

$$E_p \pm H_{\pi\pi}$$

have two-fold degeneracy. The s, σ 4×4 block reduces into two 2×2 subblocks

$$\begin{pmatrix} E_s + H_{ss} & H_{s\sigma} \\ H_{\sigma s} & E_p + H_{\sigma\sigma} \end{pmatrix}, \quad \begin{pmatrix} E_s - H_{ss} & H_{s\sigma} \\ H_{\sigma s} & E_p - H_{\sigma\sigma} \end{pmatrix},$$

whose solutions are

$$\frac{E_s + E_p \pm (H_{ss} + H_{\sigma\sigma})}{2} \pm \sqrt{\left(\frac{E_s - E_p \pm (H_{ss} - H_{\sigma\sigma})}{2}\right)^2 + H_{s\sigma}^2} .$$

Numerical values of these levels with Harrison's parameters can be found in [4.29], as can a comparison with more sophisticated calculations.

4.2 The dispersion curve of the p_z band of graphite is like the one which would be obtained for an s band on the same lattice. With two atoms per unit cell one obtains two bands given by

$$E_z(\boldsymbol{k}) = E_p \pm H_{\pi\pi} \left| \sum_{i=1}^{3} \mathrm{e}^{i\boldsymbol{k}\cdot\boldsymbol{\delta}_i} \right| ,$$

where the $\boldsymbol{\delta}_i$ are vectors joining one atom to its three neighbors.

For the other bands the simplest approach is to use a method such as that developed in Sect. 1.1.3 reducing the problem to the one of a "s" band. One writes the wave function as

$$\psi = \sum a_{ij} \Phi_{ij} ,$$

where Φ_{ij} is the sp^2 orbital pointing from atom i towards atom j. One can write

$$(E - \bar{E}) a_{ij} = \Delta \sum_{j' \neq j} a_{ij'} + \beta a_{ji}$$

where $\bar{E}$ is the sp^2 atomic energy. It is interesting to write $S_i = \sum_j a_{ij}$ and to rewrite the above equation as

$$(E - \bar{E} + \Delta) a_{ij} = \Delta S_i + \beta a_{ji} .$$

Writing the same equation for a_{ji} and inserting a_{ji} into the above equation one gets

$$\{(E - \bar{E} + \Delta)^2 - \beta^2\} a_{ij} = (E - \bar{E} + \Delta) \Delta S_i + \beta \Delta S_j .$$

Summing this over j gives

$$\left[\left(E-\bar{E}-\frac{\Delta}{2}\right)^2-\frac{9}{4}\Delta^2-\beta^2\right]S_i=\beta\Delta\sum_j S_j\,.$$

This is identical to the problem of an s band on the same lattice allowing one to write

$$E=\bar{E}+\frac{\Delta}{2}\pm\sqrt{\frac{9}{4}\Delta^2+\beta^2+\beta\Delta\varepsilon}$$

with

$$\varepsilon=\pm\left|\sum_{i=1}^{3}e^{ik\cdot\delta_i}\right|\,.$$

These are the broad bands. However, there still remain other solutions with all $S_i=0$ but $a_{ij}\neq 0$ given by

$$E=\bar{E}-\Delta\pm\beta$$

which is the equation of flat bands representing one extreme of the broad bands.

4.3 According to expression (4.14) the self-energy of a dangling bond $|n\rangle$ is

$$\sum_n(E)=\sum_\alpha\frac{|H_{n\alpha}|^2}{E-E_\alpha}$$

and the effective interaction between $|n\rangle$ and $|n'\rangle$ is

$$H^*_{nn'}=H_{nn'}+\sum_\alpha\frac{H_{n\alpha}H_{\alpha n'}}{E-E_\alpha}\,.$$

If one approximates the states $|\alpha\rangle$ by the bonding and anti-bonding states one gets

$$\sum_n(E)=\sum_B\frac{|\langle n|H|B\rangle|^2}{E-\beta}+\sum_A\frac{|\langle n|H|A\rangle|^2}{E+\beta}$$

$$H^*_{nn'}=\sum_B\frac{\langle n|H|B\rangle\langle B|H|n'\rangle}{E-\beta}+\sum_A\frac{\langle n|H|A\rangle\langle A|H|n'\rangle}{E+\beta}$$

where B and A denote the bonding and anti-bonding states. Both quantities can be evaluated in detail in a nearest neighbor tight-binding approximation. The main result is that the effective interaction between nearest neighbor dangling bonds at the Si(111) surface becomes positive as discussed in [4.7].

4.4 With nearest neighbor effective interactions β^*, the system of dangling bonds at the Si(111) surface becomes equivalent to a two-dimensional s band. The resulting dispersion relation is

$$E(\boldsymbol{k})=\sum+2\beta^*\left[\cos\boldsymbol{k}\cdot\boldsymbol{a}_1+\cos\boldsymbol{k}\cdot\boldsymbol{a}_2+\cos\boldsymbol{k}\cdot(\boldsymbol{a}_1-\boldsymbol{a}_2)\right]\,,$$

where $\boldsymbol{a}_1$ and $\boldsymbol{a}_2$ are the two basis vectors of exercise 1.3.

4.5 As shown in [4.30], classical image theory states that, in the case of an interface between two materials 1 and 2, of dielectric constants ε_1 and ε_2, the potential produced by a charge e at a distance r in material 1 is given by

$$\Phi_1 = \frac{1}{\varepsilon_1}\left(\frac{e}{r} + \frac{e'}{r'}\right)$$

where e' is the image charge $e' = (\varepsilon_1 - \varepsilon_2)/(\varepsilon_1 + \varepsilon_2)e$. Near the interface $r = r'$ and this potential becomes

$$\Phi_1 = \frac{2}{\varepsilon_1 + \varepsilon_2}\frac{e}{r}$$

showing that the effective dielectric constant is $\varepsilon = (\varepsilon_1 + \varepsilon_2)/2$.

4.6 In the molecular model of silicon there are four electrons per atom. Applying the considerations of exercise 2.1 one can write the cohesive energy per atom as

$$E_c = \left(1 - \frac{q}{p}\right)\beta_0 e^{-qR_e} .$$

The creation of the neutral vacancy requires the breaking of four covalent bonds which costs an energy equal to E_c. However one regains an energy E_c by placing the ejected atom on the surface. In the absence of atomic relaxation the formation energy of the vacancy is thus given by

$$E_{FV} = E_c .$$

As $E_c \sim 4.5\,\text{eV}$ for silicon one thus predicts the same value for E_{FV} which is precisely the result of local density calculations.

The same reasoning applied to the breaking of one covalent bond (i.e. the creation of two dangling bonds) leads us to a cost in energy equal to $E_c/2 \sim 2.25\,\text{eV}$, i.e. about 1.1 eV per dangling bond.

4.7 Assuming that the major contribution to U is the sp^3 contribution, delocalization effects will reduce the sp^3 atomic value by a multiplicative factor equal to $(0.7)^2$, i.e. ~ 0.5. Furthermore, the addition of one electron to the dangling bond will lead to screening, mainly via the polarization of the back-bonds. In view of the short-range nature of the screening in these materials this will reduce the excess charge from $-e$ to $-e/\varepsilon$ practically within the trivalent atom. In other words, this will reduce U by an extra factor of ε. Starting from an atomic value of 10 eV this leads to $U \sim 0.5\,\text{eV}$. Such a reasoning is only valid for a dangling bond in the bulk material.

4.8 The perfect crystal is simulated by a central atom coupled to four sp^3 hybrids. This leads to an 8×8 matrix which can be factorized by symmetry. The s state of the central atom couples to the symmetric combination of the sp^3 hybrids, the p_x, p_y, p_z to other combinations. This leads to four 2×2 matrices, three of which are identical. The levels are

$$\frac{E_s + E_{sp^3}}{2} \pm \sqrt{\left(\frac{E_s - E_{sp^3}}{2}\right)^2 + \left(H_{ss} + \sqrt{3}H_{s\sigma}\right)^2}$$

$$\frac{E_p + E_{sp^3}}{2} \pm \sqrt{\left(\frac{E_p - E_{sp^3}}{2}\right)^2 + \left(H_{\sigma s} + \sqrt{3}H_{\sigma\sigma}\right)^2} \quad \left(\begin{matrix}\text{threefold} \\ \text{degenerate}\end{matrix}\right) .$$

One fills the levels with 8 electrons. Summing their energies one gets the band structure energy versus the interatomic distance R if all $H_{\alpha\beta}$ are assumed to vary as $\exp(-qR)$. Adding four repulsive potentials that vary as $\exp(-pR)$, one can write the total energy and demand that it be minimum for $R = R_e$.

To simulate the trivalent atom one simply removes an sp^3 hybrid and performs a similar calculation, again using symmetry and repulsive potentials deduced as above. It is then possible to take the derivatives of the total energy with respect to an axial displacement of the trivalent atom. This allows one to obtain F_0 and F_1 numerically.

Chapter 5

5.1 The electron density n of a free electron gas in a potential $V = 0$ can be expressed in terms of its Fermi energy E_F by

$$n = \frac{1}{3\pi^2}\left(\frac{2mE_F}{\hbar^2}\right)^{3/2} .$$

In the Thomas–Fermi approximation this becomes

$$n(\boldsymbol{r}) = \frac{1}{3\pi^2}\left\{\frac{2m}{\hbar^2}\left[E_F - V(\boldsymbol{r})\right]\right\}^{3/2} .$$

Linearizing this with respect to V gives the change in electron density

$$\delta n(\boldsymbol{r}) = -\frac{mk_F}{\hbar^2\pi^2}V(\boldsymbol{r}) .$$

This potential V comprises a bare part V_b plus a self-consistent part $\delta V(\boldsymbol{r})$ induced by $\delta n(\boldsymbol{r})$. Using Poisson's equation and taking Fourier transforms one has

$$\delta n(\boldsymbol{q}) = -\frac{mk_F}{\hbar^2\pi}V(\boldsymbol{q}) , \quad \delta V(\boldsymbol{q}) = \frac{4\pi e^2}{q^2}\delta n(\boldsymbol{q}) .$$

Writing $V(\boldsymbol{q}) = V_b(\boldsymbol{q})/\varepsilon(\boldsymbol{q})$, one readily obtains

$$\varepsilon(\boldsymbol{q}) = 1 + \frac{\lambda^2}{q^2} \quad \lambda^2 = \frac{4e^2mk_F}{\hbar^2\pi} .$$

5.2 Denoting the sp^3 energies of atoms i and j by E_i and E_j, with β as the covalent coupling between two sp^3 hybrids, the electron population N_i on atom

i is

$$N_i = 4 + \sum_j \frac{\frac{1}{2}(E_j - E_i)}{\sqrt{\left(\frac{E_j - E_i}{2}\right)^2 + \beta^2}},$$

where j labels nearest neighbors. Expanding this to first order in δE_i and δE_j and denoting the unperturbed value of $(E_j - E_i)/2$ by δ_0 one finds

$$\delta N_i = \frac{\beta^2}{(\delta_0^2 + \beta^2)^{3/2}} \sum_j \frac{\delta E_j - \delta E_i}{2}$$

directly giving the susceptibilities of (5.13).

5.3 One writes

$$S = \sum_{j=1}^{n} \frac{e^2}{\sqrt{R_j^2 + d^2}} + \frac{2\pi}{s} \int_{R_{\min}}^{R_{\max}} \frac{e^2 R dR}{\sqrt{R^2 + d^2}},$$

where s is the surface area per atom in the plane. The radius $R_{\min}$ is determined in such a way that

$$\pi R_{\min}^2 = ns$$

while $R_{\max}$ is taken to remain finite to avoid electrostatic divergences. Performing the integral leads to

$$S = \sum_{j=1}^{n} \frac{e^2}{\sqrt{R_j^2 + d^2}} + \frac{2\pi e^2}{s} \left(R_{\max} - \sqrt{\frac{ns}{\pi} + d^2} \right).$$

For a (111) plane with $d = 0$, one has $s = a^2\sqrt{3}/2$ where a is the nearest neighbor separation in the plane. With no discrete corrections one gets

$$S - \frac{2\pi e^2}{s} R_{\max} = -\frac{e^2}{a} 2\sqrt{\frac{2\pi}{\sqrt{3}}} = -3.81 \frac{e^2}{a}$$

whereas with the first shell of neighbors (6 at distance a) one obtains

$$S - \frac{2\pi e^2}{s} R_{\max} = \frac{e^2}{a} \left(6 - 2\sqrt{\frac{14\pi}{\sqrt{3}}} \right) = -4.08 \frac{e^2}{a}$$

showing that the two estimates are fairly close.

5.4 Core level shifts

	GaP	GaAs	GaSb	InP	InAs	InSb	ZnS	ZnSe	ZnTe	AlSb	CdTe
f	0.61	0.56	0.51	0.65	0.60	0.57	0.79	0.82	0.74	0.65	0.82
$\Delta E_A = -\Delta E_c$ [eV]	0.3	0.35	0.39	0.28	0.32	0.34	0.17	0.14	0.21	0.28	0.14

5.5 For a (111) face the reasoning of Sect. 5.3.1 leads to

$$\Delta V = \frac{4\pi eq}{s}\frac{R}{R+R'}L$$

where L is the total thickness, q the net charge per atom, s the surface area per atom. If d is the interatomic distance one obtains

$$\Delta V = \frac{4\pi eq}{(4/\sqrt{3})d^2}\frac{d/3}{4d/3}L = \frac{\pi\sqrt{3}}{4}eq\frac{L}{d^2}\,.$$

With $q = 0.2e$, $L = 1\,\mu$m and $d = 2.35$ Å, this yields $\Delta V = 8.1 \times 10^3$ eV.

Chapter 6

6.1 To calculate the donor levels one must use (6.13) with the help of the self-consistency condition (6.11). The results deduced from the numerical values of Table 6.1 must correspond to Fig. 6.6
The slope 0.2 of the straight line corresponds to surface screening with an effective dielectric constant of 5, which is quite reasonable.

6.2 From exercise 5.3 one can write directly

$$S = \sum_{j<j_M}\left(\frac{1}{R_j} - \frac{1}{\sqrt{R_j^2+4d'^2}}\right) + \frac{4\pi Nd'}{\frac{1}{2d'}\sqrt{\frac{n}{\pi N}} + \sqrt{1+\frac{n}{4\pi Nd'^2}}}$$

where n is the number of atoms considered explicitly. The different levels of approximations are thus:

- no discrete sum for which one recovers (6.24)
- first shell of neighbors for which one gets the formula above with $n = 4$ and the discrete contribution given by

$$4\left(\frac{1}{a} - \frac{1}{\sqrt{a^2+4d'^2}}\right)$$

where a is the square lattice parameter (note that $s = a^2$).

6.3 The free holes are repelled by the positive surface charge density. This leaves negatively charged ionized acceptors in the vicinity of the surface. The simplest way to account for this is to use the "depletion approximation" which assumes that ionized acceptors exist in a region of thickness X where the charge

density if thus $\varrho = -eN_A$. Integrating Poisson's equation between the surface ($x = 0$) and the point ($x = X$) where the electric field is zero leads to equation (6.15).
The depletion approximation is valid when X is much larger than the typical screening length L_D (Debye length) of the material. This can be understood easily by noticing that a macroscopic perturbative potential $V(\boldsymbol{r})$ induces a change in hole concentration $N_A[\exp(-V/kT) - 1]$ according to Boltzmann statistics. Linearizing this leads to a charge density $-(eN_A/kT)V$ which, when introduced into Poisson's equation, gives a Debye screening length $L_D^{-2} = e^2N_A/\varepsilon kT$. From (6.15) the condition $X/L_D \gg 1$ is thus equivalent to $\Delta V/kT \gg 1$.

6.4 The monolayer problem corresponds to a half filled electronic system. With symmetrical coupling of the "s" band with the anion and cation dangling bonds, symmetry allows one to state that local neutrality is obtained for ε_s lying midway between ε_{Ga} and ε_{As}. Taking $\varepsilon_s = 0$ one can write $\varepsilon_{As} = -\delta$ and $\varepsilon_{Ga} = +\delta$. The eigenstates, for a given $\boldsymbol{k}$ wavevector, will be solutions of

$$\begin{vmatrix} \varepsilon & V & \beta f & 0 \\ V & \varepsilon+\delta & 0 & 0 \\ \beta f & 0 & \varepsilon & V \\ 0 & 0 & V & \varepsilon-\delta \end{vmatrix} = 0$$

with $f(\boldsymbol{k}) = 2(\cos k_x a + \cos k_y a)$, giving

$$\varepsilon = V^2 + \frac{\delta^2 + \beta^2 f^2}{2} \pm \sqrt{\left(V^2 + \frac{\delta^2 + \beta^2 f^2}{2}\right)^2 - V^4 - \delta^2\beta^2 f^2}\,.$$

Chapter 7

7.1 Equation (7.9) becomes

$$E_F - E_d = -\frac{\delta V^2}{\delta^2 + W^2}\,.$$

This is negligible if V and δ are weak compared to W. From Chap. 6 we know that $V \sim 0.5\,\text{eV}$; free-electron theory leads to a value of W of several eV and, as given in Table 6.1 $2\delta \sim 0.7\,\text{eV}$. Taking $W = 2\,\text{eV}$ this yields $E_F - E_d \simeq -0.02\,\text{eV}$.

7.2 Using the relation $\tan^{-1} x + \tan^{-1} y = \tan^{-1}(x + y/1 - xy)$, one can easily show that, if the functions R and I are the same for the anion and the cation, one obtains (7.10).

7.3 Using the fact that $g_{00} = 1/E - E_d + \mathrm{i}\eta$ and $g_{kk} = 1/E - \varepsilon_k + \mathrm{i}\eta$ one can show that (7.6) reads

$$\delta N(E) = -\frac{1}{\pi}\mathrm{Im}\left\{\ln\left(1 - \frac{1}{E - E_d + \mathrm{i}\eta}\sum_k \frac{|V_{0k}|^2}{E - \varepsilon_k + \mathrm{i}\eta}\right)\right\}$$

which can easily be transformed into (7.7). We can factor $1/E - E_d + \mathrm{i}\eta$ in this expression and then take the derivative $d\delta N(E)/dE$ which gives the change in density of states $\delta n(E)$ in the form

$$\delta n(E) = -\frac{1}{\pi}\,\mathrm{Im}\left\{\frac{1-R'-\mathrm{i}\Gamma'}{E-E_\mathrm{d}-R-\mathrm{i}\Gamma} - \frac{1}{E-E_\mathrm{d}+\mathrm{i}\eta}\right\}.$$

The second term corresponds to $-\delta(E-E_\mathrm{d})$, i.e. to the removal of the atomic state at E_d and to its replacement by the first term. If $V_{\mathrm{d}k}$ is small, this first term is important in the vicinity of E_d. It can thus be rewritten approximately as

$$-\frac{1}{\pi}\,\frac{\Gamma(E_\mathrm{d})[1-R'(E_\mathrm{d})]-\Gamma'(E_\mathrm{d})[E-E_\mathrm{d}-R(E_\mathrm{d})]}{[E-E_\mathrm{d}-R(E_\mathrm{d})]^2+\Gamma^2(E_\mathrm{d})}$$

which represents a Lorentzien curve.

7.4 The dipole layer is given by

$$\Delta V = 4\pi e^2 q\frac{d}{s}$$

where d is the distance between the two planes and s the surface area per atom within the plane. In terms of the nearest neighbor separation we find:

$$\begin{aligned}
&(111)\ \text{face}: d=\frac{R}{3} && s=\frac{4}{\sqrt{3}}R^2 && \Delta V=\frac{\pi}{\sqrt{3}}\frac{qe^2}{R}\\
&(110)\ \text{face}: d=\sqrt{\frac{2}{3}}R && s=\frac{8R^2}{3\sqrt{2}} && \Delta V=\pi\sqrt{3}\frac{qe^2}{R}\\
&(100)\ \text{face}: d=\frac{R}{\sqrt{3}} && s=\frac{8R^2}{3} && \Delta V=\frac{\pi\sqrt{3}}{2}\frac{qe^2}{R}.
\end{aligned}$$

For $q=1$, one obtains $\Delta V(111) = 11.4\,\mathrm{eV}$, $\Delta V(110) = 34.2\,\mathrm{eV}$ and $\Delta V(100) = 17.1\,\mathrm{eV}$.

7.5 From (7.23) and (7.24) one can show that

$$\delta = \frac{\delta_0 + \frac{U}{2}\left(3+n_\mathrm{E}+\frac{n_\mathrm{A}}{2}-n_{\mathrm{d}0}\right)}{1+\frac{U}{2\Delta}(3-n_\mathrm{A})}.$$

For very large U this simplifies to

$$\delta \sim \Delta\frac{3+n_\mathrm{E}+\frac{n_\mathrm{A}}{2}-n_{\mathrm{d}0}}{3-n_\mathrm{A}},$$

which, in (7.24), exactly corresponds to $n_\mathrm{d}=0$, i.e. to charge neutrality on the impurity atom.

7.6 We have seen in exercise 4.7 that for the bulk dangling bond, U takes the atomic value $\sim 10\,\mathrm{eV}$, multiplied by the delocalization factor $(0.7)^2$ and divided by the bulk dielectric constant $\varepsilon = 10$. This gives $(U_\mathrm{db})_\mathrm{bulk} \sim 0.5\,\mathrm{eV}$.

Exactly the same reasoning applies to the Si-SiO_2 interface replacing ε by its appropriate interface value. From exercise 4.5 this is equal to $\varepsilon + 1/2 \simeq 5$ so that $(U_{\mathrm{db}})_{\mathrm{surface}} \sim 1\,\mathrm{eV}$.

Chapter 8

8.1 In this case it is not necessary to specify the coordinate α which is along the chain. Dropping this index we get for the force constant matrix elements

$$A_{ii} = 2k \quad A_{i,i\pm1} = -k \ .$$

The problem is strictly analogous to a tight-binding electronic band structure calculation. One thus gets

$$M\omega^2 = k(2 - 2\cos ka) \ ,$$

i.e.

$$\omega = 2\sqrt{\frac{k}{M}}\left|\sin\frac{ka}{2}\right| \ .$$

8.2 Consider a surface wave propagating along x and decaying exponentially along z (with $z < 0$). From the basic equation $(\partial^2 u/\partial t^2) - c^2 u = 0$, one immediately obtains

$$u \propto \mathrm{e}^{\mathrm{i}(kx-\omega t)}\mathrm{e}^{Kz}$$

with

$$K = \sqrt{k^2 - \omega^2/c^2} \ .$$

There are two possible waves of this kind: a longitudinal wave $\boldsymbol{u}_l$ and a transverse wave $\boldsymbol{u}_t$. The general solution is the sum of these two. As shown in [8.14] the boundary conditions are such that $\boldsymbol{u}$ has only two components u_x and u_z, corresponding to four unknown amplitudes u_{tx}, u_{tz}, u_{lx}, u_{lz}. The two general conditions div $\boldsymbol{u}_t = 0$ and curl $\boldsymbol{u}_l = 0$ give two relations. The other two conditions

$$\frac{\partial u_x}{\partial z} + \frac{\partial u_z}{\partial x} = 0$$

$$c_l^2\frac{\partial u_z}{\partial z} + (c_l^2 - 2c_t^2)\frac{\partial u_x}{\partial x} = 0$$

complete these to form a system of homogeneous equations which has a solution only if (8.63) is satisfied.

8.3 The simplest approach is to consider two types of deformation since there are two parameters to determine. The first type is a uniform compression along [100] such that $d_{i,i+1} = -u$ (R is the interatomic distance). From (8.64) the bulk energy per atom becomes

$$\delta E = \left(\frac{\beta}{2} + \gamma\right)u^2 \ .$$

The same energy can be evaluated in the valence force field model and turns out to be

$$\delta E = \left(\tfrac{1}{3}k_r + 4k_\theta\right) u^2 .$$

The second deformation consists in displacing one plane over another by an amount u. This gives, per atom of each displaced plane,

$$\delta E' = (\beta - 2\gamma)u^2$$

or, in terms of k_r and k_θ,

$$\delta E' = \left(\tfrac{2}{3}k_r + \tfrac{16}{3}k_\theta\right) u^2 .$$

Identification of δE and $\delta E'$ directly gives

$$\beta = \tfrac{2}{3}k_r + \tfrac{20}{3}k_\theta , \quad \gamma = \tfrac{2}{3}k_\theta .$$

From the numerical values of k_r and k_θ given in Chap. 4 the ratio $2\gamma/\beta$ will be small and the long range relaxation will be severely damped.

8.4 The results are given in Sect. 8.3.2. To calculate the contribution in k_θ, the best method is to use the following expression for $d\theta_{jik}$ the change in bond angle $\widehat{jik}$

$$R d\theta_{jik} = -\frac{3}{\sqrt{8}}\left[\boldsymbol{u}_{ij}\left(\boldsymbol{n}_{ik} + \frac{1}{3}\boldsymbol{n}_{ij}\right) + \boldsymbol{u}_{ik}\left(\boldsymbol{n}_{ij} + \frac{1}{3}\boldsymbol{n}_{ik}\right)\right]$$

where $\boldsymbol{u}_{ij} = \boldsymbol{u}_j - \boldsymbol{u}_i$ ($\boldsymbol{u}$ displacement vector) and $\boldsymbol{n}_{ij}$ is the unit vector of the direction $i \to j$.

8.5 The Fermi golden rule gives the probability $w(h\nu)$ for transitions at frequency ν:

$$w(h\nu) = \sum_n p_n |\langle\psi_i|\Omega|\psi_f\rangle|^2 |\langle n|n'\rangle|^2 \delta(h\nu - h\nu_0 - (n' - n)\hbar\omega)$$

where $h\nu_0$ is the electronic threshold and $|\langle\psi_i|\Omega|\psi_f\rangle|^2$ is the electronic optical matrix element. Since $\sum_n p_n = 1$, the moments of the normalized curve are:

$$\mu_k = \sum_n p_n \langle n|n'\rangle\langle n'|n\rangle [h\nu_0 + (n' - n)\hbar\omega]^k .$$

Writing the vibrational Hamiltonian in the ground state as $H = T_N + 1/2kQ^2$ and in the excited state as $H' = T_N + 1/2k(Q - Q_e)^2$ where T_N is the nuclear kinetic energy and Q_e the equilibrium position in the excited state, we get

$$\mu_k = \sum_n p_n \langle n| \left[H' - \left(n + \tfrac{1}{2}\right)\hbar\omega\right]^k |n\rangle .$$

For $k = 1, 2$ one can replace $(n + 1/2)\hbar\omega$ by H in this expression and $H' - H$ can be expressed as in (8.80) taking into account the fact that $-kQ_eQ = -FQ$. This directly leads to (8.81) and (8.83).

References

Chapter 1

1.1 H. Lüth: In *Surfaces and Interfaces of Solids*, Springer Ser. Surf. Sci. Vol. 15 (Springer, Berlin, Heidelberg 1991)
1.2 N.W. Ashcroft, N.D. Mermin: In *Solid State Physics* (Holt-Saunders, New York 1976)
1.3 J.C. Phillips: In *Bonds and Bands in Semiconductors* (Wiley, New York 1973)
1.4 G.F. Koster: In *Solid State Physics*, Vol. 5. ed. by F. Seitz, D. Turnbull (Academic, New York 1957) p.174
1.5 R.J. Meyer, C.B. Duke, A. Paton, A. Kahn, E. So, J.L. Yeh, P. Mark: Phys. Rev. B **19**, 5194 (1979)
1.6 D.J. Chadi: J. Vac. Sci. Technol. **16**, 1290 (1979)
1.7 J.A. Appelbaum, G.A. Baraff, D.R. Hamann: Phys. Rev. B **14**, 588 (1976)
1.8 P.K. Larsen, J.F. van der Veen, A. Mazur, J. Pollmann, J.H. Neave, B.A. Joyce: Phys. Rev. B **26**, 3222 (1982)
1.9 M.D. Pashley, W. Haberern, J.W. Woodall: J. Vac. Sci. Technol. B **6**, 1468 (1988)
1.10 R.W. Nosker, P. Mark, D. Levine: Surf. Sci. **19**, 291 (1970)
1.11 D. Haneman: Phys. Rev. **121**, 1093 (1961)
1.12 F. Himpsel: Appl. Phys. A **38**, 205 (1985)
1.13 K.C. Pandey: Phys. Rev. Lett. **47**, 1913 (1981)
1.14 R.E. Schlier, H.E. Farnsworth: J. Chem. Phys. **30**, 917 (1959)
1.15 K. Takayanagi, Y. Tanishiro, M. Takahashi, S. Takahashi: J. Vac. Sci. Technol. A **3**, 1502 (1985)
1.16 J.J. Lander, J. Morrison: Surf. Sci. **2**, 553 (1964)
1.17 J.E. Northrup: Phys. Rev. Lett. **53**, 683 (1984)
1.18 J.M. Nicholls, B. Reihl, J.E. Northrup: Phys. Rev. B **35**, 4137 (1987)
1.19 G.V. Hansson, R.Z. Bachrach, R.S. Bauer, P. Chiaradia: Phys. Rev. Lett. **46**, 1033 (1981)
1.20 M. Alff, W. Moritz: Surf. Sci. **80**, 24 (1979)
1.21 J.N. Andersen, H.B. Nielsen, L. Petersen, D.L. Adams: J. Phys. C **17**, 17 (1984)
1.22 D.L. Adams, H.B. Nielsen, J.N. Andersen: Surf. Sci. **128**, 294 (1983)
1.23 D.L. Adams, L. Petersen, C.E. Sorensen: J. Phys. C **18**, 1753 (1984)
1.24 H.D. Shih, F. Jona, D.W. Jepsen, P.M. Marcus: Surf. Sci. **104**, 39 (1981)
1.25 R.P. Gupta: Phys. Rev. B **23**, 6265 (1981)
1.26 R. Smoluchowski: Phys. Rev. **60**, 661 (1941); M.W. Finnis, V. Heine: J. Phys. F **4**, L37 (1974)
1.27 H.L. Davis, J.R. Noonan: Surf. Sci. **126**, 245 (1983)
1.28 Y. Gauthier, R. Baudoing, Y. Joly, C. Gaubert, J. Rundgren: J. Phys. C **17**, 4547 (1984)
1.29 D.L. Adams, C.S. Sorensen: Surf. Sci. **166**, 495 (1986)
1.30 J. Sokolov, H.D. Shih, U. Bardi, F. Jona, P.M. Marcus: J. Phys. C **17**, 371 (1984)
1.31 G. Allan, M. Lannoo: Phys. Rev. B **37**, 2678 (1988)
1.32 G. Allan: Prog. in Surf. Sci. **25**, 43 (1987)
1.33 P. Jiang, P.M. Marcus, F. Jona: Solid State Commun. **59**, 275 (1986)
1.34 M.K. Debe, D.A. King: Surf. Sci. **81**, 193 (1979)
1.35 J.A. Walker, M.K. Debe, D.A. King: Surf. Sci. **104**, 405 (1981)

1.36 R.A. Barker, S. Semancik, P.J. Estrup: Surf. Sci. **94**, L162 (1980)
1.37 M. Lannoo, J.N. Decarpigny: Phys. Rev. B **8**, 5704 (1973)
1.38 M. Lannoo: In *Handbook of Surfaces and Interfaces*, Vol. 1, ed. by L. Dobrzynski (Garland, New York 1978)

Chapter 2

2.1 P. Hohenberg, W. Kohn: Phys. Rev. B **136**, 864 (1964)
2.2 W. Kohn, L.J. Sham: Phys. Rev. A **140**, 1133 (1965)
2.3 G.B. Bachelet, D.R. Hamann, M. Schluter: Phys. Rev. B **26**, 4199 (1982)
2.4 M.L. Cohen: Proc. Enrico Fermi Summer School, Varenna (1983)
2.5 M. Schluter: Proc. Enrico Fermi Summer School, Varenna (1983)
2.6 D.R. Hamann: Phys. Rev. Lett. **42**, 662 (1979)
2.7 M.S. Hybertsen, S.G. Louie: Solid State Commun. **51**, 451 (1984); Phys. Rev. B **30**, 5777 (1984)
2.8 G.A. Baraff, M. Schluter: Phys. Rev. B **30**, 3460 (1984)
2.9 L.J. Sham, M. Schluter: Phys. Rev. Lett. **51**, 1888 (1983), Phys. Rev. B **32**, 3883 (1985); J. Perdew, M. Levy: Phys. Rev. Lett. **51**, 1884 (1983)
2.10 M. Lannoo, M. Schluter, L.J. Sham: Phys. Rev. B **32**, 3890 (1985)
2.11 M.S. Hybertsen, S.G. Louie: Phys. Rev. Lett. **55**, 1418 (1985); Phys. Rev. B **34**, 5390 (1986)
2.12 R.W. Godby, M. Schluter, L.J. Sham: Phys. Rev. Lett. **56**, 2415 (1986); Phys. Rev. B **35**, 4170 (1987)
2.13 L. Hedin, S. Lundquist: Solid State Physics **23**, 1 (1969)
2.14 A. Gigy: Submitted to Phys. Rev. B
2.15 J.C. Slater, G.J. Koster: Phys. Rev. **94**, 1498 (1954)
2.16 W.A. Harrison: In *Electronic Structure and the Properties of Solids. The Physics of the Chemical Bond* (Freeman, New York 1980)
2.17 S.G. Louie: Phys. Rev. B **22**, 1933 (1980)
2.18 R.C. Chaney, C. Lin, E.E. Lafon: Phys. Rev. B **3**, 459 (1971)
2.19 E.O. Kane: Phys. Rev. B **13**, 3478 (1976)
2.20 D.J. Chadi: Phys. Rev. B **16**, 3572 (1977)
2.21 G. Leman, J. Friedel: J. Appl. Phys., Supp. **33**, 281 (1962)
2.22 D. Weaire, M.F. Thorpe: Phys. Rev. B **4**, 2508 (1971)
2.23 M. Lannoo, M. Bensoussan: Phys. Rev. B **16**, 3546 (1977)
2.24 M. Bensoussan, M. Lannoo: J. de Phys. (Paris) **40**, 749 (1979)
2.25 P.O. Lowdin: J. Chem. Phys. **18**, 365 (1950)
2.26 M. Lannoo: J. de Phys. (Paris) **40**, 461 (1979)
2.27 G.Allan: In *Electronic Structure of Crystal Defects and Disordered Systems*, (Les Editions de Physique, Paris 1981) p.37
2.28 See the review by M. Schluter: In *Festkörperprobleme* (Advances in Solid State Physics), Vol. XVIII, p.155, ed. by J. Treusch, (Vieweg, Braunschweig 1978)
2.29 J.R. Chelikowsky, M. Schluter, S.G. Louie, M.L. Cohen: Sol. State Comm. **17**, 1103 (1975); M.L. Cohen, M. Schluter, J.R. Chelikowsky, S.G. Louie: Phys. Rev. B **12**, 5575 (1975)
2.30 G.P. Alldredge, L. Kleinmann: Phys. Rev. B **10**, 559 (1974)
2.31 See the review by J.A. Appelbaum, D.R. Hamann: Rev. Mod. Phys. **48**, 479 (1976)
2.32 J.A. Appelbaum, D.R. Hamann: Phys. Rev. B **6**, 2166 (1972)
2.33 J.Pollmann, R. Kalla, P. Kruger, A. Mazur, G. Wolfgarten: Appl. Phys. A **41**, 21 (1986)
2.34 M. Lannoo, J. Bourgoin: In *Point Defects in Semiconductors* I, Springer Ser. Solid State Sci. Vol. 22 (Springer, Berlin, Heidelberg 1981)
2.35 G.A. Baraff, M. Schluter: Phys. Rev. Lett. **41**, 892 (1978); Phys. Rev. B **19**, 4965 (1979)
2.36 J. Bernholc, N.O. Lipari, S.T. Pantelides: Phys. Rev. Lett. **41**, 895 (1978)
2.37 F. Cyrot-Lackmann: Adv. Phys. **16**, 393 (1967); J. Phys. Chem. Sol. **29**, 1235 (1968)
2.38 R.C. Gordon: J. Math. Phys. **9**, 655 (1968)

2.39 R. Haydock, V. Heine, M.J. Kelly: J. Phys. C **5**, 2845 (1972)
2.40 P. Turchi, F. Ducastelle, G. Treglia: J. Phys. C **15**, 1891 (1982)
2.41 G. Allan: J. Phys. C **17**, 3945 (1984)
2.42 G. Allan: Thesis, Ann. de Phys. **5**, 169 (1970)
2.43 F. Guinea, J. Sanchez-Dehesa, F. Flores: J. Phys. C **16**, 6499 (1983); F. Guinea, C. Tejedor, F. Flores, E. Louis: Phys. Rev. B **28**, 4397 (1983)
2.44 N.D. Lang, W. Kohn: Phys. Rev. B **12**, 4555 (1970); Phys. Rev. B **4**, 1215 (1971)
2.45 P.A. Serena, J.M. Loler, N. Garcia: Phys. Rev. B **37**, 8701 (1988)

Chapter 3

3.1 J. Friedel: In *The Physics of Metals*, ed. by J.M Ziman (Cambridge University, Cambridge 1969) p.340–408
3.2 K.A. Gschneider: In *Solid State Phys.*, ed. by F. Seitz, D. Turnbull Vol. 16 (Academic, New York 1965) p.275
3.3 F. Ducastelle: J. Phys. **31**, 1055 (1970)
3.4 F. Ducastelle, F. Cyrot-Lackmann: J. Phys. Chem. Sol. **32**, 285 (1971)
3.5 W.A. Harrison: In *Electronic Structure and the Properties of Solids, The Physics of the Chemical Bond*, (Freemann, New York 1980)
3.6 G.Allan, J. Lopez: In *Vibrations at Surfaces*, ed. by R. Caudano, J.M. Gilles, A.A. Lucas, (Plenum, New York 1982)
3.7 B. Legrand: Private communication
3.8 The first unambiguous measurements were reported by Tran Minh Duc, C. Guillot, Y. Lassailly, J. Lecante, Y. Jugnet, J.C. Vedrine: Phys. Rev. Lett. **43**, 789 (1979)
3.9 E. Clementi, D.L. Raimondi, W.P. Reinhardt: J. Chem. Phys. **47**, 1300 (1967)
3.10 C.E. Moore: In *Atomic Energy Levels* (National Bureau of Standards, U.S. Department of Commerce 1949)
3.11 L.Hodges, H.E. Watson, H. Ehrenreich: Phys. Rev. B **5**, 3953 (1972)
3.12 N.F. Mott, H. Jones: In *The Theory of the Properties of Metals and Alloys* (Dover, New York 1936) p.318
3.13 This correction is due to the fact that in the atom the spatial extension of the d-wave function has to be taken into account
3.14 J.C. Riviere: In *Solid State Surface Science*, ed. by M. Green (M. Dekker, New York 1969) pp.179–290
3.15 G. Allan, M. Lannoo: Le Vide, Les Couches Minces, **30A**, 1 (1975); Le Vide, **30**, 48 (1975)
3.16 G.Allan: In *Handbook of Surfaces and Interfaces*, ed. by L. Dobrzynski, (Garland, New York 1978)
3.17 J.A. Pople, G.A. Segal: J. Chem. Phys. **44**, 3289 (1966)
3.18 M.P. Tosi: In *Solid State Physics* Vol. 16, ed. by F. Seitz, D. Turnbull, (Academic, New York 1964) p.107
3.19 L.H. Thomas: Proc. Cambridge Phil. Soc. **23**, 542 (1927), J. Chem. Phys. **22**, 1758 (1954) E. Fermi, Atti. Accad. Nazl. Lincei **6**, 602 (1927), **7**, 342 (1928), Z. Physik **48**, 73 (1928), **49**, 550 (1928)
3.20 J. Friedel: Il Nuovo Cimento, Supp. **7**, 287 (1968)
3.21 M.C. Desjonqueres, F. Cyrot-Lackmann: J. de Phys. **36**, L45 (1975)
3.22 D. Spanjaard, C. Guillot, M.C. Desjonqueres, G. Treglia, J. Lecante: Surf. Sci. Reports **5**, (1-2) (1985)
3.23 J.C. Slater: In *The Self Consistent Field for Molecules and Solids*, (Mc Graw-Hill, New York 1974)
3.24 G. Allan: Annales Phys. (Paris), **5**, 169 (1970)
3.25 G. Allan, M. Lannoo: Phys. Stat. Sol. (b) **74**, 409 (1976)
3.26 G. Allan, M. Lannoo: Surf. Sci. **40**, 375 (1973)

3.27 For a more detailed discussion see G. Allan: In *Progress in Surface Science*, to be published; G. Allan, M. Lannoo: Phys. Rev. B **37**, 2678 (1988) and Proceedings of the 2nd Int. Conf. on the structure of surfaces, Amsterdam, to be published
3.28 J.W. Frenken, F. van der Veen, G. Allan: Phys. Rev. Lett. **51**, 1876 (1983); J. van der Veen, R.G. Smeenk, R.M. Tromp, F.W. Saris Surf. Sci. **79**, 212 (1979)
3.29 J. van der Veen, R.G. Smeenk, R.M. Tromp, F.W. Saris: Surf. Sci. **79**, 219 (1979)
3.30 L.E. Klebanoff, R.H. Victoria, L.M. Falicov, D.A. Shirley: Phys. Rev. B **32**, 1977 (1985)
3.31 G. Allan: Phys. Rev. B **19**, 4774 (1979); In *Handbook of Surfaces and Interfaces* Vol.4, ed. by L. Dobrzynski (Garland, New York 1978)
3.32 L.Gunther: Phys. Lett. **25A**, 649 (1967); **26A**, 216 (1968)
3.33 L. Lajzerowicz, L. Dobrzynski: Phys. Rev. B **14**, 2695 (1976)

Chapter 4

4.1 W.A. Harrison: In *Electronic Structure and the Properties of Solids, The Physics of the Chemical Bond* (Freeman, New York 1980)
4.2 C.A. Coulson: In *Valence* (Oxford University Press, Oxford 1961)
4.3 J.C. Slater: In *Quantum Theory of Molecules and Solids*, Vol. 1 (McGraw Hill, New York 1963)
4.4 C. Priester, G. Allan, J. Conard: Phys. Rev. B **26**, 4680 (1982)
4.5 K.C. Pandey, J.C. Phillips: Solid State Commun. **14**, 439 (1974); Phys. Rev. Lett. **32**, 1433 (1974)
4.6 K. Hirabayashi: J. Phys. Soc. Japan **27**, 1475 (1969)
4.7 M. Lannoo: In *Handbook of Surfaces and Interfaces*, Vol.1, ed. by L. Dobrzynski (Garland, New York 1978)
4.8 M. Lannoo, J. Bourgoin: In *Point Defects in Semiconductors I, Theoretical Aspects*, Springer Ser. Solid State Sci. Vol. 22 (Springer, Berlin, Heidelberg 1981)
4.9 J. Petit, M. Lannoo, G. Allan: Solid State Commun. **60**, 861 (1986)
4.10 K.C. Pandey: Phys. Rev. Lett. **47**, 1913 (1981) and **49**, 223 (1982)
4.11 G. Allan, M. Lannoo: Surf. Sci. **63**, 11 (1977)
4.12 C.B. Duke, W.K. Ford: Surf. Sci **111**, L685 (1981)
4.13 R. del Sole, D.J. Chadi: Phys. Rev. B **24**, 7430 (1981)
4.14 G. Allan, M. Lannoo: Phys. Rev. B **26**, 5279 (1982)
4.15 B.N. Lee, J.D. Joannopoulos: Phys. Rev. B **29**, 1473 (1984); G. Allan, M. Lannoo: Phys. Rev. B **29**, 1474 (1984)
4.16 J.E. Northrup, M.L. Cohen: J. Vac. Sci. Technol. **21**, 333 (1982); Phys. Rev. Lett. **49**, 1349 (1982)
4.17 P.N. Keating: Phys. Rev. **145**, 637 (1966)
4.18 W.A. Harrison: Surf. Sci. **55**, 1 (1976)
4.19 Y.Bar-Yam, J.D. Joannopoulos: Phys. Rev. Lett. **56**, 2203 (1986)
4.20 M. Lannoo, G. Allan: Phys. Rev. B **25**, 4089 (1982)
4.21 P. Vogl, H.P. Hjalmarson, J.D Dow: J. Phys. Chem. Sol. **44**, 365 (1983)
4.22 D.N. Talwar, C.S. Ting: Phys. Rev. B **25**, 2660 (1982)
4.23 D. Haneman: Phys. Rev. **121**, 1093 (1961)
4.24 M. Lannoo, G. Allan: Solid State Commun. **47**, 153 (1983)
4.25 G.X. Qian, D.J. Chadi: Phys. Rev. B **35**, 1288 (1987)
4.26 G. Binnig, H. Rohrer, C. Gerber, E. Weibel: Phys. Rev. Lett. **50**, 120 (1983)
4.27 R.J. Hamers, R.M. Tromp, J.E. Demuth: Phys. Rev. B **56**, 1972 (1986)
4.28 J. Pollmann, R. Kalla, P. Kruger, A. Mazur, G. Wolfgarten: Appl. Phys. A **41**, 21 (1986)
4.29 W.A. Harrison: In *Electronic Structure and the Properties of Solids, The Physics of the Chemical Bond* (Freemann, New York 1980)
4.30 L. Landau, E. Lifshitz: In *Electrodynamics of Continuous Media*, 3rd ed. (Pergamon, Oxford 1978)

Chapter 5

5.1 S.L. Adler: Phys. Rev. **126**, 413 (1962)
5.2 N. Wiser: Phys. Rev. **129**, 62 (1963)
5.3 M. Lannoo: J. de Phys. **38**, 473 (1977)
5.4 G. Srinivasan: Phys. Rev. **178**, 1244 (1969)
5.5 P. Vogl, H.P. Hjalmarson, J.D. Dow: J. Phys. Chem. Sol. **44**, 365 (1983)
5.6 D.N. Talwar, C.S. Ting: Phys. Rev. B **25**, 2660 (1982)
5.7 M. Lannoo: Phys. Rev. B **10**, 2544 (1974)
5.8 C.E. Moore: In *Atomic Energy Levels* (National Bureau of Standards, U.S. Department of Commerce 1949)
5.9 J.B. Mann: In *Atomic Structure Calculations I* (distributed by Clearinghouse for Technical Information, Springfield, VA 22125 (1967)
5.10 M. Lannoo, J.N. Decarpigny: Phys. Rev. B **8**, 5704 (1973)
5.11 W.A. Harrison: Phys. Rev. B **31**, 2121 (1985)
5.12 M.P. Tosi: In *Sol. State Phys.* Vol.16, ed. by F. Seitz and D. Turnbull (Academic Press New York 1964) p.107
5.13 J. Petit, M. Lannoo, G. Allan: Solid State Commun. **60**, 861 (1986)
5.14 M. Lannoo, J. Bourgoin: In *Point Defects in Semiconductors*, Vol. I, Springer Series in Solid State Sciences, Vol. 22, (Springer, Berlin, Heidelberg 1981)
5.15 I. Lefebvre: unpublished
5.16 J. Menendez: Phys. Rev. B **38**, 6305 (1988)
5.17 J. Tersoff: Phys. Rev. Lett. **52**, 465 (1984); Phys. Rev. B **30**, 4874 (1984); J. Vac. Sci. Technol. B **3**(4), 1157 (1985)
5.18 D.E. Eastman, J.L. Freeouf: Phys. Rev. Lett. **33**, 1601 (1974) and **34**, 1624 (1975)
5.19 P.E. Gregory, W.E. Spicer, W.A. Harrison: Appl. Phys. Lett. **25**, 511 (1974)
5.20 J. van Laar, J.J. Scheer: Surf. Sci. **8**, 342 (1967); A. Huijser, J. van Laar: Surf. Sci. **52**, 202 (1975)
5.21 J. van Laar, A. Huijser: J. Vac. Sci. Tech. **13**, 769 (1976); J. van Laar, A. Huijser, T. van Rooy: J. Vac. Sci. Tech. **14**, 893 (1977)
5.22 W.E. Spicer, I. Lindau, P.E. Gregory, C. M. Garner, P. Pianetta, P.W. Chye: J. Vac. Sci. Tech. **13**, 233 (1976); P.E. Gregory, W.E. Spicer: Phys. Rev. B **13**, 725 (1976)
5.23 W.E. Spicer, P. Pianetta, I. Lindau, P.W. Chye: J. Vac. Sci. Tech. **14**, 885 (1977)
5.24 W. Gudat, D. E. Eastman: J. Vac. Sci. Tech. **13**, 831 (1976)
5.25 D.J. Chadi: Phys. Rev. Lett. **43**, 43 (1979)
5.26 J.R. Chelikowsky, M.L. Cohen: Solid State Commun. **29**, 267 (1979)
5.27 J.R. Chelikowsky, M.L. Cohen: Phys. Rev. B **13**, 826 (1976)
5.28 D. Lohez, P. Masri, M. Lannoo, L Soonckindt, L. Lassabatere: Surf. Sci. **99**, 132 (1980)
5.29 C.B. Duke, C. Mailhiot, A. Paton, D.J. Chadi, A. Kahn: J. Vac. Sci. and Tech. B **3** (4), 1087 (1985)
5.30 Guo-Xin Quian, R.M. Martin, D.J. Chadi: Phys. Rev. B **37**, 1303 (1988)
5.31 S. Brennan, J. Stohr, R. Jaeger, J.E. Rowe: Phys. Rev. Lett. **45**, 1414 (1980)
5.32 F.J. Himpsel, P. Heinmann, T.C. Chiang, D.E. Eastmann: Phys. Rev. Lett. **45**, 1112, (1980)
5.33 D.E. Eastman, T.C. Chiang, P. Heinmann, F.J. Himpsel: Phys. Rev. Lett. **45**, 656 (1980)
5.34 M. Taniguchi, S. Suga, M. Seki, B. Stin, K.L.I. Kobayashi, M. Kanzaki: J. Phys. C **16**, L45 (1983)
5.35 In *Photoemission in Solids II*, Topics Appl. Phys. **27** ed. by L. Ley, M. Cardona (Springer, Berlin, Heidelberg 1979)
5.36 C. Priester, G. Allan, M. Lannoo: Phys. Rev. Lett. **58**, 1989 (1987)
5.37 W.A. Harrison: Phys. Rev. B **8**, 4487 (1973)
5.38 M. Lannoo, J.N. Decarpigny: Phys. Rev. B **8**, 5704 (1973)
5.39 D. Spanjaard, C. Guillot, M.C. Desjonqueres, G. Treglia, J. Lecante: Surf. Sci. Rep. **5**, 1 (1985)
5.40 W. Monch: Solid State Commun. **58**, 215 (1986)

5.41 W.A. Harrison: In *The Physics of Solid State Chemistry* – Festkörperprobleme XVII. Springer Tracts Mod. Phys. (Springer, Berlin, Heidelberg 1977)
5.42 A. Kahn: Surf. Sci. **168**, 1 (1986)
5.43 C. Priester, G. Allan, M. Lannoo: Phys. Rev. B **33**, 7386 (1986)
5.44 R.W. Nosker, P. Mark, J.D. Levine: Surf. Sci. **19**, 291 (1970)
5.45 P. Masri, M. Lannoo: Surf. Sci **52**, 377 (1975)
5.46 W.A. Harrison, E.A. Kraut, J.R. Waldrop, R.W. Grant: Phys. Rev. B **18**, 4402 (1978)
5.47 K. Jacobi: Surf. Sci **132**, 1 (1983)
5.48 K. Haneman: Phys. Rev. **121**, 1093 (1961)
5.49 S.Y. Iong, G. Xu, W.N. Mei: Phys. Rev. Lett. **52**, 1963 (1984)
5.50 J. Bohr, R. Feidenhans'l, M. Neielsen, M. Toney, R.L. Johnson, Y.K. Robinson: Phys. Rev. Lett. **54**, 1275 (1985)
5.51 D.J. Chadi: Phys. Rev. Lett. **52**, 1911 (1984)
5.52 D.J. Chadi: Phys. Rev. B **29**, 785 (1984)
5.53 E. Kaxiras, Y. Bar-Yam, J.D. Joannopoulos: Phys. Rev. B **35**, 9625 (1987)
5.54 E. Kaxiras, Y. Bar-Yam, J.D. Joannopoulos, K.C. Pandey: Phys. Rev. B **35**, 9636 (1987)
5.55 P.K. Larsen, J.F. van der Veen, A. Mazur, J. Pollmann, J.H. Neave, B.A. Joyce: Phys. Rev. B **26**, 3222 (1982)
5.56 P.K. Larsen, D.J. Chadi: Phys. Rev. B **37**, 8282 (1988)
5.57 R. Ludeke, T.C. Chiang, D.E. Eastman: Physica B + C **117** & **118** B, 819 (1983)
5.58 R.Z. Bachrach, R.S. Bauer, P. Chiaradia, G.V. Harrison: J. Vac. Sci. Technol. **18**, 797 (1981) and ibidem p.335
5.59 J.H. Neave, B.A. Joyce, P.J. Dobson, N. Norton: Appl. Phys. A **31**, 1 (1983)
5.60 J. Massies, P. Etienne, F. Dezaly, N.T. Link: Surf. Sci **99**, 121 (1980)
5.61 M.D. Pashley, W. Haberern, J.W. Woodall: J. Vac. Sci. Technol. B **6**, 1468 (1988)
5.62 D.J. Chadi, C. Tanner, J. Ihm: Surf. Sci. **120**, L425 (1982)
5.63 B.A. Joyce, J.H. Neave, P.J. Dolson, P.K. Larsen, J. Zhang: J. Vac. Sci. Technol. B **3**, 562 (1985)
5.64 B.A. Joyce, J.H. Neave, P.J. Dolson, P.K. Larsen: Phys. Rev. B **29**, 814 (1984)
5.65 K.C. Pandey: Phys. Rev. Lett. **47**, 1913 (1981)
5.66 J.R. Arthur: Surf. Sci. **43**, 449 (1974)
5.67 P.Friedel, P.K. Larsen, S. Gourrier, J.P. Cabanie, W. Gerits: J. Vac. Sci. Technol. B **2**, 675 (1984)
5.68 T. Carette, M. Lannoo, G. Allan, P. Friedel: Surf. Sci. **164**, 260 (1985)
5.69 G.X. Qian, R.M. Martin, D.J. Chadi: Phys. Rev. B **37**, 1303 (1988)
5.70 D.J. Chadi: J. Vac. Sci. Technol. A **5**, 834 (1987)

Chapter 6

6.1 K. Stiles, D. Mao, S.F. Horng, A. Kahn, J. McKinley, D.G. Kilday, G. Margaritondo: in *Metallization and Metal Semiconductor Interfaces*, NATO ASI Series, ed. by I.P. Batra, Plenum Press (1989)
6.2 W.E. Spicer, R. Cao, K. Miyano, C. McCants, T.T. Chiang, C.J. Spindt, N. Newman, T. Kendelewicz, I. Lindau: Ibidem
6.3 W.E. Spicer, P.W. Chye, P.R. Skeath, C.Y. Su, I. Lindau: J. Vac. Sci. Technol. **16**, 1427 (1979); **17**, 1019 (1980)
6.4 K. Stiles, A. Kahn, D.G. Kilday, G. Margaritondo: J. Vac. Sci. Technol. B **5**, 987 (1987)
6.5 K. Stiles, S.F. Horng, A. Kahn, J. McKinley, D.G. Kilday, G. Margaritondo: J. Vac. Sci. Technol. A **6**, 1462 (1988) and B **6** (1988), in press
6.6 R. Cao, K. Miyano, T. Kendelewicz, K.K. Chin, I. Lindau, W.E. Spicer: J. Vac. Sci. Technol. B **5**, 998 (1987)
6.7 K. Stiles, A. Kahn: Phys. Rev. Lett. **60**, 440 (1988)
6.8 R. Cao, K. Miyano, T. Kendelewicz, I. Lindau, W.E. Spicer: J. Vac. Sci. A **6**, 1571 (1988)

6.9 W. Monch: Europhys. Lett. **7** (3), 275 (1988); J. Vac. Sci. Technol. B **6** (4), 1270 (1988)
6.10 T. Kendelewicz, P. Soukiassian, M.H. Bacshi, Z. Hurych, I. Lindau, W.E. Spicer: Phys. Rev. B **38**, 7568 (1988); J. Vac. Sci. Technol. B **6**, 1331 (1988)
6.11 G. Picoli, A. Chomette, M. Lannoo: Phys. Rev. B **30**, 7138 (1984)
6.12 D.M. Newns: J. Chem. Phys. **50**, 4572 (1969)
6.13 For a proof in LDA see J.F. Janak: Phys. Rev. B **18**, 7165 (1978)
6.14 E.Clementi, D.L. Raimondi, W.P. Reinhardt: J. Chem. Phys. **17**, 1300 (1967)
6.15 I. Lefebvre, M. Lannoo, G. Allan, A. Ibanez, J. Fourcade, J.C. Jumas, E. Beaurepaire: Phys. Rev. Lett. **59**, 2471 (1987); I. Lefebvre, M. Lannoo, G. Allan, L. Martinage: Phys. Rev. B **38**, 8593 (1988)
6.16 W.A. Harrison: Phys. Rev. B **24**, 5835 (1981)
6.17 D.N. Talwar, C.S. Ting: Phys. Rev. B **25**, 2660 (1982)
6.18 W.A. Harrison: Phys. Rev. B **31**, 2121 (1985)
6.19 J.B. Mann: In *Atomic Structure Calculations I* (Clearing House for Technical Information, Springfield, VA 1967)
6.20 A. Zur, McGill, D.L. Smith: Phys. Rev. B **28**, 2060 (1983)
6.21 I. Lefebvre, M. Lannoo, G. Allan: Europhysics Letters **10**, 359 (1989)
6.22 J.E. Klepeis, W.A. Harrison: J. Vac. Sci. Technol. B **7**, 964 (1989)
6.23 M.S. Hybertsen, S.G. Louie: Phys. Rev. B **38**, 4033 (1988)
6.24 O. Madelung: *Introduction to Solid State Theory*, Springer Ser. Solid-State Sci. Vol.2 (Springer, Berlin, Heidelberg 1978)

Chapter 7

7.1 W. Schottky: Naturwissenschaften **26**, 843 (1938); Z. Phys. **113**, 367 (1939)
7.2 S. Kurtin, T.C. McGill, C.A. Mead: Phys. Rev. Lett. **22**, 1433 (1969)
7.3 M. Schluter: In *Festkörperprobleme* Vol.18. Advances in Solid State Physics, (Vieweg, Braunschweig, 1978) p.155
7.4 J. Bardeen: Phys. Rev. **71**, 717 (1947)
7.5 V. Heine: Phys. Rev. **138**, A 1689 (1965)
7.6 W.E. Spicer, I. Lindau, P. Skeath, C.Y. Yu: J. Vac. Sci. Technol. **17**, 1019 (1980)
7.7 W.E. Spicer, T. Kendelewicz, N. Newman, K.K. Chin, I. Lindau: Surf. Sci. **168**, 240 (1986)
7.8 C. Tejedor, F. Flores, E. Louis: J. Phys. C **10**, 2163 (1977)
7.9 J. Tersoff: Phys. Rev. Lett. **52**, 465 (1984); Phys. Rev. B **30**, 4874 (1984); J. Vac. Sci. Technol. B 3 (4), 1157 (1985)
7.10 J. Tersoff: In *Heterojunction Band Discontinuities, Physics and Device Applications*, ed. by F. Capasso, G. Margaritondo (North Holland, Amsterdam 1987)
7.11 E.J. Mele, J.D. Joannopoulos: Phys. Rev. B **17**, 1528 (1978)
7.12 I. Lefebvre, M. Lannoo, C. Priester, G. Allan, C. Delerue: Phys. Rev. B **36**, 1336 (1987)
7.13 G. Allan: In *Handbook of Surfaces and Interfaces*, ed. by L. Dobrzynski (Garland STPM, 368 1978)
7.14 P. Vogl. H.P. Hjalmarson, J.D. Dow: J. Phys. Chem. Sol. **44**, 365 (1983)
7.15 D.N. Talwar, C.S. Ting: Phys. Rev. B **25**, 2660 (1982)
7.16 O.F. Sankey, R.E. Allen, S.F. Ren, J.D. Dow: J. Vac. Sci. Technol. B **3**, 1182 (1985)
7.17 A. Zur, T.C. McGill, D.L. Smith: Phys. Rev. B **28**, 2060 (1983)
7.18 F. Flores, C. Tejedor: J. Phys. C. Sol. State Phys. **20**, 145 (1987)
7.19 S.G. Louie, J.R. Chelikowsky, M.L. Cohen: Phys. Rev. B **15**, 2154 (1977)
7.20 R.H. Williams: Proc. of 17th Int. Conf. Phys. Semicond. ed. by J.D. Chadi, W.A. Harrison (Springer, Berlin, Heidelberg 1985) p.175
7.21 L.J. Brillson: In *Handbook of Synchrotron Radiation* Vol.II, ed. by G.V. Marr (North Holland, Amsterdam 1985)
7.22 D.R. Hamann: Phys. Rev. Lett. **60**, 313 (1988)
7.23 G.P. Das, P. Blochl, N.E. Christensen, O.K. Andersen: in *Metallization and Metal-Semiconductor Interfaces*, NATO ASI Series, ed. by I.P. Batra, Plenum Press (1988)

7.24 R.L. Anderson: Solid State Electron **5**, 341 (1972)
7.25 W.A. Harrison: J. Vac. Sci. Technol. **14**, 1016 (1977)
7.26 R. Martin, C. van de Walle: J. Vac. Sci. Technol. B **3**, 1256 (1985) and B **4**, 1055 (1986); Phys. Rev. B **34**, 5621 (1986) and B **35**, 8154 (1987)
7.27 N. Christensen: Phys. Rev. B **37**, 4528 (1988)
7.28 A. Munoz, J.C. Duran, F. Flores: Phys. Rev. B **35**, 7121 (1987)
7.29 B. Haussy, C. Priester, G. Allan, M. Lannoo: 18th Intl Conference on the Physics of Semiconductors, Stockholm, 1986 (World Scientific, Singapore 1987)
7.30 C. Priester, G. Allan, M. Lannoo: J. Vac. Sci. Technol. B **6**, 1290 (1988)
7.31 N. Christensen, M. Cardona: Phys. Rev. B **35**, 6182, (1987)
7.32 S.B. Zhang, D. Tomanek, S. Louie, M.L. Cohen, M.S. Hybertsen: Solid State Commun. **66**, 585 (1988)
7.33 C. Priester, G. Allan, M. Lannoo: Phys. Rev. B, to be published
7.34 F.H. Pollack, M. Cardona: Phys. Rev. **172**, 816 (1968)
7.35 C. Priester, G. Allan, M. Lannoo: Phys. Rev. B **37**, 8519 (1987); Phys. Rev. B **38**, 13451 (1988)
7.36 S.P. Kowalczyk, W.J. Schaffer, E.A. Kraut, R.W. Grant: J. Vac. Sci. Technol. **20**, 705 (1982)
7.37 J. Menendez, A. Pinczuck, D.J. Werder, S.K. Sputz, R.C. Miller, D.L. Sivco, A.Y. Cho: Phys. Rev. B **36**, 8165 (1987)
7.38 J.Y. Marzin, M.N. Charasse, B. Sermage: Phys. Rev. B **31**, 8298 (1985)
7.39 J.Y. Marzin: Thesis, Paris (1987) p.198, unpublished
7.40 T.P. Pearsall, J. Bevk, L.C. Feldman, J.M. Bonar, J.P. Mannaerts, A. Ourmazd: Phys. Rev. Lett. **58**, 729 (1987)
7.41 J. Bevk, A. Ourmazd, L.C. Feldman, T.P. Pearsall, J.M. Bonar, B.A. Davidson, J.P. Mannaerts; Appl. Phys. Lett. **50**, 760 (1987)
7.42 L. Brey, C. Tejedor: Phys. Rev. Lett. **59**, 1022 (1987)
7.43 S. Froyen, D.M. Wood, A. Zunger: Phys. Rev. B **36**, 4547 (1987); **37**, 6893 (1988)
7.44 M.S. Hybertsen, M. Schlüter: Phys. Rev. B **36**, 9683 (1987)
7.45 S. Satpathy, R.M. Martin, C.G. van de Walle: Phys. Rev. B **38**, 13237 (1988)
7.46 L. Morrison, M. Jaros, K.B. Wong: Phys. Rev. B **35**, 9693 (1987)
7.47 M.S. Hybertsen, M. Schlüter, R. People, S.A. Jackson, D.V. Lang T.P. Pearsall, J.C. Bean, J.M. Vandenberg, J. Bevk: Phys. Rev. B **37**, 10195 (1988)
7.48 P. Friedel, M.S. Hybertsen, M. Schlüter: Phys. Rev. B **39**, 7974 (1989)
7.49 P. Friedel, M.S. Hybertsen, M. Schlüter: To be published
7.50 R. Zachai, E. Friess, G. Abstreiter, E. Kasper, H. Kibbel: Proc. 19th ICPS, ed. by W. Zawadski (Polish Academy of Sciences, Warsaw, 1988)
7.51 T.P. Pearsall, J.C. Bean, R.H. Hull, J.M Bonar: Proc. of the Euro-MRS, Strasbourg, France (1989)
7.52 E.A. Montie, G.F.A. van de Walle, D.J. Gravesteijn, A.A. van Gorkum, C.W.T. Bulle-Lieuwma: Appl. Phys. Lett., to be published
7.53 J.M Langer, H. Heinrich: Phys. Rev. Lett. **55**, 1414, (1985) and Physica B **134**, 444 (1985)
7.54 A. Zunger: Ann. Rev. Mater. Sci. **15**, 411 (1985); Solid State Physics **39**, 275 (1986)
7.55 J. Tersoff, W.A. Harrison: Phys. Rev. Lett. **58**, 2367 (1987); J. Vac. Sci. Technol. B **5**, 1221 (1987)
7.56 G. Picoli, A. Chomette, M. Lannoo: Phys. Rev. B **30**, 7138 (1984)
7.57 L.A. Ledebo, B.K. Ridley: J. Phys. C **15**, L 961 (1982)
7.58 M.J. Caldas, A. Fazzio, A. Zunger: Appl. Phys. Lett. **45**, 671 (1984); A. Zunger: Phys. Rev. Lett. **54**, 849 (1985)
7.59 J.M. Langer, C. Delerue, M. Lannoo, H. Heinrich: Phys. Rev. B **38**, 7723 (1988)
7.60 H. Heinrich, J.M. Langer: In *Festkörperprobleme*, Vol.26, ed. by P. Grosse (Vieweg, Braunschweig, 1986) p.251; H. Heinrich: In *Lecture Notes in Physics* (Springer, Berlin, Heidelberg 1989)
7.61 D.W. Niles, G. Margaritondo: Phys. Rev. B **34**, 2923 (1986)
7.62 C. Tejedor, F. Flores: J. Phys. C **11**, L19 (1978); F. Flores, C. Tejedor: ibid. **12**, 731 (1979)

7.63 C. Delerue, G. Allan, M. Lannoo: Mat. Sci. Forum, **10–12**, 37 (1986); Z. Liro, C. Delerue, M. Lannoo: Phys. Rev. B **36**, 17 (1987); C. Delerue, M. Lannoo, G. Allan: Phys. Rev. B **39**, 1669 (1989)
7.64 W.A. Harrison: In *Electronic Structure and the Properties of Solids* (Freeman, New York 1980)
7.65 R.D.M. Haldane, P.W. Anderson: Phys. Rev. B **13**, 2553 (1976)
7.66 F. Herman, S. Skillman: In *Atomic Structure Calculations* (Prentice Hall, New York 1963)
7.67 O.L. Krivanek, T.T. Scheng, D.C. Tsui: Appl. Phys. Lett. **32**, 439 (1978)
7.68 G. Hollinger, F.J. Himpsel: Appl. Phys. Lett. **44**, 93 (1984)
7.69 F.J. Grunthaner, P.J. Grunthaner, R.P. Vasquez, B.F. Lewis, J. Maserjian, A. Madhukar: Phys. Rev. Lett. **43**, 1683 (1979)
7.70 G. Hollinger, Y. Jugnet, Tran Minh Duc: Solid State Commun. **22**, 277 (1977)
7.71 G. Hollinger, E. Bergignat, H. Chermette, F. Himpsel, D. Lohez, M. Lannoo, M. Bensoussan: Philos. Mag. **55**, 735 (1987)
7.72 C.A. Coulson: In *Valence* (Oxford University Press, Oxford 1961)
7.73 S.T. Pantelides, W.A. Harrison: Phys. Rev. B **13**, 2667 (1976)
7.74 T.H. di Stefano, D.E. Eastman: Phys. Rev. Lett. **29**, 1088 (1972)
7.75 H. Ibach, J.E. Rowe: Phys. Rev. B **10**, 710 (1974)
7.76 J.R. Chelikowsky, M. Schluter: Phys. Rev. B **15**, 4020 (1977)
7.77 E.P. O'Reilly, J. Robertson: Phys. Rev. B **27**, 3780 (1983)
7.78 M. Lannoo, G. Allan: Sol. State Comm. **28**, 733 (1978)
7.79 E. Martinez, F. Yndurain: Phys. Rev. B **24**, 5718 (1981)
7.80 K. Hubner, A. Stern, E.D. Klinkenberg: Phys. Stat. Sol. **136**, 211 (1986)
7.81 Y.P. Li, W.Y. Ching: Phys. Rev. B **31**, 2172 (1985)
7.82 M. Bensoussan, M. Lannoo: J. Phys. Paris **40**, 749 (1979)
7.83 G. Hollinger, S.J. Sferco, M. Lannoo: Phys. Rev. B **37**, 7149 (1988)
7.84 M. Lannoo: J. Phys. **40**, 461 (1979)
7.85 E.H. Poindexter, P.J. Caplan: Prog. Surf. Sci. **14**, 201 (1983)
7.86 P.J. Caplan, E.H. Poindexter, B.E. Deal, R.R. Razouk: J. Appl. Phys. **50**, 5847 (1983)
7.87 N.M Johnson, D.K. Biegelsen, M.D. Moyer, S.T. Chang, E.H. Poindexter, P.J. Caplan: Appl. Phys. Lett. **43**, 563 (1983)
7.88 K.L. Brower: Appl Phys. Lett. **43**, 1111 (1983)
7.89 B. Henderson: Appl. Phys. Lett. **44**, 228 (1984)
7.90 J.D. Cohen, D.V. Lang: Phys. Rev. B **25**, 5285 (1982)
7.91 W.B. Jackson: Solid State Commun. **44**, 477 (1982)
7.92 N.M. Johnson, W.B. Jackson, M.D. Moyer: Phys. Rev. B **31**, 1194 (1985)
7.93 J. Petit, M. Lannoo, G. Allan: Sol. State Commun. **60**, 861 (1986)
7.94 J. Bourgoin, M. Lannoo: In *Point Defects in Semiconductors II*, ed. by M. Cardona (Springer, New York 1983)
7.95 S. Loualiche, A. Nouailhat, G. Guillot, M. Lannoo: Phys. Rev. B **30**, 5822 (1984)
7.96 M. Lannoo: Unpublished
7.97 Y. Bar-Yam, J.D. Toannopoulos: Phys. Rev. Lett. **56**, 2203 (1986)

Chapter 8

8.1 M.J.P. Musgrave, J.A. Pople: Proc. R. Soc. (London) A **268**, 474 (1962)
8.2 M. Lannoo, G. Allan: Phys. Rev. B **25**, 4089 (1982)
8.3 P.N. Keating: Phys. Rev. **145**, 637 (1966)
8.4 R.M. Martin: Phys. Rev. B **1**, 4005 (1970)
8.5 M. Lannoo: J. de Phys. **40**, 461 (1979)
8.6 D.J. Chadi: Phys. Rev. Lett. **41**, 1062 (1978); Phys. Rev. Lett. **43**, 43 (1979); J. Vac. Sci. Technol. **16**, 1290 (1979)
8.7 D.C. Allan, E.J. Mele: Phys. Rev. Lett. **53**, 826 (1984)
8.8 O.L. Alerhand, E.J. Mele: Phys. Rev. Lett. **59**, 657 (1987); Phys. Rev. B **37**, 2536 (1988)

8.9 J.M. Ziman: In *Models of Disorder* (Cambridge University Press, Cambridge 1979)
8.10 R.F. Wallis: In *The Structure and Chemistry of Solid Surfaces* (Wiley, New York 1968)
8.11 J.B. Theeten, L. Dobrzynski: Phys. Rev. B **5**, 1529 (1973)
8.12 R.A. Swalin: In *Thermodynamics of Solids* (Wiley, New York 1962)
8.13 L. Rayleigh: Proc. London Math. Soc. **17**, 4 (1885)
8.14 L.D. Landau, E.M. Lifshitz: In *Theory of Elasticity* (Pergamon, Oxford 1959) p.105
8.15 U. Harten, J.P. Toennies, Ch. Woll: Phys. Rev. Lett. **57**, 2947 (1986)
8.16 G. Allan, M. Lannoo: Surf. Sci. **40**, 375 (1973) and Phys. Stat. Sol. (b) **74**, 409 (1976)
8.17 J. Sokolov, H.D. Shih, U. Bardi, F. Jona, P.M. Marcus: Solid State Commun. **48**, 739 (1983) and J. Phys. C **17**, 371 (1984)
8.18 J. Feidenhans'l, J.E. Sorensen, I. Stengaard: Surf. Sci. **134**, 329 (1983)
8.19 J. Sokolov, F. Jona, P.M. Marcus: Solid State Commun. **49**, 307 (1984)
8.20 Y. Gauthier, R. Baudoing, Y. Joly, C. Gaubert, J. Rundgren: J. Phys. C **17**, 4547 (1984)
8.21 R.N Barnett, Uzi Landman, C.L. Cleveland: Phys. Rev. Lett. **51**, 1359 (1983); Phys. Rev. B **28**, 1685 (1983)
8.22 C.L. Fu, S. Ohnishi, E. Wimmer, A.J. Freeman: Phys. Rev. Lett. **53**, 675 (1984)
8.23 C.L. Fu, S. Ohnishi, H.J.F. Jansen, A.J. Freeman: Phys. Rev. B **31**, 1168 (1985)
8.24 B. Legrand, G. Treglia, M.C. Desjonqueres, D. Spanjaard: J. Phys. C **19**, 4463 (1986)
8.25 W. Kohn: Phys. Rev. **115**, 809 (1959)
8.26 D. Castiel, L. Dobrzynski, D. Spanjaard: Surf. Sci. **59**, 252 (1976); J.E. Black, D.A. Campbell, R.F. Wallis: Surf. Sci. **105**, 629 (1981)
8.27 M.L. Xu, S.Y. Tong: Phys. Rev. B **31**, 6332 (1985)
8.28 J. Sokolov, F. Jona, P.M. Marcus: Phys. Rev. B **33**, 1397 (1986)
8.29 F. Ciccacci, S. Selci, G. Chiarotti, P. Chiaradia: Phys. Rev. Lett. **56**, 2411 (1986)
8.30 J. Bourgoin, M. Lannoo: In *Point Defects in Semiconductors II*, Springer Ser. Solid State Sci. Vol. 35 (Springer, Berlin Heidelberg 1983)

Subject Index

GPSR Compliance

The European Union's (EU) General Product Safety Regulation (GPSR) is a set of rules that requires consumer products to be safe and our obligations to ensure this.

If you have any concerns about our products, you can contact us on ProductSafety@springernature.com

In case Publisher is established outside the EU, the EU authorized representative is:

Springer Nature Customer Service Center GmbH
Europaplatz 3
69115 Heidelberg, Germany

Batch number: 10607214

Printed by Printforce, the Netherlands